D1452271

4G LTE/LTE-Advanced for Mobile Broadband

4G LTE/LTE-Advanced for Mobile Broadband

Erik Dahlman, Stefan Parkvall, and Johan Sköld

ELSEVIER

AMSTERDAM • BOSTON • HEIDELBERG • LONDON • NEW YORK • OXFORD
PARIS • SAN DIEGO • SAN FRANCISCO • SINGAPORE • SYDNEY • TOKYO
Academic Press is an imprint of Elsevier

Academic Press is an imprint of Elsevier
The Boulevard, Langford Lane, Kidlington, Oxford, OX5 1GB, UK
30 Corporate Drive, Suite 400, Burlington, MA 01803, USA

First published 2011

British Library Cataloguing-in-Publication Data
A catalogue record for this book is available from the British Library

Library of Congress Control Number: 2011921244

ISBN: 978-0-12-385489-6

For information on all Academic Press publications
visit our website at www.elsevierdirect.com

Typeset by MPS Limited, a Macmillan Company, Chennai, India
www.macmillansolutions.com

Printed and bound in the UK

11 12 13 14   10 9 8 7 6 5 4 3 2 1

Contents

Preface

During the past years, there has been a quickly rising interest in radio access technologies for providing mobile as well as nomadic and fixed services for voice, video, and data. The difference in design, implementation, and use between telecom and datacom technologies is also becoming more blurred. One example is cellular technologies from the telecom world being used for broadband data and wireless LAN from the datacom world being used for voice-over IP.

Today, the most widespread radio access technology for mobile communication is digital cellular, with the number of users passing 5 billion by 2010, which is more than half of the world's population. It has emerged from early deployments of an expensive voice service for a few car-borne users, to today's widespread use of mobile-communication devices that provide a range of mobile services and often include camera, MP3 player, and PDA functions. With this widespread use and increasing interest in mobile communication, a continuing evolution ahead is foreseen.

This book describes LTE, developed in 3GPP (*Third Generation Partnership Project*) and providing true 4G broadband mobile access, starting from the first version in release 8 and through the continuing evolution to release 10, the latest version of LTE. Release 10, also known as LTE-Advanced, is of particular interest as it is the major technology approved by the ITU as fulfilling the IMT-Advanced requirements. The description in this book is based on LTE release 10 and thus provides a complete description of the LTE-Advanced radio access from the bottom up.

Chapter 1 gives the background to LTE and its evolution, looking also at the different standards bodies and organizations involved in the process of defining 4G. It also gives a discussion of the reasons and driving forces behind the evolution.

Chapters 2–6 provide a deeper insight into some of the technologies that are part of LTE and its evolution. Because of its generic nature, these chapters can be used as a background not only for LTE as described in this book, but also for readers who want to understand the technology behind other systems, such as WCDMA/HSPA, WiMAX, and CDMA2000.

Chapters 7–17 constitute the main part of the book. As a start, an introductory technical overview of LTE is given, where the most important technology components are introduced based on the generic technologies described in previous chapters. The following chapters provide a detailed description of the protocol structure, the downlink and uplink transmission schemes, and the associated mechanisms for scheduling, retransmission and interference handling. Broadcast operation and relaying are also described. This is followed by a discussion of the spectrum flexibility and the associated requirements from an RF perspective.

Finally, in Chapters 18–20, an assessment is made on LTE. Through an overview of similar technologies developed in other standards bodies, it will be clear that the technologies adopted for the evolution in 3GPP are implemented in many other systems as well. Finally, looking into the future, it will be seen that the evolution does not stop with LTE-Advanced but that new features are continuously added to LTE in order to meet future requirements.

Acknowledgements

We thank all our colleagues at Ericsson for assisting in this project by helping with contributions to the book, giving suggestions and comments on the contents, and taking part in the huge team effort of developing LTE.

The standardization process involves people from all parts of the world, and we acknowledge the efforts of our colleagues in the wireless industry in general and in 3GPP RAN in particular. Without their work and contributions to the standardization, this book would not have been possible.

Finally, we are immensely grateful to our families for bearing with us and supporting us during the long process of writing this book.

Abbreviations and Acronyms

3GPP	Third Generation Partnership Project
3GPP2	Third Generation Partnership Project 2
ACIR	Adjacent Channel Interference Ratio
ACK	Acknowledgement (in ARQ protocols)
ACLR	Adjacent Channel Leakage Ratio
ACS	Adjacent Channel Selectivity
AM	Acknowledged Mode (RLC configuration)
AMC	Adaptive Modulation and Coding
A-MPR	Additional Maximum Power Reduction
AMPS	Advanced Mobile Phone System
AQPSK	Adaptive QPSK
ARI	Acknowledgement Resource Indicator
ARIB	Association of Radio Industries and Businesses
ARQ	Automatic Repeat-reQuest
AS	Access Stratum
ATIS	Alliance for Telecommunications Industry Solutions
AWGN	Additive White Gaussian Noise
BC	Band Category
BCCH	Broadcast Control Channel
BCH	Broadcast Channel
BER	Bit-Error Rate
BLER	Block-Error Rate
BM-SC	Broadcast Multicast Service Center
BPSK	Binary Phase-Shift Keying
BS	Base Station
BSC	Base Station Controller
BTS	Base Transceiver Station
CA	Carrier Aggregation
CC	Convolutional Code (in the context of coding), or Component Carrier (in the context of carrier aggregation)
CCCH	Common Control Channel
CCE	Control Channel Element
CCSA	China Communications Standards Association
CDD	Cyclic-Delay Diversity
CDF	Cumulative Density Function
CDM	Code-Division Multiplexing
CDMA	Code-Division Multiple Access

CEPT	European Conference of Postal and Telecommunications Administrations
CN	Core Network
CoMP	Coordinated Multi-Point transmission/reception
CP	Cyclic Prefix
CPC	Continuous Packet Connectivity
CQI	Channel-Quality Indicator
C-RAN	Centralized RAN
CRC	Cyclic Redundancy Check
C-RNTI	Cell Radio-Network Temporary Identifier
CRS	Cell-specific Reference Signal
CS	Circuit Switched (or Cyclic Shift)
CS	Capability Set (for MSR base stations)
CSA	Common Subframe Allocation
CSG	Closed Subscriber Group
CSI	Channel-State Information
CSI-RS	CSI reference signals
CW	Continuous Wave
DAI	Downlink Assignment Index
DCCH	Dedicated Control Channel
DCH	Dedicated Channel
DCI	Downlink Control Information
DFE	Decision-Feedback Equalization
DFT	Discrete Fourier Transform
DFTS-OFDM	DFT-Spread OFDM (DFT-precoded OFDM, see also SC-FDMA)
DL	Downlink
DL-SCH	Downlink Shared Channel
DM-RS	Demodulation Reference Signal
DRX	Discontinuous Reception
DTCH	Dedicated Traffic Channel
DTX	Discontinuous Transmission
DwPTS	The downlink part of the special subframe (for TDD operation).
EDGE	Enhanced Data rates for GSM Evolution, Enhanced Data rates for Global Evolution
EGPRS	Enhanced GPRS
eNB	eNodeB
eNodeB	E-UTRAN NodeB
EPC	Evolved Packet Core
EPS	Evolved Packet System
ETSI	European Telecommunications Standards Institute
E-UTRA	Evolved UTRA
E-UTRAN	Evolved UTRAN
EV-DO	Evolution-Data Only (of CDMA2000 1x)
EV-DV	Evolution-Data and Voice (of CDMA2000 1x)
EVM	Error Vector Magnitude

FACH	Forward Access Channel
FCC	Federal Communications Commission
FDD	Frequency Division Duplex
FDM	Frequency-Division Multiplex
FDMA	Frequency-Division Multiple Access
FEC	Forward Error Correction
FFT	Fast Fourier Transform
FIR	Finite Impulse Response
FPLMTS	Future Public Land Mobile Telecommunications Systems
FRAMES	Future Radio Wideband Multiple Access Systems
FSTD	Frequency Switched Transmit Diversity
GERAN	GSM/EDGE Radio Access Network
GGSN	Gateway GPRS Support Node
GP	Guard Period (for TDD operation)
GPRS	General Packet Radio Services
GPS	Global Positioning System
GSM	Global System for Mobile communications
HARQ	Hybrid ARQ
HII	High-Interference Indicator
HLR	Home Location Register
HRPD	High Rate Packet Data
HSDPA	High-Speed Downlink Packet Access
HSPA	High-Speed Packet Access
HSS	Home Subscriber Server
HS-SCCH	High-Speed Shared Control Channel
ICIC	Inter-Cell Interference Coordination
ICS	In-Channel Selectivity
ICT	Information and Communication Technologies
IDFT	Inverse DFT
IEEE	Institute of Electrical and Electronics Engineers
IFDMA	Interleaved FDMA
IFFT	Inverse Fast Fourier Transform
IMT-2000	International Mobile Telecommunications 2000 (ITU's name for the family of 3G standards)
IMT-Advanced	International Mobile Telecommunications Advanced (ITU's name for the family of 4G standards)
IP	Internet Protocol
IR	Incremental Redundancy
IRC	Interference Rejection Combining
ITU	International Telecommunications Union
ITU-R	International Telecommunications Union-Radiocommunications Sector

J-TACS	Japanese Total Access Communication System
LAN	Local Area Network
LCID	Logical Channel Index
LDPC	Low-Density Parity Check Code
LTE	Long-Term Evolution
MAC	Medium Access Control
MAN	Metropolitan Area Network
MBMS	Multimedia Broadcast/Multicast Service
MBMS-GW	MBMS gateway
MBS	Multicast and Broadcast Service
MBSFN	Multicast-Broadcast Single Frequency Network
MC	Multi-Carrier
MCCH	MBMS Control Channel
MCE	MBMS Coordination Entity
MCH	Multicast Channel
MCS	Modulation and Coding Scheme
MDHO	Macro-Diversity Handover
MIB	Master Information Block
MIMO	Multiple-Input Multiple-Output
ML	Maximum Likelihood
MLSE	Maximum-Likelihood Sequence Estimation
MME	Mobility Management Entity
MMS	Multimedia Messaging Service
MMSE	Minimum Mean Square Error
MPR	Maximum Power Reduction
MRC	Maximum Ratio Combining
MSA	MCH Subframe Allocation
MSC	Mobile Switching Center
MSI	MCH Scheduling Information
MSP	MCH Scheduling Period
MSR	Multi-Standard Radio
MSS	Mobile Satellite Service
MTCH	MBMS Traffic Channel
MU-MIMO	Multi-User MIMO
MUX	Multiplexer or Multiplexing
NAK, NACK	Negative Acknowledgement (in ARQ protocols)
NAS	Non-Access Stratum (a functional layer between the core network and the terminal that supports signaling and user data transfer)
NDI	New-data indicator
NSPS	National Security and Public Safety
NMT	Nordisk MobilTelefon (Nordic Mobile Telephony)

NodeB	NodeB, a logical node handling transmission/reception in multiple cells. Commonly, but not necessarily, corresponding to a base station.
NS	Network Signaling
OCC	Orthogonal Cover Code
OFDM	Orthogonal Frequency-Division Multiplexing
OFDMA	Orthogonal Frequency-Division Multiple Access
OI	Overload Indicator
OOB	Out-Of-Band (emissions)
PAPR	Peak-to-Average Power Ratio
PAR	Peak-to-Average Ratio (same as PAPR)
PARC	Per-Antenna Rate Control
PBCH	Physical Broadcast Channel
PCCH	Paging Control Channel
PCFICH	Physical Control Format Indicator Channel
PCG	Project Coordination Group (in 3GPP)
PCH	Paging Channel
PCRF	Policy and Charging Rules Function
PCS	Personal Communications Systems
PDA	Personal Digital Assistant
PDC	Personal Digital Cellular
PDCCH	Physical Downlink Control Channel
PDCP	Packet Data Convergence Protocol
PDSCH	Physical Downlink Shared Channel
PDN	Packet Data Network
PDU	Protocol Data Unit
PF	Proportional Fair (a type of scheduler)
P-GW	Packet-Data Network Gateway (also PDN-GW)
PHICH	Physical Hybrid-ARQ Indicator Channel
PHS	Personal Handy-phone System
PHY	Physical layer
PMCH	Physical Multicast Channel
PMI	Precoding-Matrix Indicator
POTS	Plain Old Telephony Services
PRACH	Physical Random Access Channel
PRB	Physical Resource Block
P-RNTI	Paging RNTI
PS	Packet Switched
PSK	Phase Shift Keying
PSS	Primary Synchronization Signal
PSTN	Public Switched Telephone Networks
PUCCH	Physical Uplink Control Channel

PUSC	Partially Used Subcarriers (for WiMAX)
PUSCH	Physical Uplink Shared Channel
QAM	Quadrature Amplitude Modulation
QoS	Quality-of-Service
QPP	Quadrature Permutation Polynomial
QPSK	Quadrature Phase-Shift Keying
RAB	Radio Access Bearer
RACH	Random Access Channel
RAN	Radio Access Network
RA-RNTI	Random Access RNTI
RAT	Radio Access Technology
RB	Resource Block
RE	Reseource Element
RF	Radio Frequency
RI	Rank Indicator
RIT	Radio Interface Technology
RLC	Radio Link Control
RNC	Radio Network Controller
RNTI	Radio-Network Temporary Identifier
RNTP	Relative Narrowband Transmit Power
ROHC	Robust Header Compression
R-PDCCH	Relay Physical Downlink Control Channel
RR	Round-Robin (a type of scheduler)
RRC	Radio Resource Control
RRM	Radio Resource Management
RS	Reference Symbol
RSPC	IMT-2000 radio interface specifications
RSRP	Reference Signal Received Power
RSRQ	Reference Signal Received Quality
RTP	Real Time Protocol
RTT	Round-Trip Time
RV	Redundancy Version
RX	Receiver
S1	The interface between eNodeB and the Evolved Packet Core.
S1-c	The control-plane part of S1
S1-u	The user-plane part of S1
SAE	System Architecture Evolution
SCM	Spatial Channel Model
SDMA	Spatial Division Multiple Access
SDO	Standards Developing Organization
SDU	Service Data Unit
SEM	Spectrum Emissions Mask

SF	Spreading Factor
SFBC	Space-Frequency Block Coding
SFN	Single-Frequency Network (in general, see also MBSFN) or System Frame Number (in 3GPP)
SFTD	Space–Frequency Time Diversity
SGSN	Serving GPRS Support Node
S-GW	Serving Gateway
SI	System Information message
SIB	System Information Block
SIC	Successive Interference Combining
SIM	Subscriber Identity Module
SINR	Signal-to-Interference-and-Noise Ratio
SIR	Signal-to-Interference Ratio
SI-RNTI	System Information RNTI
SMS	Short Message Service
SNR	Signal-to-Noise Ratio
SOHO	Soft Handover
SORTD	Spatial Orthogonal-Resource Transmit Diversity
SR	Scheduling Request
SRS	Sounding Reference Signal
SSS	Secondary Synchronization Signal
STBC	Space–Time Block Coding
STC	Space–Time Coding
STTD	Space-Time Transmit Diversity
SU-MIMO	Single-User MIMO
TACS	Total Access Communication System
TCP	Transmission Control Protocol
TC-RNTI	Temporary C-RNTI
TD-CDMA	Time-Division Code-Division Multiple Access
TDD	Time-Division Duplex
TDM	Time-Division Multiplexing
TDMA	Time-Division Multiple Access
TD-SCDMA	Time-Division-Synchronous Code-Division Multiple Access
TF	Transport Format
TIA	Telecommunications Industry Association
TM	Transparent Mode (RLC configuration)
TR	Technical Report
TS	Technical Specification
TSG	Technical Specification Group
TTA	Telecommunications Technology Association
TTC	Telecommunications Technology Committee
TTI	Transmission Time Interval
TX	Transmitter

UCI	Uplink Control Information
UE	User Equipment, the 3GPP name for the mobile terminal
UL	Uplink
UL-SCH	Uplink Shared Channel
UM	Unacknowledged Mode (RLC configuration)
UMB	Ultra Mobile Broadband
UMTS	Universal Mobile Telecommunications System
UpPTS	The uplink part of the special subframe (for TDD operation).
US-TDMA	US Time-Division Multiple Access standard
UTRA	Universal Terrestrial Radio Access
UTRAN	Universal Terrestrial Radio Access Network
VAMOS	Voice services over Adaptive Multi-user channels
VoIP	Voice-over-IP
VRB	Virtual Resource Block
WAN	Wide Area Network
WARC	World Administrative Radio Congress
WCDMA	Wideband Code-Division Multiple Access
WG	Working Group
WiMAX	Worldwide Interoperability for Microwave Access
WLAN	Wireless Local Area Network
WMAN	Wireless Metropolitan Area Network
WP5D	Working Party 5D
WRC	World Radiocommunication Conference
X2	The interface between eNodeBs.
ZC	Zadoff-Chu
ZF	Zero Forcing

Background of LTE

1.1 INTRODUCTION

Mobile communications has become an everyday commodity. In the last decades, it has evolved from being an expensive technology for a few selected individuals to today's ubiquitous systems used by a majority of the world's population. From the first experiments with radio communication by Guglielmo Marconi in the 1890s, the road to truly mobile radio communication has been quite long. To understand the complex mobile-communication systems of today, it is important to understand where they came from and how cellular systems have evolved. The task of developing mobile technologies has also changed, from being a national or regional concern, to becoming an increasingly complex task undertaken by global standards-developing organizations such as the *Third Generation Partnership Project* (3GPP) and involving thousands of people.

Mobile communication technologies are often divided into generations, with 1G being the analog mobile radio systems of the 1980s, 2G the first digital mobile systems, and 3G the first mobile systems handling broadband data. The *Long-Term Evolution* (LTE) is often called "4G", but many also claim that LTE release 10, also referred to as *LTE-Advanced*, is the true 4G evolution step, with the first release of LTE (release 8) then being labeled as "3.9G". This continuing race of increasing sequence numbers of mobile system generations is in fact just a matter of labels. What is important is the actual system capabilities and how they have evolved, which is the topic of this chapter.

In this context, it must first be pointed out that LTE and LTE-Advanced is the same technology, with the "Advanced" label primarily being added to highlight the relation between LTE release 10 (LTE-Advanced) and ITU/IMT-Advanced, as discussed later. This does not make LTE-Advanced a different system than LTE and it is not in any way the final evolution step to be taken for LTE. Another important aspect is that the work on developing LTE and LTE-Advanced is performed as a continuing task within 3GPP, the same forum that developed the first 3G system (WCDMA/HSPA).

This chapter describes the background for the development of the LTE system, in terms of events, activities, organizations and other factors that have played an important role. First, the technologies and mobile systems leading up to the starting point for 3G mobile systems will be discussed. Next, international activities in the ITU that were part of shaping 3G and the 3G evolution and the market and technology drivers behind LTE will be discussed. The final part of the chapter describes the standardization process that provided the detailed specification work leading to the LTE systems deployed and in operation today.

4G LTE/LTE-Advanced for Mobile Broadband.

1.2 EVOLUTION OF MOBILE SYSTEMS BEFORE LTE

The US *Federal Communications Commission* (FCC) approved the first commercial car-borne telephony service in 1946, operated by AT&T. In 1947 AT&T also introduced the cellular concept of reusing radio frequencies, which became fundamental to all subsequent mobile-communication systems. Similar systems were operated by several monopoly telephone administrations and wire-line operators during the 1950s and 1960s, using bulky and power-hungry equipment and providing car-borne services for a very limited number of users.

The big uptake of subscribers and usage came when mobile communication became an international concern involving several interested parties, in the beginning mainly the operators. The first international mobile communication systems were started in the early 1980s; the best-known ones are NMT that was started up in the Nordic countries, AMPS in the USA, TACS in Europe, and J-TACS in Japan. Equipment was still bulky, mainly car-borne, and voice quality was often inconsistent, with "cross-talk" between users being a common problem. With NMT came the concept of "roaming", giving a service also for users traveling outside the area of their "home" operator. This also gave a larger market for mobile phones, attracting more companies into the mobile-communication business.

The analog first-generation cellular systems supported "plain old telephony services" (POTS) – that is, voice with some related supplementary services. With the advent of digital communication during the 1980s, the opportunity to develop a second generation of mobile-communication standards and systems, based on digital technology, surfaced. With digital technology came an opportunity to increase the capacity of the systems, to give a more consistent quality of the service, and to develop much more attractive and truly mobile devices.

In Europe, the GSM (originally *Groupe Spécial Mobile*, later *Global System for Mobile communications*) project to develop a pan-European mobile-telephony system was initiated in the mid 1980s by the telecommunication administrations in CEPT[1] and later continued within the new *European Telecommunication Standards Institute* (ETSI). The GSM standard was based on *Time-Division Multiple Access* (TDMA), as were the US-TDMA standard and the Japanese PDC standard that were introduced in the same time frame. A somewhat later development of a *Code-Division Multiple Access* (CDMA) standard called IS-95 was completed in the USA in 1993.

All these standards were "narrowband" in the sense that they targeted "low-bandwidth" services such as voice. With the second-generation digital mobile communications came also the opportunity to provide data services over the mobile-communication networks. The primary data services introduced in 2G were text messaging (*Short Message Services*, SMS) and circuit-switched data services enabling e-mail and other data applications, initially at a modest peak data rate of 9.6 kbit/s. Higher data rates were introduced later in evolved 2G systems by assigning multiple time slots to a user and through modified coding schemes.

Packet data over cellular systems became a reality during the second half of the 1990s, with *General Packet Radio Services* (GPRS) introduced in GSM and packet data also added to other cellular technologies such as the Japanese PDC standard. These technologies are often referred to as 2.5G. The success of the wireless data service iMode in Japan, which included a complete "ecosystem"

---

[1] The European Conference of Postal and Telecommunications Administrations (CEPT) consists of the telecom administrations from 48 countries.

for service delivery, charging etc., gave a very clear indication of the potential for applications over packet data in mobile systems, in spite of the fairly low data rates supported at the time.

With the advent of 3G and the higher-bandwidth radio interface of UTRA (*Universal Terrestrial Radio Access*) came possibilities for a range of new services that were only hinted at with 2G and 2.5G. The 3G radio access development is today handled in 3GPP. However, the initial steps for 3G were taken in the early 1990s, long before 3GPP was formed.

What also set the stage for 3G was the internationalization of cellular standards. GSM was a pan-European project, but it quickly attracted worldwide interest when the GSM standard was deployed in a number of countries outside Europe. A global standard gains in economy of scale, since the market for products becomes larger. This has driven a much tighter international cooperation around 3G cellular technologies than for the earlier generations.

1.2.1 The First 3G Standardization

Work on a third-generation mobile communication started in ITU (*International Telecommunication Union*) in the 1980s, first under the label *Future Public Land Mobile Telecommunications Systems* (FPLMTS), later changed to IMT-2000 [1]. The World Administrative Radio Congress WARC-92 identified 230 MHz of spectrum for IMT-2000 on a worldwide basis. Of these 230 MHz, 2×60 MHz was identified as paired spectrum for FDD (*Frequency-Division Duplex*) and 35 MHz as unpaired spectrum for TDD (*Time-Division Duplex*), both for terrestrial use. Some spectrum was also set aside for satellite services. With that, the stage was set to specify IMT-2000.

In parallel with the widespread deployment and evolution of 2G mobile-communication systems during the 1990s, substantial efforts were put into 3G research activities worldwide. In Europe, a number of partially EU-funded projects resulted in a multiple access concept that included a Wideband CDMA component that was input to ETSI in 1996. In Japan, the *Association of Radio Industries and Businesses* (ARIB) was at the same time defining a 3G wireless communication technology based on Wideband CDMA and also in the USA a Wideband CDMA concept called WIMS was developed within the T1.P1[2] committee. South Korea also started work on Wideband CDMA at this time.

When the standardization activities for 3G started in ETSI in 1996, there were WCDMA concepts proposed both from a European research project (FRAMES) and from the ARIB standardization in Japan. The Wideband CDMA proposals from Europe and Japan were merged and came out as part of the winning concept in early 1998 in the European work on Universal Mobile Telecommunication Services (UMTS), which was the European name for 3G. Standardization of WCDMA continued in parallel in several standards groups until the end of 1998, when the *Third Generation Partnership Project* (3GPP) was formed by standards-developing organizations from all regions of the world. This solved the problem of trying to maintain parallel development of aligned specifications in multiple regions. The present organizational partners of 3GPP are ARIB (Japan), CCSA (China), ETSI (Europe), ATIS (USA), TTA (South Korea), and TTC (Japan).

At this time, when the standardization bodies were ready to put the details into the 3GPP specifications, work on 3G mobile systems had already been ongoing for some time in the international arena within the ITU-R. That work was influenced by and also provided a broader international framework for the standardization work in 3GPP.

---

[2]The T1.P1 committee was part of T1, which presently has joined the ATIS standardization organization.

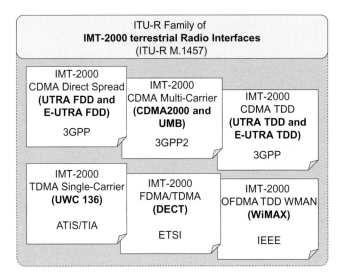

FIGURE 1.1

The definition of IMT-2000 in ITU-R.

1.3 **ITU ACTIVITIES**

1.3.1 **IMT-2000 and IMT-Advanced**

ITU-R *Working Party 5D* (WP5D) has the responsibility for IMT systems, which is the umbrella name for 3G (IMT-2000) and 4G (IMT-Advanced). WP5D does not write technical specifications for IMT, but has kept the role of defining IMT in cooperation with the regional standardization bodies and to maintain a set of recommendations for IMT-2000 and IMT-Advanced.

The main IMT-2000 recommendation is ITU-R M.1457 [2], which identifies the IMT-2000 *radio interface specifications* (RSPC). The recommendation contains a "family" of radio interfaces, all included on an equal basis. The family of six terrestrial radio interfaces is illustrated in Figure 1.1, which also shows the *Standards Developing Organizations* (SDO) or Partnership Projects that produce the specifications. In addition, there are several IMT-2000 satellite radio interfaces defined, not illustrated in Figure 1.1.

For each radio interface, M.1457 contains an overview of the radio interface, followed by a list of references to the detailed specifications. The actual specifications are maintained by the individual SDOs and M.1457 provides references to the specifications maintained by each SDO.

With the continuing development of the IMT-2000 radio interfaces, including the evolution of UTRA to Evolved UTRA, the ITU recommendations also need to be updated. ITU-R WP5D continuously revises recommendation M.1457 and at the time of writing it is in its ninth version. Input to the updates is provided by the SDOs and Partnership Projects writing the standards. In the latest revision of ITU-R M.1457, LTE (or E-UTRA) is included in the family through the 3GPP family members for UTRA FDD and TDD, as shown in the figure.

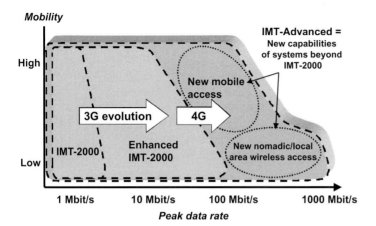

FIGURE 1.2

Illustration of capabilities of IMT-2000 and IMT-Advanced, based on the framework described in ITU -R Recommendation M.1645 [4].

In addition to maintaining the IMT-2000 specifications, a main activity in ITU-R WP5D is the work on systems beyond IMT-2000, now called IMT-Advanced. The term IMT-Advanced is used for systems that include new radio interfaces supporting the new capabilities of systems beyond IMT-2000, as demonstrated with the "van diagram" in Figure 1.2. The step into IMT-Advanced capabilities is seen by ITU-R as the step into 4G, the next generation of mobile technologies after 3G.

The process for defining IMT-Advanced was set by ITU-R WP5D [3] and was quite similar to the process used in developing the IMT-2000 recommendations. ITU-R first concluded studies for IMT-Advanced of services and technologies, market forecasts, principles for standardization, estimation of spectrum needs, and identification of candidate frequency bands [4]. Evaluation criteria were agreed, where proposed technologies were to be evaluated according to a set of minimum technical requirements. All ITU members and other organizations were then invited to the process through a circular letter [5] in March 2008. After submission of six candidate technologies in 2009, an evaluation was performed in cooperation with external bodies such as standards-developing organizations, industry forums, and national groups.

An evolution of LTE as developed by 3GPP was submitted as one candidate to the ITU-R evaluation. While actually being a new release (release 10) of the LTE system and thus an integral part of the continuing LTE development, the candidate was named LTE-Advanced for the purpose of ITU submission. 3GPP also set up its own set of technical requirements for LTE-Advanced, with the ITU-R requirements as a basis. The specifics of LTE-Advanced will be described in more detail as part of the description of LTE later in this book. The performance evaluation of LTE-Advanced for the ITU-R submission is described further in Chapter 18.

The target of the process was always harmonization of the candidates through consensus building. ITU-R determined in October 2010 that two technologies will be included in the first release of IMT-Advanced, those two being LTE release 10 ("LTE-Advanced") and WirelessMAN-Advanced [6] based on the IEEE 802.16m specification. The two can be viewed as the "family" of IMT-Advanced

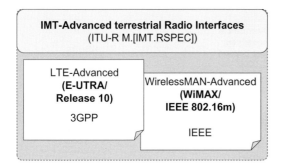

FIGURE 1.3

Radio interface technologies for IMT-Advanced.

technologies as shown in Figure 1.3. The main IMT-Advanced recommendation, identifying the IMT-Advanced radio interface specifications, is presently named ITU-R.[IMT.RSPEC] [7] and will be completed during 2011. As for the corresponding IMT-2000 specification, it will contain an overview of each radio interface, followed by a list of references to the detailed specifications.

1.3.2 Spectrum for IMT Systems

Another major activity within ITU-R concerning IMT-Advanced has been to identify globally available spectrum, suitable for IMT systems. The spectrum work has involved sharing studies between IMT and other technologies in those bands. Adequate spectrum availability and globally harmonized spectrum are identified as essential for IMT-Advanced.

Spectrum for 3G was first identified at the *World Administrative Radio Congress* WARC-92, where 230 MHz was identified as intended for use by national administrations that want to implement IMT-2000. The so-called IMT-2000 "core band" at 2 GHz is in this frequency range and was the first band where 3G systems were deployed.

Additional spectrum was identified for IMT-2000 at later World Radio communication conferences. WRC-2000 identified the existing 2G bands at 800/900 MHz and 1800/1900 MHz plus an additional 190 MHz of spectrum at 2.6 GHz, all for IMT-2000. As additional spectrum for IMT-2000, WRC'07 identified a band at 450 MHz, the so-called "digital dividend" at 698–806 MHz, plus an additional 300 MHz of spectrum at higher frequencies. The applicability of these new bands varies on a regional and national basis.

The worldwide frequency arrangements for IMT-2000 are outlined in ITU-R recommendation M.1036 [8], which is presently being updated with the arrangements for the most recent frequency bands added at WRC'07. The recommendation outlines the regional variations in how the bands are implemented and also identifies which parts of the spectrum are paired and which are unpaired. For the paired spectrum, the bands for uplink (mobile transmit) and downlink (base-station transmit) are identified for *Frequency-Division Duplex* (FDD) operation. The unpaired bands can, for example, be used for *Time-Division Duplex* (TDD) operation. Note that the band that is most globally deployed for 3G is still 2 GHz.

1.4 **DRIVERS FOR LTE**

The evolution of 3G systems into 4G is driven by the creation and development of new services for mobile devices, and is enabled by advancement of the technology available for mobile systems. There has also been an evolution of the environment in which mobile systems are deployed and operated, in terms of competition between mobile operators, challenges from other mobile technologies, and new regulation of spectrum use and market aspects of mobile systems.

The rapid evolution of the technology used in telecommunication systems, consumer electronics, and specifically mobile devices has been remarkable in the last 20 years. Moore's law illustrates this and indicates a continuing evolution of processor performance and increased memory size, often combined with reduced size, power consumption, and cost for devices. High-resolution color displays and megapixel camera sensors are also coming into all types of mobile devices. Combined with a high-speed internet backbone often based on optical fiber networks, we see that a range of technology enablers are in place to go hand-in-hand with advancement in mobile communications technology such as LTE.

The rapid increase in use of the internet to provide all kinds of services since the 1990s started at the same time as 2G and 3G mobile systems came into widespread use. The natural next step was that those internet-based services also moved to the mobile devices, creating what is today know as *mobile broadband*. Being able to support the same *Internet Protocol* (IP)-based services in a mobile device that people use at home with a fixed broadband connection is a major challenge and a prime driver for the evolution of LTE. A few services were already supported by the evolved 2.5G systems, but it is not until the systems are designed primarily for IP-based services that the real mobile IP revolution can take off. An interesting aspect of the migration of broadband services to mobile devices is that a mobile "flavor" is also added. The mobile position and the mobility and roaming capabilities do in fact create a whole new range of services tailored to the mobile environment.

Fixed telephony (POTS) and earlier generations of mobile technology were built for circuit switched services, primarily voice. The first data services over GSM were circuit switched, with packet-based GPRS coming in as a later addition. This also influenced the first development of 3G, which was based on circuit switched data, with packet-switched services as an add-on. It was not until the 3G evolution into HSPA and later LTE/LTE-Advanced that packet-switched services and IP were made the primary design target. The old circuit-switched services remain, but will on LTE be provided over IP, with Voice-over IP (VoIP) as an example.

IP is in itself service agnostic and thereby enables a range of services with different requirements. The main service-related design parameters for a radio interface supporting a variety of services are:

- **Data rate.** Many services with lower data rates such as voice services are important and still occupy a large part of a mobile network's overall capacity, but it is the higher data rate services that drive the design of the radio interface. The ever increasing demand for higher data rates for web browsing, streaming and file transfer pushes the peak data rates for mobile systems from kbit/s for 2G, to Mbit/s for 3G and getting close to Gbit/s for 4G.
- **Delay.** Interactive services such as real-time gaming, but also web browsing and interactive file transfer, have requirements for very low delay, making it a primary design target. There are, however, many applications such as e-mail and television where the delay requirements are not as strict. The delay for a packet sent from a server to a client and back is called *latency*.

- **Capacity.** From the mobile system operator's point of view, it is not only the peak data rates provided to the end-user that are of importance, but also the total data rate that can be provided on average from each deployed base station site and per hertz of licensed spectrum. This measure of capacity is called *spectral efficiency*. In the case of capacity shortage in a mobile system, the Quality-of-Service (QoS) for the individual end-users may be degraded.

How these three main design parameters influenced the development of LTE is described in more detail in Chapter 7 and an evaluation of what performance is achieved for the design parameters above is presented in Chapter 18.

The demand for new services and for higher peak bit rates and system capacity is not only met by evolution of the technology to 4G. There is also a demand for more spectrum resources to expand systems and the demand also leads to more competition between an increasing number of mobile operators and between alternative technologies to provide mobile broadband services. An overview of some technologies other than LTE is given in Chapter 19.

With more spectrum coming into use for mobile broadband, there is a need to operate mobile systems in a number of different frequency bands, in spectrum allocations of different sizes and sometimes also in fragmented spectrum. This calls for high spectrum flexibility with the possibility for a varying channel bandwidth, which was also a driver and an essential design parameter for LTE.

The demand for new mobile services and the evolution of the radio interface to LTE have served as drivers to evolve the core network. The core network developed for GSM in the 1980s was extended to support GPRS, EDGE, and WCDMA in the 1990s, but was still very much built around the circuit-switched domain. A *System Architecture Evolution* (SAE) was initiated at the same time as LTE development started and has resulted in an *Evolved Packet Core* (EPC), developed to support HSPA and LTE/LTE-Advanced, focusing on the packet-switched domain. For more details on SAE/EPC, please refer to [9].

1.5 STANDARDIZATION OF LTE

With a framework for IMT systems set up by the ITU-R, with spectrum made available by the WRC and with an ever increasing demand for better performance, the task of specifying the LTE system that meets the design targets falls on 3GPP. 3GPP writes specifications for 2G, 3G, and 4G mobile systems, and 3GPP technologies are the most widely deployed in the world, with more than 4.5 billion connections in 2010. In order to understand how 3GPP works, it is important to understand the process of writing standards.

1.5.1 The Standardization Process

Setting a standard for mobile communication is not a one-time job, it is an ongoing process. The standardization forums are constantly evolving their standards trying to meet new demands for services and features. The standardization process is different in the different forums, but typically includes the four phases illustrated in Figure 1.4:

1. *Requirements*, where it is decided what is to be achieved by the standard.
2. *Architecture*, where the main building blocks and interfaces are decided.

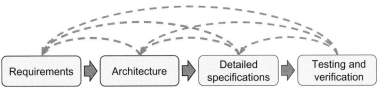

FIGURE 1.4

The standardization phases and iterative process.

3. *Detailed specifications*, where every interface is specified in detail.
4. *Testing and verification*, where the interface specifications are proven to work with real-life equipment.

These phases are overlapping and iterative. As an example, requirements can be added, changed, or dropped during the later phases if the technical solutions call for it. Likewise, the technical solution in the detailed specifications can change due to problems found in the testing and verification phase.

Standardization starts with the *requirements* phase, where the standards body decides what should be achieved with the standard. This phase is usually relatively short.

In the *architecture* phase, the standards body decides about the architecture – that is, the principles of how to meet the requirements. The architecture phase includes decisions about reference points and interfaces to be standardized. This phase is usually quite long and may change the requirements.

After the architecture phase, the *detailed specification* phase starts. It is in this phase the details for each of the identified interfaces are specified. During the detailed specification of the interfaces, the standards body may find that previous decisions in the architecture or even in the requirements phases need to be revisited.

Finally, the *testing and verification* phase starts. It is usually not a part of the actual standardization in the standards bodies, but takes place in parallel through testing by vendors and interoperability testing between vendors. This phase is the final proof of the standard. During the testing and verification phase, errors in the standard may still be found and those errors may change decisions in the detailed standard. Albeit not common, changes may also need to be made to the architecture or the requirements. To verify the standard, products are needed. Hence, the implementation of the products starts after (or during) the detailed specification phase. The testing and verification phase ends when there are stable test specifications that can be used to verify that the equipment is fulfilling the standard.

Normally, it takes one to two years from the time when the standard is completed until commercial products are out on the market. However, if the standard is built from scratch, it may take longer since there are no stable components to build from.

1.5.2 The 3GPP Process

The *Third-Generation Partnership Project* (3GPP) is the standards-developing body that specifies the LTE/LTE-Advanced, as well as 3G UTRA and 2G GSM systems. 3GPP is a partnership project formed by the standards bodies ETSI, ARIB, TTC, TTA, CCSA, and ATIS. 3GPP consists of four *Technical Specifications Groups* (TSGs) – see Figure 1.5.

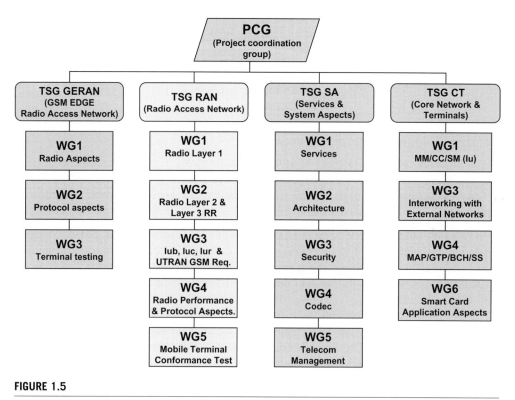

FIGURE 1.5

3GPP organization.

A parallel partnership project called 3GPP2 was formed in 1999. It also develops 3G specifications, but for CDMA2000, which is the 3G technology developed from the 2G CDMA-based standard IS-95. It is also a global project, and the organizational partners are ARIB, CCSA, TIA, TTA, and TTC.

3GPP TSG RAN (*Radio Access Network*) is the technical specification group that has developed WCDMA, its evolution HSPA, as well as LTE/LTE-Advanced, and is in the forefront of the technology. TSG RAN consists of five working groups (WGs):

1. RAN WG1, dealing with the physical layer specifications.
2. RAN WG2, dealing with the layer 2 and layer 3 radio interface specifications.
3. RAN WG3, dealing with the fixed RAN interfaces – for example, interfaces between nodes in the RAN – but also the interface between the RAN and the core network.
4. RAN WG4, dealing with the *radio frequency* (RF) and *radio resource management* (RRM) performance requirements.
5. RAN WG 5, dealing with the terminal conformance testing.

The scope of 3GPP when it was formed in 1998 was to produce global specifications for a 3G mobile system based on an evolved GSM core network, including the WCDMA-based radio access of the UTRA FDD and the TD-CDMA-based radio access of the UTRA TDD mode. The task to maintain and develop the GSM/EDGE specifications was added to 3GPP at a later stage and the work now also includes LTE (E-UTRA). The UTRA, E-UTRA and GSM/EDGE specifications are developed, maintained, and approved in 3GPP. After approval, the organizational partners transpose them into appropriate deliverables as standards in each region.

In parallel with the initial 3GPP work, a 3G system based on TD-SCDMA was developed in China. TD-SCDMA was eventually merged into release 4 of the 3GPP specifications as an additional TDD mode.

The work in 3GPP is carried out with relevant ITU recommendations in mind and the result of the work is also submitted to ITU. The organizational partners are obliged to identify regional requirements that may lead to options in the standard. Examples are regional frequency bands and special protection requirements local to a region. The specifications are developed with global roaming and circulation of terminals in mind. This implies that many regional requirements in essence will be global requirements for all terminals, since a roaming terminal has to meet the strictest of all regional requirements. Regional options in the specifications are thus more common for base stations than for terminals.

The specifications of all releases can be updated after each set of TSG meetings, which occur four times a year. The 3GPP documents are divided into releases, where each release has a set of features added compared to the previous release. The features are defined in Work Items agreed and undertaken by the TSGs. The releases from release 8 and onwards, with some main features listed for LTE, are shown in Figure 1.6. The date shown for each release is the day the content of the release was frozen. Release 10 of LTE is the version approved by ITU-R as an IMT-Advanced technology and is therefore also named *LTE-Advanced*.

1.5.3 The 3G Evolution to 4G

The first release of WCDMA Radio Access developed in TSG RAN was called release 99[3] and contained all features needed to meet the IMT-2000 requirements as defined by the ITU. This included circuit-switched voice and video services, and data services over both packet-switched and circuit-switched bearers. The first major addition of radio access features to WCDMA was HSPA, which was added in release 5 with *High Speed Downlink Packet Access* (HSDPA) and release 6 with *Enhanced Uplink*. These two are together referred to as HSPA and an overview of HSPA is given in Chapter 19 of this book. With HSPA, UTRA goes beyond the definition of a 3G mobile system and also encompasses broadband mobile data.

The 3G evolution continued in 2004, when a workshop was organized to initiate work on the 3GPP *Long-Term Evolution* (LTE) radio interface. The result of the LTE workshop was that a study item in 3GPP TSG RAN was created in December 2004. The first 6 months were spent on defining the requirements, or design targets, for LTE. These were documented in a 3GPP technical report [10] and approved in June 2005. Most notable are the requirements on high data rate at the cell edge and the importance of low delay, in addition to the normal capacity and peak data rate requirements. Furthermore, spectrum flexibility and maximum commonality between FDD and TDD solutions are pronounced.

---

[3]For historical reasons, the first 3GPP release is named after the year it was frozen (1999), while the following releases are numbered 4, 5, 6, etc.

During the fall of 2005, 3GPP TSG RAN WG1 made extensive studies of different basic physical layer technologies and in December 2005 the TSG RAN plenary decided that the LTE radio access should be based on OFDM in the downlink and DFT-precoded OFDM in the uplink. TSG RAN and its working groups then worked on the LTE specifications and the specifications were approved in December 2007. Work has since then continued on LTE, with new features added in each release, as shown in Figure 1.6. Chapters 7–17 will go through the details of the LTE radio interface in more detail.

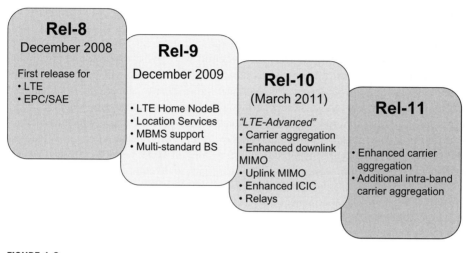

FIGURE 1.6

Releases of 3GPP specifications for LTE.

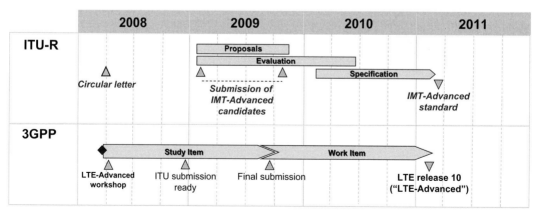

FIGURE 1.7

3GPP time schedule for LTE-Advanced in relation to ITU time-schedule on IMT-Advanced.

The work on IMT-Advanced within ITU-R WP5D came in 2008 into a phase where the detailed requirements and process were announced through a circular letter [5]. Among other things, this triggered activities in 3GPP, where a study item on *LTE-Advanced* was started. The task was to define requirements and investigate and propose technology components to be part of LTE-Advanced. The work was turned into a Work Item in 2009 in order to develop the detailed specifications.

Within 3GPP, LTE-Advanced is seen as the next major step in the evolution of LTE. LTE-Advanced is therefore not a new technology; it is an evolutionary step in the continuing development of LTE. As shown in Figure 1.6, the features that form LTE-Advanced are part of release 10 of 3GPP LTE specifications. Wider bandwidth through aggregation of multiple carriers and evolved use of advanced antenna techniques in both uplink and downlink are the major components added in LTE release 10 to reach the IMT-Advanced targets.

The work on LTE-Advanced within 3GPP is planned with the ITU-R time frame in mind, as shown in Figure 1.7. LTE-Advanced was submitted as a candidate to the ITU-R in 2009 and is now included in the set of radio interface technologies announced by ITU-R [6] in October 2010 to be included as a part of the IMT-Advanced radio interface specifications. This is very much aligned with what was from the start stated as a goal for LTE, namely that *LTE should provide the starting point for a smooth transition to 4G (= IMT-Advanced) radio access*.

Since LTE-Advanced is an integral part of 3GPP LTE release 10, it is described in detail together with the corresponding components of LTE in Chapters 7–17 of this book.

High Data Rates in Mobile Communication

As discussed in Chapter 1, one main target for the evolution of mobile communication is to provide the possibility for significantly higher end-user data rates compared to what is achievable with, for example, the first releases of the 3G standards. This includes the possibility for higher peak data rates but, as pointed out in the previous chapter, even more so the possibility for significantly higher data rates over the entire cell area, also including, for example, users at the cell edge. The initial part of this chapter will briefly discuss some of the more fundamental constraints that exist in terms of what data rates can actually be achieved in different scenarios. This will provide a background to subsequent discussions in the later part of the chapter, as well as in the subsequent chapters, concerning different means to increase the achievable data rates in different mobile-communication scenarios.

2.1 HIGH DATA RATES: FUNDAMENTAL CONSTRAINTS

In Ref. [11], Shannon provided the basic theoretical tools needed to determine the maximum rate, also known as the *channel capacity*, by which information can be transferred over a given communication channel. Although relatively complicated in the general case, for the special case of communication over a channel, for example a radio link, only impaired by additive white Gaussian noise, the channel capacity C is given by the relatively simple expression [12]:

$$C = BW \cdot \log_2 \left(1 + \frac{S}{N} \right),$$

(2.1)

where BW is the bandwidth available for the communication, S denotes the received signal power, and N denotes the power of the white noise impairing the received signal.

Already from Eqn (2.1) it should be clear that the two fundamental factors limiting the achievable data rate are the available received signal power, or more generally the available signal-power-to-noise-power ratio S/N, and the available bandwidth. To further clarify how and when these factors limit the achievable data rate, assume communication with a certain information rate R. The received signal power can then be expressed as $S = E_b \cdot R$, where E_b is the received energy per information bit. Furthermore, the noise power can be expressed as $N = N_0 \cdot BW$, where N_0 is the constant noise power spectral density measured in W/Hz.

Clearly, the information rate can never exceed the channel capacity. Together with the above expressions for the received signal power and noise power, this leads to the inequality:

$$R \leq C = BW \cdot \log_2\left(1 + \frac{S}{N}\right) = BW \cdot \log_2\left(1 + \frac{E_b \cdot R}{N_0 \cdot BW}\right) \tag{2.2}$$

or, by defining the radio-link *bandwidth utilization* $\gamma = R/BW$,

$$\gamma \leq \log_2\left(1 + \gamma \cdot \frac{E_b}{N_0}\right). \tag{2.3}$$

This inequality can be reformulated to provide a lower bound on the required received energy per information bit, normalized to the noise power density, for a given bandwidth utilization γ:

$$\frac{E_b}{N_0} \geq \min\left\{\frac{E_b}{N_0}\right\} = \frac{2^\gamma - 1}{\gamma}. \tag{2.4}$$

The rightmost expression – that is, the minimum required E_b/N_0 at the receiver as a function of the bandwidth utilization – is illustrated in Figure 2.1. As can be seen, for bandwidth utilizations significantly less than 1 – that is, for information rates substantially smaller than the utilized bandwidth – the minimum required E_b/N_0 is relatively constant, regardless of γ. For a given noise power density, any increase of the information data rate then implies a similar relative increase in the minimum required signal power $S = E_b \cdot R$ at the receiver. On the other hand, for bandwidth utilizations larger than 1, the minimum required E_b/N_0 increases rapidly with γ. Thus, in the case of data rates of the same order as or larger than the communication bandwidth, any further increase of the information data rate, without a corresponding increase in the available bandwidth, implies a larger, eventually much larger, relative increase in the minimum required received signal power.

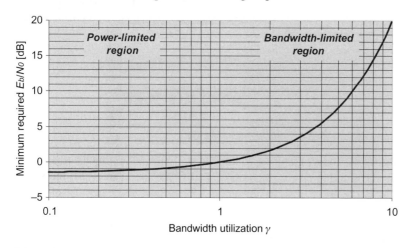

FIGURE 2.1

Minimum required E_b/N_0 at the receiver as a function of bandwidth utilization.

2.1.1 **High Data Rates in Noise-Limited Scenarios**

From the discussion above, some basic conclusions can be drawn regarding the provisioning of higher data rates in a mobile-communication system when noise is the main source of radio-link impairment (a *noise-limited* scenario):

- The data rates that can be provided in such scenarios are always limited by the available received signal power or, in the general case, the received signal-power-to-noise-power ratio. Furthermore, any increase of the achievable data rate within a given bandwidth will require *at least* the same relative increase of the received signal power. At the same time, if sufficient received signal power can be made available, basically any data rate can, at least in theory, be provided within a given limited bandwidth.
- In the case of low-bandwidth utilization – that is, as long as the radio-link data rate is substantially lower than the available bandwidth – any further increase of the data rate requires *approximately the same* relative increase in the received signal power. This can be referred to as *power-limited* operation (in contrast to *bandwidth-limited* operation; see below) as, in this case, an increase in the available bandwidth does not substantially impact what received signal power is required for a certain data rate.
- On the other hand, in the case of high-bandwidth utilization – that is, in the case of data rates of the same order as or exceeding the available bandwidth – any further increase in the data rate requires a *much larger* relative increase in the received signal power unless the bandwidth is increased in proportion to the increase in data rate. This can be referred to as *bandwidth-limited operation* as, in this case, an increase in the bandwidth will reduce the received signal power required for a certain data rate.

Thus, to make efficient use of the available received signal power or, in the general case, the available signal-to-noise ratio, the transmission bandwidth should at least be of the same order as the data rates to be provided.

Assuming a constant transmit power, the received signal power can always be increased by reducing the distance between the transmitter and the receiver, thereby reducing the attenuation of the signal as it propagates from the transmitter to the receiver. Thus, in a noise-limited scenario it is, at least in theory, always possible to increase the achievable data rates, assuming that one is prepared to accept a reduction in the transmitter/receiver distance – that is, a reduced range. In a mobile-communication system this would correspond to a reduced cell size and thus the need for more cell sites to cover the same overall area. In particular, providing data rates of the same order as or larger than the available bandwidth – that is, with a high-bandwidth utilization – would require a significant cell-size reduction. Alternatively, one has to accept that the high data rates are only available for terminals in the center of the cell and not over the entire cell area.

Another means to increase the overall received signal power for a given transmit power is the use of additional antennas at the receiver side, also known as *receive-antenna diversity*. Multiple receive antennas can be applied at the base station (that is, for the uplink) or at the terminal (that is, for the downlink). By proper combination of the signals received at the different antennas, the signal-to-noise ratio after the antenna combination can be increased in proportion to the number of receive antennas, thereby allowing for higher data rates for a given transmitter/receiver distance.

Multiple antennas can also be applied at the transmitter side, typically at the base station, and be used to focus a given total transmit power in the direction of the receiver – that is, toward the target terminal. This will increase the received signal power and thus, once again, allow for higher data rates for a given transmitter/receiver distance.

However, providing higher data rates by the use of multiple transmit or receive antennas is only efficient up to a certain level – that is, as long as the data rates are power limited rather than bandwidth limited. Beyond this point, the achievable data rates start to saturate and any further increase in the number of transmit or receive antennas, although leading to a correspondingly improved signal-to-noise ratio at the receiver, will only provide a marginal increase in the achievable data rates. This saturation in achievable data rates can be avoided though, by the use of multiple antennas at both the transmitter *and* the receiver, enabling what can be referred to as *spatial multiplexing*, often also referred to as *MIMO* (Multiple-Input Multiple-Output). Different types of multi-antenna techniques, including spatial multiplexing, will be discussed in more detail in Chapter 5. Multi-antenna techniques for the specific case of LTE are discussed in Chapters 10 and 11.

An alternative to increasing the received signal power is to reduce the noise power, or more exactly the noise power density, at the receiver. This can, at least to some extent, be achieved by more advanced receiver RF design, allowing for a reduced receiver noise figure.

2.1.2 Higher Data Rates in Interference-Limited Scenarios

The discussion above assumed communication over a radio link only impaired by noise. However, in actual mobile-communication scenarios, interference from transmissions in neighboring cells, also referred to as *inter-cell interference*, is often the dominant source of radio-link impairment, more so than noise. This is especially the case in small-cell deployments with a high traffic load. Furthermore, in addition to inter-cell interference there may in some cases also be interference from other transmissions *within the current cell*, also referred to as *intra-cell interference*.

In many respects the impact of interference on a radio link is similar to that of noise. In particular, the basic principles discussed above apply also to a scenario where interference is the main radio-link impairment:

- The maximum data rate that can be achieved in a given bandwidth is limited by the available signal-power-to-interference-power ratio.
- Providing data rates larger than the available bandwidth (high-bandwidth utilization) is costly in the sense that it requires a disproportionately high signal-to-interference ratio.

Also, similar to a scenario where noise is the dominant radio-link impairment, reducing the cell size as well as the use of multi-antenna techniques are key means to increase the achievable data rates in an interference-limited scenario:

- Reducing the cell size will obviously reduce the number of users, and thus also the overall traffic, per cell. This will reduce the relative interference level and thus allow for higher data rates.
- Similar to the increase in signal-to-noise ratio, proper combination of the signals received at multiple antennas will also increase the signal-to-interference ratio after the antenna combination.
- The use of beam-forming by means of multiple transmit antennas will focus the transmit power in the direction of the target receiver, leading to reduced interference to other radio links and thus improving the overall signal-to-interference ratio in the system.

One important difference between interference and noise is that interference, in contrast to noise, typically has a certain structure which makes it, at least to some extent, predictable and thus possible to further suppress or even remove completely. As an example, a dominant interfering signal may arrive from a certain direction, in which case the corresponding interference can be further suppressed, or even completely removed, by means of *spatial processing* using multiple antennas at the receiver. This will be further discussed in Chapter 5. Also, any differences in the spectral properties between the target signal and an interfering signal can be used to suppress the interferer and thus reduce the overall interference level.

2.2 HIGHER DATA RATES WITHIN A LIMITED BANDWIDTH: HIGHER-ORDER MODULATION

As discussed in the previous section, providing data rates larger than the available bandwidth is fundamentally inefficient in the sense that it requires disproportionately high signal-to-noise and signal-to-interference ratios at the receiver. Still, bandwidth is often a scarce and expensive resource and, at least in some mobile-communication scenarios, high signal-to-noise and signal-to-interference ratios can be made available, for example in small-cell environments with a low traffic load or for terminals close to the cell site. Mobile-communication systems should preferably be designed to be able to take advantage of such scenarios – that is, they should be able to offer very high data rates within a limited bandwidth when the radio conditions so allow.

A straightforward means to provide higher data rates within a given transmission bandwidth is the use of *higher-order modulation*, implying that the modulation alphabet is extended to include additional signaling alternatives and thus allowing for more bits of information to be communicated per modulation symbol.

In the case of QPSK modulation, which is the modulation scheme used for the downlink in the first releases of the 3G mobile-communication standards (WCDMA and CDMA2000), the modulation alphabet consists of four different signaling alternatives. These four signaling alternatives can be illustrated as four different points in a two-dimensional plane (see Figure 2.2a). With four different signaling alternatives, QPSK allows for up to 2 bits of information to be communicated during each modulation-symbol interval. By extending to 16QAM modulation (Figure 2.2b), 16 different signaling alternatives are available. The use of 16QAM thus allows for up to 4 bits of information to be communicated per symbol interval. Further extension to 64QAM (Figure 2.2c), with 64 different signaling alternatives, allows for up to 6 bits of information to be communicated per symbol interval. At the same time, the bandwidth of the transmitted signal is, at least in principle, independent of the size of the modulation alphabet and mainly depends on the modulation rate – that is, the number of modulation symbols per second. The maximum bandwidth utilization, expressed in bits/s/Hz, of 16QAM and 64QAM are thus, at least in principle, two and three times that of QPSK respectively.

It should be pointed out that there are many other possible modulation schemes, in addition to those illustrated in Figure 2.2. One example is 8PSK, consisting of eight signaling alternatives and thus providing up to 3 bits of information per modulation symbol. Readers are referred to [12] for a more thorough discussion on different modulation schemes.

The use of higher-order modulation provides the possibility for higher bandwidth utilization – that is, the possibility to provide higher data rates within a given bandwidth. However, the higher bandwidth utilization comes at the cost of reduced robustness to noise and interference. Alternatively expressed, higher-order modulation schemes, such as 16QAM or 64QAM, require a higher E_b/N_0 at the receiver for a given bit-error probability, compared to QPSK. This is in line with the discussion in

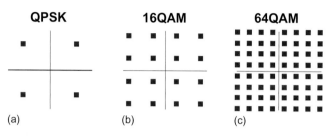

FIGURE 2.2

Signal constellations for: (a) QPSK; (b) 16QAM; (c) 64QAM.

the previous section, where it was concluded that high-bandwidth utilization – that is, a high informa-tion rate within a limited bandwidth – in general requires a higher receiver E_b/N_0.

2.2.1 Higher-Order Modulation in Combination with Channel Coding

Higher-order modulation schemes such as 16QAM and 64QAM require, in themselves, a higher receiver E_b/N_0 for a given error rate, compared to QPSK. However, in combination with channel cod-ing the use of higher-order modulation will sometimes be more efficient – that is, require a lower receiver E_b/N_0 for a given error rate – compared to the use of lower-order modulation such as QPSK. This may, for example, occur when the target bandwidth utilization implies that, with lower-order modulation, no or very little channel coding can be applied. In such a case, the additional channel coding that can be applied by using a higher-order modulation scheme such as 16QAM may lead to an overall gain in power efficiency compared to the use of QPSK.

As an example, if a bandwidth utilization of close to 2 information bits per modulation symbol is required, QPSK modulation would allow for very limited channel coding (channel-coding rate close to 1). On the other hand, the use of 16QAM modulation would allow for a channel-coding rate of the order of one-half. Similarly, if a bandwidth efficiency close to 4 information bits per modulation symbol is required, the use of 64QAM may be more efficient than 16QAM modulation, taking into account the possibility for lower-rate channel coding and corresponding additional coding gain in the case of 64QAM. It should be noted that this does not contradict the general discussion in Section 2.1, where it was concluded that transmission with high-bandwidth utilization is inherently power ineffi-cient. The use of rate 1/2 channel coding for 16QAM obviously reduces the information data rate, and thus also the bandwidth utilization, to the same level as uncoded QPSK.

From the discussion above it can be concluded that, for a given signal-to-noise/interference ratio, a certain combination of modulation scheme and channel-coding rate is optimal in the sense that it can deliver the highest-bandwidth utilization (the highest data rate within a given bandwidth) for that signal-to-noise/interference ratio.

2.2.2 Variations in Instantaneous Transmit Power

A general drawback of higher-order modulation schemes such as 16QAM and 64QAM, where infor-mation is also encoded in the instantaneous amplitude of the modulated signal, is that the modulated signal will have larger variations, and thus also larger peaks, in its instantaneous power. This can be

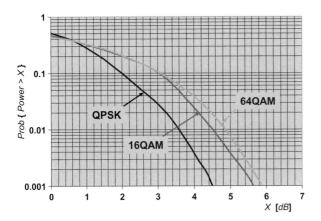

FIGURE 2.3

Distribution of instantaneous power for different modulation schemes. Average power is the same in all cases.

seen in Figure 2.3, which illustrates the distribution of the instantaneous power, more specifically the probability that the instantaneous power is above a certain value, for QPSK, 16QAM, and 64QAM respectively. Clearly, the probability for large peaks in the instantaneous power is higher in the case of higher-order modulation.

Larger peaks in the instantaneous signal power imply that the transmitter power amplifier must be over-dimensioned to avoid power-amplifier nonlinearities, occurring at high instantaneous power levels, causing corruption to the signal to be transmitted. As a consequence, the power-amplifier efficiency will be reduced, leading to increased power consumption. In addition, there will be a negative impact on the power-amplifier cost. Alternatively, the average transmit power must be reduced, implying a reduced range for a given data rate. High power-amplifier efficiency is especially important for the terminal – that is, in the uplink direction – due to the importance of low mobile-terminal power consumption and cost. For the base station, high power-amplifier efficiency, although far from irrelevant, is still somewhat less important. Thus, large peaks in the instantaneous signal power are less of an issue for the downlink compared to the uplink and, consequently, higher-order modulation is more suitable for the downlink compared to the uplink.

2.3 WIDER BANDWIDTH INCLUDING MULTI-CARRIER TRANSMISSION

As was shown in Section 2.1, transmission with a high-bandwidth utilization is fundamentally power inefficient in the sense that it will require disproportionately high signal-to-noise and signal-to-interference ratios for a given data rate. Providing very high data rates within a limited bandwidth, for example by means of higher-order modulation, is thus only possible in situations where relatively high signal-to-noise and signal-to-interference ratios can be made available, for example in small-cell environments with low traffic load or for terminals close to the cell site.

Instead, to provide high data rates as efficiently as possible in terms of required signal-to-noise and signal-to-interference ratios, implying as good coverage as possible for high data rates, the transmission bandwidth should be at least of the same order as the data rates to be provided.

Having in mind that the provisioning of higher data rates with good coverage is one of the main targets for mobile communication, it can thus be concluded that support for even wider transmission bandwidth is an important part of this evolution.

However, there are several critical issues related to the use of wider transmission bandwidths in a mobile-communication system:

- Spectrum is, as already mentioned, often a scarce and expensive resource, and it may be difficult to find spectrum allocations of sufficient size to allow for very wideband transmission, especially at lower-frequency bands.
- The use of wider transmission and reception bandwidths has an impact on the complexity of the radio equipment, both at the base station and at the terminal. As an example, a wider transmission bandwidth has a direct impact on the transmitter and the receiver sampling rates, and thus on the complexity and power consumption of digital-to-analog and analog-to-digital converters, as well as front-end digital signal processing. RF components are also, in general, more complicated to design and more expensive to produce, the wider the bandwidth they have to handle.

The two issues above are mainly outside the scope of this book. However, a more specific technical issue related to wider-band transmission is the increased corruption of the transmitted signal due to time dispersion on the radio channel. Time dispersion occurs when the transmitted signal propagates to the receiver via multiple paths with different delays (see Figure 2.4a). In the frequency domain, a time-dispersive channel corresponds to a non-constant channel frequency response, as illustrated in Figure 2.4b. This radio-channel *frequency selectivity* will corrupt the frequency-domain structure of the transmitted signal and lead to higher error rates for given signal-to-noise/interference ratios. Every radio channel is subject to frequency selectivity, at least to some extent. However, the extent to which the frequency selectivity impacts the radio communication depends on the bandwidth of the transmitted signal with, in general, larger impact for wider-band transmission. The amount of radio-channel frequency selectivity also depends on the environment, with typically less frequency selectivity (less time dispersion) in the case of small cells and in environments with few obstructions and potential reflectors, such as rural environments.

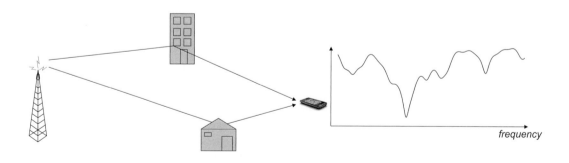

FIGURE 2.4

Multi-path propagation causing time dispersion and radio-channel frequency selectivity.

It should be noted that Figure 2.4b illustrates a "snapshot" of the channel frequency response. As a terminal is moving through the environment, the detailed structure of the multi-path propagation, and thus also the detailed structure of the channel frequency response, may vary rapidly with time. The rate of the variations in the channel frequency response is related to the channel *Doppler spread*, f_D, defined as $f_D = v/c \cdot fc$, where v is the speed of the terminal, fc is the carrier frequency (for example, 2 GHz), and c is the speed of light.

Receiver-side *equalization* [12] has for many years been used to counteract signal corruption due to radio-channel frequency selectivity. Equalization has been shown to provide satisfactory performance with reasonable complexity at least up to bandwidths corresponding to the WCDMA bandwidth of 5 MHz (see, for example, [13]). However, if the transmission bandwidth is further increased up to, for example, 20 MHz, as is done in LTE, the complexity of straightforward high-performance equalization starts to become a serious issue. One option is then to apply less optimal equalization, with a corresponding negative impact on the equalizer capability to counteract the signal corruption due to radio-channel frequency selectivity and thus a corresponding negative impact on the radio-link performance.

An alternative approach is to consider specific transmission schemes and signal designs that allow for good radio-link performance also in the case of substantial radio-channel frequency selectivity without a prohibitively large receiver complexity. In the following, two such approaches to wider-band transmission will be discussed:

1. The use of different types of *multi-carrier transmission* – that is, transmitting an overall wider-band signal as several more narrowband frequency-multiplexed signals (see below). One special case of multi-carrier transmission is *OFDM transmission*, to be discussed in more detail in Chapter 3.
2. The use of specific *single-carrier* transmission schemes, especially designed to allow for efficient but still reasonably low-complexity equalization. This is further discussed in Chapter 4.

2.3.1 Multi-Carrier Transmission

One way to increase the overall transmission bandwidth, without suffering from increased signal corruption due to radio-channel frequency selectivity, is the use of so-called *multi-carrier transmission*. As illustrated in Figure 2.5, multi-carrier transmission implies that, instead of transmitting a single more wideband signal, multiple more narrowband signals, often referred to as *subcarriers*, are frequency multiplexed and jointly transmitted over the same radio link to the same receiver. By transmitting M signals in parallel over the same radio link, the overall data rate can be increased up to M times. At the same time, the impact in terms of signal corruption due to radio-channel frequency selectivity depends on the bandwidth of each subcarrier. Thus, the impact from a frequency-selective channel is essentially the same as for a more narrowband transmission scheme with a bandwidth that corresponds to the bandwidth of each subcarrier.

A drawback of the kind of multi-carrier evolution outlined in Figure 2.5, where an existing more narrowband radio-access technology is extended to a wider overall transmission bandwidth by the parallel transmission of M more narrowband carriers, is that the spectrum of each subcarrier typically does not allow for very tight subcarrier "packing". This is illustrated by the "valleys" in the overall multi-carrier spectrum outlined in the lower part of Figure 2.5. This has a somewhat negative impact on the overall bandwidth efficiency of this kind of multi-carrier transmission.

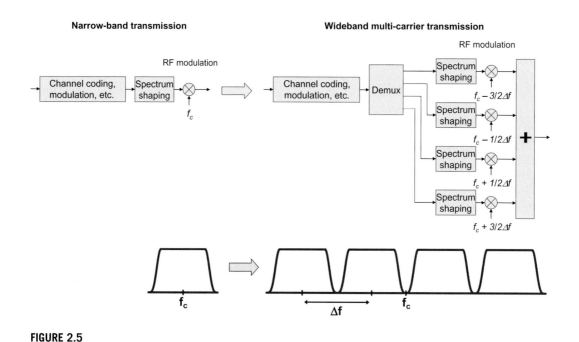

FIGURE 2.5

Extension to wider transmission bandwidth by means of multi-carrier transmission.

As an example, consider the multi-carrier evolution of WCDMA/HSPA towards wider bandwidth. WCDMA has a modulation rate, also referred to as the *WCDMA chip rate*, of $f_{cr} = 3.84$ Mchips/s. However, due to spectrum shaping, even the theoretical WCDMA spectrum, not including spectrum widening due to transmitter imperfections, has a bandwidth that significantly exceeds 3.84 MHz. More specifically, as can be seen in Figure 2.6, the theoretical WCDMA spectrum has a *raised-cosine* shape with roll-off $\alpha = 0.22$. As a consequence, the bandwidth outside of which the WCDMA theoretical spectrum equals zero is approximately 4.7 MHz (see right part of Figure 2.6).

For a straightforward multi-carrier extension of WCDMA, the subcarriers must thus be spaced approximately 4.7 MHz from each other to completely avoid inter-subcarrier interference. It should be noted, though, that a smaller subcarrier spacing can be used with only limited inter-subcarrier interference.

A second drawback of multi-carrier transmission is that, similar to the use of higher-order modulation, the parallel transmission of multiple carriers will lead to larger variations in the instantaneous transmit power. Thus, similar to the use of higher-order modulation, multi-carrier transmission will have a negative impact on the transmitter power-amplifier efficiency, implying increased transmitter power consumption and increased power-amplifier cost. Alternatively, the average transmit power must be reduced, implying a reduced range for a given data rate. For this reason, similar to the use of higher-order modulation, multi-carrier transmission is more suitable for the downlink (base-station transmission), compared to the uplink (mobile-terminal transmission), due to the higher importance of high power-amplifier efficiency at the terminal.

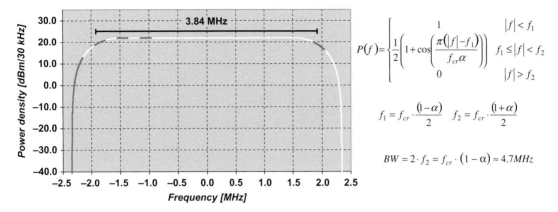

FIGURE 2.6

Theoretical WCDMA spectrum. Raised-cosine shape with roll-off $\alpha\ =\ 0.22$.

The main advantage with the kind of multi-carrier extension outlined in Figure 2.5 is that it provides a very smooth evolution, in terms of both radio equipment and spectrum, of an already existing radio-access technology to wider transmission bandwidth and a corresponding possibility for higher data rates, especially for the downlink. In essence this kind of multi-carrier evolution to wider bandwidth can be designed so that, for legacy terminals not capable of multi-carrier reception, each downlink "subcarrier" will appear as an original, more narrowband carrier, while, for a multi-carrier-capable terminal, the network can make use of the full multi-carrier bandwidth to provide higher data rates.

The next chapter will discuss, in more detail, a different approach to multi-carrier transmission, based on the so-called OFDM technique.

OFDM Transmission

In this chapter, a more detailed overview of OFDM or *Orthogonal Frequency-Division Multiplexing* will be given. OFDM is the transmission scheme used for 3GPP LTE and is also used for several other radio-access technologies, for example WiMAX [14] and the DVB broadcast technologies [15].

3.1 BASIC PRINCIPLES OF OFDM

Transmission by means of OFDM can be seen as a kind of multi-carrier transmission. The basic characteristics of OFDM transmission, which distinguish it from a straightforward multi-carrier extension of a more narrowband transmission scheme as outlined in Figure 2.5 in the previous chapter, are:

- The use of a typically very large number of relatively narrowband subcarriers. In contrast, a straightforward multi-carrier extension as outlined in Figure 2.5 would typically consist of only a few subcarriers, each with a relatively wide bandwidth. As an example, the HSPA multi-carrier evolution to a 20 MHz overall transmission bandwidth consists of four (sub)carriers, each with a bandwidth of the order of 5 MHz. In comparison, OFDM transmission may imply that several hundred subcarriers are transmitted over the same radio link to the same receiver.
- Simple rectangular pulse shaping is illustrated in Figure 3.1a. This corresponds to a sinc-square-shaped per-subcarrier spectrum, as illustrated in Figure 3.1b.
- Tight frequency-domain packing of the subcarriers with a subcarrier spacing $\Delta f = 1/T_{\mathrm{u}}$, where T_{u} is the per-subcarrier modulation-symbol time (see Figure 3.2). The subcarrier spacing is thus equal to the per-subcarrier modulation rate $1/T_{\mathrm{u}}$.[1]

An illustrative description of a basic OFDM modulator is provided in Figure 3.3. It consists of a bank of N_c complex modulators, where each modulator corresponds to one OFDM subcarrier.

In complex baseband notation, a basic OFDM signal $x(t)$ during the time interval $mT_{\mathrm{u}} \leq t < (m + 1)T_{\mathrm{u}}$ can thus be expressed as:

$$x(t) = \sum_{k=0}^{N_c-1} x_k(t) = \sum_{k=0}^{N_c-1} a_k^{(m)} e^{j2\pi k\Delta ft}, \tag{3.1}$$

where $x_k(t)$ is the kth modulated subcarrier with frequency $f_k = k \cdot \Delta f$ and $a_k^{(m)}$ is the, generally complex, modulation symbol applied to the kth subcarrier during the mth OFDM symbol interval – that is, during the time interval $mT_{\mathrm{u}} \leq t < (m + 1)T_{\mathrm{u}}$. OFDM transmission is thus block based, implying

---

[1] This ignores a possible *cyclic prefix*, see Section 3.4.

4G LTE/LTE-Advanced for Mobile Broadband.

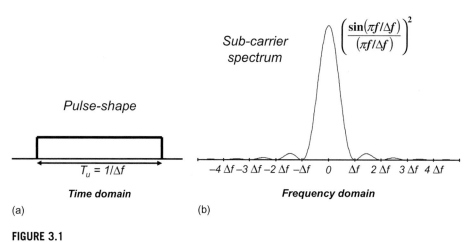

FIGURE 3.1

(a) Per-subcarrier pulse shape. (b) Spectrum for basic OFDM transmission.

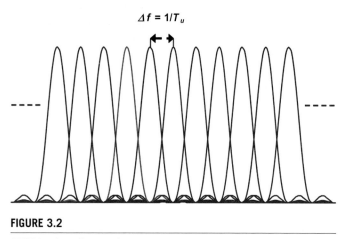

FIGURE 3.2

OFDM subcarrier spacing.

that, during each OFDM symbol interval, N_c modulation symbols are transmitted in parallel. The modulation symbols can be from any modulation alphabet, such as QPSK, 16QAM, or 64QAM.

The number of OFDM subcarriers can range from less than hundred to several thousand, with the subcarrier spacing ranging from several hundred kHz down to a few kHz. What subcarrier spacing to use depends on what types of environments the system is to operate in, including such aspects as the maximum expected radio-channel frequency selectivity (maximum expected time dispersion) and the maximum expected rate of channel variations (maximum expected Doppler spread). Once the subcarrier spacing has been selected, the number of subcarriers can be decided based on the assumed overall transmission bandwidth, taking into account acceptable out-of-band emission, etc.

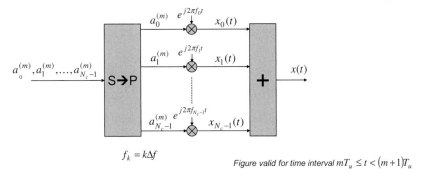

$$f_k = k\Delta f$$

Figure valid for time interval $mT_u \leq t < (m+1)T_u$

FIGURE 3.3

Principles for OFDM modulation.

The selection of OFDM subcarrier spacing and number of subcarriers is discussed in somewhat more detail in Section 3.8.

As an example, for 3GPP LTE the basic subcarrier spacing equals 15 kHz. On the other hand, the number of subcarriers depends on the transmission bandwidth, with of the order of 600 subcarriers in the case of operation in a 10 MHz spectrum allocation and correspondingly fewer/more subcarriers in the case of smaller/larger overall transmission bandwidths.

The term *Orthogonal Frequency-Division Multiplex* is due to the fact that two modulated OFDM subcarriers x_{k_1} and x_{k_2} are mutually *orthogonal* over the time interval $mT_u \leq t < (m + 1)T_u$, that is:

$$\int_{mT_u}^{(m+1)T_u} x_{k_1}(t) x_{k_2}^*(t)\, dt = \int_{mT_u}^{(m+1)T_u} a_{k_1} a_{k_2}^* e^{j2\pi k_1 \Delta f t} e^{-j2\pi k_2 \Delta f t}\, dt = 0 \quad \text{for } k_1 \neq k_2. \tag{3.2}$$

Thus, basic OFDM transmission can be seen as the modulation of a set of orthogonal functions $\varphi_k(t)$, where

$$\varphi_k(t) = \begin{cases} e^{j2\pi k \Delta f t} & 0 \leq t < T_u \\ 0 & \text{otherwise} \end{cases}. \tag{3.3}$$

The "physical resource" in the case of OFDM transmission is often illustrated as a time–frequency grid according to Figure 3.4, where each "column" corresponds to one OFDM symbol and each "row" corresponds to one OFDM subcarrier.

3.2 OFDM DEMODULATION

Figure 3.5 illustrates the basic principle of OFDM demodulation consisting of a bank of correlators, one for each subcarrier. Taking into account the orthogonality between subcarriers according to Eqn (3.2), it is clear that, in the ideal case, two OFDM subcarriers do not cause any interference to

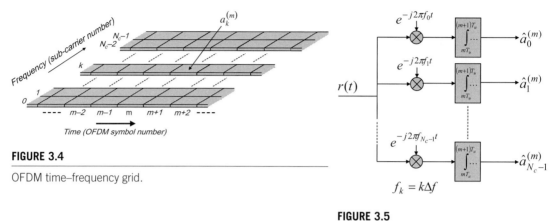

FIGURE 3.4

OFDM time–frequency grid.

FIGURE 3.5

Basic principle of OFDM demodulation.

each other after demodulation. Note that this is the case despite the fact that the spectrum of neighbor subcarriers clearly overlaps, as can be seen from Figure 3.2. Thus, the avoidance of interference between OFDM subcarriers is not simply due to a subcarrier spectrum separation, which is, for example, the case for the kind of straightforward multi-carrier extension outlined in Figure 2.5 in the previous chapter. Rather, the subcarrier orthogonality is due to the *specific* frequency-domain structure of each subcarrier in combination with the *specific* choice of a subcarrier spacing Δf equal to the per-subcarrier symbol rate $1/T_u$. However, this also implies that, in contrast to the kind of multi-carrier transmission outlined in Section 2.3.1 of the previous chapter, any corruption of the frequency-domain structure of the OFDM subcarriers, for example due to a frequency-selective radio channel, may lead to a loss of inter-subcarrier orthogonality and thus to interference between subcarriers. To handle this and to make an OFDM signal truly robust to radio-channel frequency selectivity, *cyclic-prefix insertion* is typically used, as will be further discussed in Section 3.4.

3.3 OFDM IMPLEMENTATION USING IFFT/FFT PROCESSING

Although a bank of modulators/correlators according to Figures 3.3 and 3.5 can be used to illustrate the basic principles of OFDM modulation and demodulation respectively, these are not the most appropriate modulator/demodulator structures for actual implementation. Actually, due to its specific structure and the selection of a subcarrier spacing Δf equal to the per-subcarrier symbol rate $1/T_u$, OFDM allows for low-complexity implementation by means of computationally efficient *Fast Fourier Transform* (FFT) processing.

To confirm this, consider a time-discrete (sampled) OFDM signal where it is assumed that the sampling rate f_s is a multiple of the subcarrier spacing Δf – that is, $f_s = 1/T_s = N \cdot \Delta f$. The parameter N should be chosen so that the sampling theorem [12] is sufficiently fulfilled.[2] As $N_c \cdot \Delta f$ can be seen

---

[2]An OFDM signal defined according to Eqn (3.1) in theory has an infinite bandwidth and thus the sampling theorem can never be fulfilled completely.

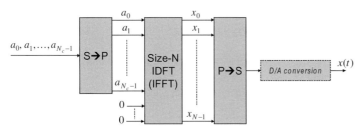

FIGURE 3.6

OFDM modulation by means of IFFT processing.

as the nominal bandwidth of the OFDM signal, this implies that N should exceed N_c with a sufficient margin.

With these assumptions, the time-discrete OFDM signal can be expressed as:[3]

$$x_n = x(nT_s) = \sum_{k=0}^{N_c-1} a_k e^{j2\pi k\Delta fnT_s} = \sum_{k=0}^{N_c-1} a_k e^{j2\pi kn/N} = \sum_{k=0}^{N-1} a'_k e^{j2\pi kn/N}, \tag{3.4}$$

where

$$a'_k = \begin{cases} a_k & 0 \le k < N_c \\ 0 & N_c \le k < N \end{cases}. \tag{3.5}$$

Thus, the sequence x_n, or, in other words, the sampled OFDM signal, is the size-N Inverse Discrete Fourier Transform (IDFT) of the block of modulation symbols a_0, a_1, ..., a_{N_c-1} extended with zeros to length N. OFDM modulation can thus be implemented by means of IDFT processing followed by digital-to-analog conversion, as illustrated in Figure 3.6. In particular, by selecting the IDFT size N equal to 2^m for some integer m, the OFDM modulation can be implemented by means of implementation-efficient radix-2 Inverse Fast Fourier Transform (IFFT) processing. It should be noted that the ratio N/N_c, which could be seen as the *over-sampling* of the time-discrete OFDM signal, could very well be, and typically is, a non-integer number. As an example and as already mentioned, for 3GPP LTE the number of subcarriers N_c is approximately 600 in the case of a 10 MHz spectrum allocation. The IFFT size can then, for example, be selected as $N = 1024$. This corresponds to a sampling rate $f_s = N \cdot \Delta f = 15.36$ MHz, where $\Delta f = 15$ kHz is the LTE subcarrier spacing.

It is important to understand that IDFT/IFFT-based implementation of an OFDM modulator, and even more so the exact IDFT/IFFT size, are just transmitter-implementation choices and not something that would be mandated by any radio-access specification. As an example, nothing forbids the implementation of an OFDM modulator as a set of parallel modulators, as illustrated in Figure 3.3. Also, nothing prevents the use of a larger IFFT size, for example a size-2048 IFFT size, even in the case of a smaller number of OFDM subcarriers.

[3]From now on the index m on the modulation symbols, indicating the OFDM symbol number, will be ignored unless especially needed.

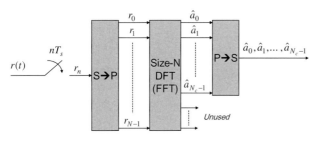

FIGURE 3.7

OFDM demodulation by means of FFT processing.

Similar to OFDM modulation, efficient FFT processing can be used for OFDM demodulation, replacing the bank of N_c parallel demodulators of Figure 3.5 with sampling with some sampling rate $f_s = 1/T_s$, followed by a size-N DFT/FFT, as illustrated in Figure 3.7.

3.4 CYCLIC-PREFIX INSERTION

As described in Section 3.2, an uncorrupted OFDM signal can be demodulated without any interference between subcarriers. One way to understand this *subcarrier orthogonality* is to recognize that a modulated subcarrier $x_k(t)$ in Eqn (3.1) consists of an integer number of periods of complex exponentials during the demodulator integration interval $T_u = 1/\Delta f$.

However, in case of a time-dispersive channel the orthogonality between the subcarriers will, at least partly, be lost. The reason for this loss of subcarrier orthogonality in the case of a time-dispersive channel is that, in this case, the demodulator correlation interval for one path will overlap with the symbol boundary of a different path, as illustrated in Figure 3.8. Thus, the integration interval will not necessarily correspond to an integer number of periods of complex exponentials of that path as the modulation symbols a_k may differ between consecutive symbol intervals. As a consequence, in the case of a time-dispersive channel there will not only be inter-symbol interference within a subcarrier, but also interference between subcarriers.

Another way to explain the interference between subcarriers in the case of a time-dispersive channel is to have in mind that time dispersion on the radio channel is equivalent to a frequency-selective channel frequency response. As clarified in Section 3.2, orthogonality between OFDM subcarriers is not simply due to frequency-domain separation but due to the *specific* frequency-domain structure of each subcarrier. Even if the frequency-domain channel is constant over a bandwidth corresponding to the main lobe of an OFDM subcarrier and only the subcarrier side lobes are corrupted due to the radio-channel frequency selectivity, the orthogonality between subcarriers will be lost with inter-subcarrier interference as a consequence. Due to the relatively large side lobes of each OFDM subcarrier, already a relatively limited amount of time dispersion or, equivalently, a relatively modest radio-channel frequency selectivity may cause substantial interference between subcarriers.

To deal with this problem and to make an OFDM signal truly insensitive to time dispersion on the radio channel, so-called *cyclic-prefix insertion* is typically used in OFDM transmission. As illustrated

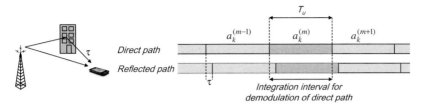

FIGURE 3.8

Time dispersion and corresponding received-signal timing.

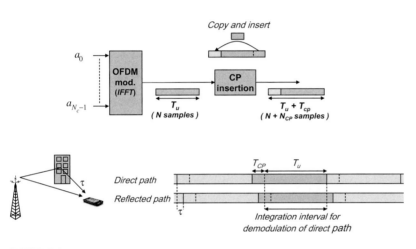

FIGURE 3.9

Cyclic-prefix insertion.

in Figure 3.9, cyclic-prefix insertion implies that the last part of the OFDM symbol is copied and inserted at the beginning of the OFDM symbol. Cyclic-prefix insertion thus increases the length of the OFDM symbol from T_u to $T_u + T_{CP}$, where T_{CP} is the length of the cyclic prefix, with a corresponding reduction in the OFDM symbol rate as a consequence. As illustrated in the lower part of Figure 3.9, if the correlation at the receiver side is still only carried out over a time interval $T_u = 1/\Delta f$, subcarrier orthogonality will then also be preserved in the case of a time-dispersive channel, as long as the span of the time dispersion is shorter than the cyclic-prefix length.

In practice, cyclic-prefix insertion is carried out on the time-discrete output of the transmitter IFFT. Cyclic-prefix insertion then implies that the last N_{CP} samples of the IFFT output block of length N are copied and inserted at the beginning of the block, increasing the block length from N to $N + N_{CP}$. At the receiver side, the corresponding samples are discarded before OFDM demodulation by means of, for example, DFT/FFT processing.

Cyclic-prefix insertion is beneficial in the sense that it makes an OFDM signal insensitive to time dispersion as long as the span of the time dispersion does not exceed the length of the cyclic prefix. The drawback of cyclic-prefix insertion is that only a fraction $T_u/(T_u + T_{CP})$ of the received signal

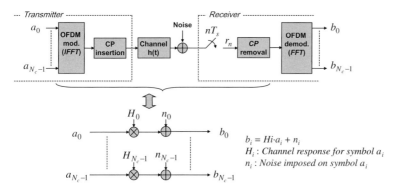

FIGURE 3.10

Frequency-domain model of OFDM transmission/reception.

power is actually utilized by the OFDM demodulator, implying a corresponding power loss in the demodulation. In addition to this power loss, cyclic-prefix insertion also implies a corresponding loss in terms of bandwidth as the OFDM symbol rate is reduced without a corresponding reduction in the overall signal bandwidth.

One way to reduce the relative overhead due to cyclic-prefix insertion is to reduce the subcarrier spacing Δf, with a corresponding increase in the symbol time T_u as a consequence. However, this will increase the sensitivity of the OFDM transmission to fast channel variations – that is, high Doppler spread – as well as different types of frequency errors (see further Section 3.8).

It is also important to understand that the cyclic prefix does not necessarily have to cover the entire length of the channel time dispersion. In general, there is a trade-off between the power loss due to the cyclic prefix and the signal corruption (inter-symbol and inter-subcarrier interference) due to residual time dispersion not covered by the cyclic prefix and, at a certain point, further reduction of the signal corruption due to further increase of the cyclic-prefix length will not justify the corresponding additional power loss. This also means that, although the amount of time dispersion typically increases with the cell size, beyond a certain cell size there is often no reason to increase the cyclic prefix further as the corresponding power loss due to a further increase of the cyclic prefix would have a larger negative impact, compared to the signal corruption due to the residual time dispersion not covered by the cyclic prefix [16].

3.5 FREQUENCY-DOMAIN MODEL OF OFDM TRANSMISSION

Assuming a sufficiently large cyclic prefix, the linear convolution of a time-dispersive radio channel will appear as a circular convolution during the demodulator integration interval T_u. The combination of OFDM modulation (IFFT processing), a time-dispersive radio channel, and OFDM demodulation (*FFT* processing) can then be seen as a *frequency-domain* channel as illustrated in Figure 3.10, where

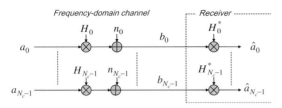

FIGURE 3.11

Frequency-domain model of OFDM transmission/
reception with "one-tap equalization" at the receiver.

the frequency-domain channel taps H_0, ..., H_{N_c-1} can be directly derived from the channel impulse response.

The demodulator output b_k in Figure 3.10 is the transmitted modulation symbol a_k scaled and phase rotated by the complex frequency-domain channel tap H_k and impaired by noise n_k. To properly recover the transmitted symbol for further processing, for example data demodulation and channel decoding, the receiver should multiply b_k by the complex conjugate of H_k, as illustrated in Figure 3.11. This is often expressed as a *one-tap equalizer* being applied to each received subcarrier.

3.6 CHANNEL ESTIMATION AND REFERENCE SYMBOLS

As described above, to demodulate the transmitted modulation symbol a_k and allow for proper decoding of the transmitted information at the receiver side, scaling with the complex conjugate of the frequency-domain channel tap H_k should be applied after OFDM demodulation (*FFT* processing) (see Figure 3.11). To be able to do this, the receiver obviously needs an estimate of the frequency-domain channel taps H_0, ..., H_{N_c-1}.

The frequency-domain channel taps can be estimated indirectly by first estimating the channel impulse response and, from that, calculating an estimate of H_k. However, a more straightforward approach is to estimate the frequency-domain channel taps directly. This can be done by inserting known *reference symbols*, sometimes also referred to as *pilot symbols*, at regular intervals within the OFDM time–frequency grid, as illustrated in Figure 3.12. Using knowledge about the reference symbols, the receiver can estimate the frequency-domain channel around the location of the reference symbol. The reference symbols should also have a sufficiently high density in both the time and the frequency domains to be able to provide estimates for the entire time–frequency grid in the case of radio channels subject to high frequency and/or time selectivity.

Different more or less advanced algorithms can be used for the channel estimation, ranging from simple averaging in combination with linear interpolation to Minimum-Mean-Square-Error (MMSE) estimation relying on more detailed knowledge of the channel time/frequency-domain characteristics. Readers are referred to, for example, [17] for a more in-depth discussion on channel estimation for OFDM.

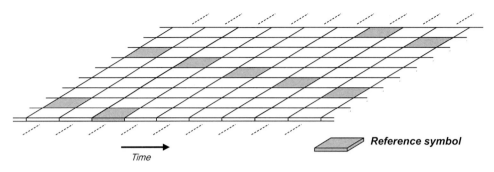

Time

Reference symbol

FIGURE 3.12

Time–frequency grid with known reference symbols.

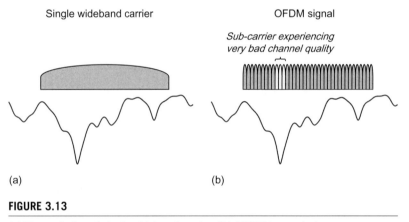

Single wideband carrier

OFDM signal

Sub-carrier experiencing
very bad channel quality

(a)   (b)

FIGURE 3.13

(a) Transmission of single wideband carrier. (b) OFDM transmission over a
frequency-selective channel.

3.7 FREQUENCY DIVERSITY WITH OFDM: IMPORTANCE OF CHANNEL CODING

As discussed in Section 2.3 in the previous chapter, a radio channel is always subject to some degree of frequency selectivity, implying that the channel quality will vary in the frequency domain. In the case of a single wideband carrier, such as a WCDMA carrier, each modulation symbol is transmitted over the entire signal bandwidth. Thus, in the case of the transmission of a single wideband carrier over a highly frequency-selective channel (see Figure 3.13a), each modulation symbol will be transmitted both over frequency bands with relatively good quality (relatively high signal strength) and frequency bands with low quality (low signal strength). Such transmission of information over multiple frequency bands with different instantaneous channel quality is also referred to as *frequency diversity*.

On the other hand, in the case of OFDM transmission each modulation symbol is mainly confined to a relatively narrow bandwidth. Thus, for OFDM transmission over a frequency-selective channel, certain modulation symbols may be fully confined to a frequency band with very low instantaneous

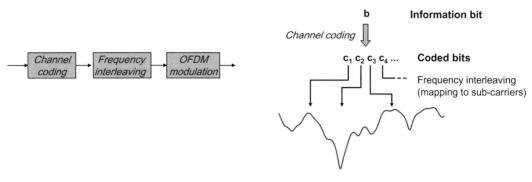

FIGURE 3.14

Channel coding in combination with frequency-domain interleaving to provide frequency diversity in the case of OFDM transmission.

signal strength, as illustrated in Figure 3.13b. Thus, the individual modulation symbols will typically not experience any substantial frequency diversity even if the channel is highly frequency selective over the overall OFDM transmission bandwidth. As a consequence, the basic error-rate performance of OFDM transmission over a frequency-selective channel is relatively poor and especially much worse than the basic error rate in the case of a single wideband carrier.

However, in practice channel coding is used in most cases of digital communication and especially in mobile communication. Channel coding implies that each bit of information to be transmitted is spread over several, often very many, code bits. If these coded bits are then, via modulation symbols, mapped to a set of OFDM subcarriers that are well distributed over the overall transmission bandwidth of the OFDM signal, as illustrated in Figure 3.14, each information bit will experience frequency diversity in the case of transmission over a radio channel that is frequency selective over the transmission bandwidth, despite the fact that the subcarriers, and thus also the code bits, will not experience any frequency diversity. Distributing the code bits in the frequency domain, as illustrated in Figure 3.14, is sometimes referred to as *frequency interleaving*. This is similar to the use of time-domain interleaving to benefit from channel coding in the case of fading that varies in time.

Thus, in contrast to the transmission of a single wideband carrier, channel coding (combined with frequency interleaving) is an essential component in order for OFDM transmission to be able to benefit from frequency diversity on a frequency-selective channel. As channel coding is typically used in most cases of mobile communication this is not a very serious drawback, especially taking into account that a significant part of the available frequency diversity can be captured already with a relatively high code rate.

3.8 SELECTION OF BASIC OFDM PARAMETERS

If OFDM is to be used as the transmission scheme in a mobile-communication system, the following basic OFDM parameters need to be decided upon:

- The subcarrier spacing Δf.
- The number of subcarriers N_c, which, together with the subcarrier spacing, determines the overall transmission bandwidth of the OFDM signal.

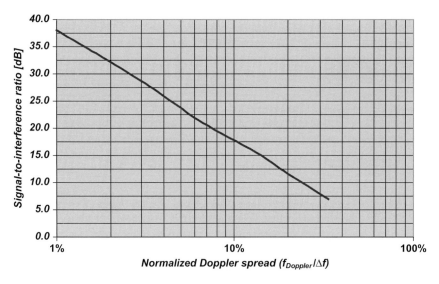

FIGURE 3.15

Subcarrier interference as a function of the normalized Doppler spread $f_{Doppler}/\Delta f$.

- The cyclic-prefix length T_{CP}. Together with the subcarrier spacing $\Delta f = 1/T_u$, the cyclic-prefix length determines the overall OFDM symbol time $T = T_{CP} + T_u$ or, equivalently, the OFDM symbol rate.

3.8.1 OFDM Subcarrier Spacing

There are two factors that constrain the selection of the OFDM subcarrier spacing:

- The OFDM subcarrier spacing should be as small as possible (T_u as large as possible) to minimize the relative cyclic-prefix overhead $T_{CP}/(T_u + T_{CP})$ – see further Section 3.8.3.
- A too small subcarrier spacing increases the sensitivity of the OFDM transmission to Doppler spread and different kinds of frequency inaccuracies.

A requirement for the OFDM subcarrier orthogonality (3.2) to hold at the receiver side – that is, after the transmitted signal has propagated over the radio channel – is that the instantaneous channel does not vary noticeably during the demodulator correlation interval T_u (see Figure 3.5). In the case of such channel variations, for example due to very high Doppler spread, the orthogonality between subcarriers will be lost, with inter-subcarrier interference as a consequence. Figure 3.15 illustrates the subcarrier signal-to-interference ratio due to inter-subcarrier interference between two neighboring subcarriers, as a function of the normalized Doppler spread. When considering Figure 3.15, it should be borne in mind that a subcarrier will be subject to interference from multiple subcarriers on both sides[4] – that is, the overall inter-subcarrier interference from all subcarriers will be higher than what is illustrated in Figure 3.15.

---

[4]Except for the subcarrier at the edge of the spectrum.

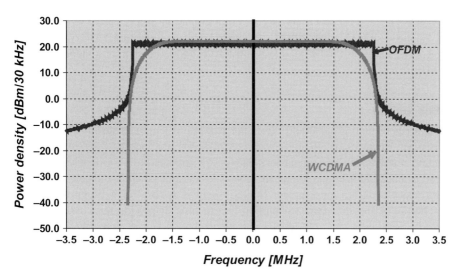

FIGURE 3.16

Spectrum of a basic 5 MHz OFDM signal compared with WCDMA spectrum.

In practice, the amount of inter-subcarrier interference that can be accepted very much depends on the service to be provided and to what extent the received signal is anyway corrupted due to noise and other impairments. As an example, on the cell border of large cells the signal-to-noise/interference ratio will anyway be relatively low, with relatively low achievable data rates as a consequence. A small amount of additional inter-subcarrier interference, for example due to Doppler spread, may then be more or less negligible. At the same time, in high signal-to-noise/interference scenarios, for example in small cells with low traffic or close to the base station, where high data rates are to be provided, the same amount of inter-subcarrier interference may have a much more negative impact.

It should also be noted that, in addition to Doppler spread, inter-subcarrier interference will also be due to different transmitter and receiver inaccuracies, such as frequency errors and phase noise.

3.8.2 Number of Subcarriers

Once the subcarrier spacing has been selected based on environment, expected Doppler spread and time dispersion, etc., the number of subcarriers can be determined based on the amount of spectrum available and the acceptable out-of-band emissions.

The basic bandwidth of an OFDM signal equals $N_c \cdot \Delta f$ – that is, the number of subcarriers multiplied by the subcarrier spacing. However, as can be seen in Figure 3.16, the spectrum of a basic OFDM signal falls off very slowly outside the basic OFDM bandwidth and especially much slower than for a WCDMA signal. The reason for the large *out-of-band emission* of a basic OFDM signal is the use of rectangular pulse shaping (Figure 3.1), leading to per-subcarrier side lobes that fall off relatively slowly. However, in practice, straightforward filtering or *time-domain windowing* [18] will be used to suppress a main part of the OFDM out-of-band emissions. Thus, in practice, typically of the order of 10% guard-band is needed for an OFDM signal implying, as an example, that in a spectrum allocation of 5 MHz, the basic OFDM bandwidth $N_c \cdot \Delta f$ could be of the order of 4.5 MHz. Assuming, for example, a subcarrier spacing of 15 kHz as selected for LTE, this corresponds to approximately 300 subcarriers in 5 MHz.

3.8.3 Cyclic-Prefix Length

In principle, the cyclic-prefix length T_{CP} should cover the maximum length of the time dispersion expected to be experienced. However, as already discussed, increasing the length of the cyclic prefix, without a corresponding reduction in the subcarrier spacing Δf, implies an additional overhead in terms of power as well as bandwidth. In particular, the power loss implies that, as the cell size grows and the system performance becomes more power limited, there is a trade-off between the loss in power due to the cyclic prefix and the signal corruption due to time dispersion not covered by the cyclic prefix. As already mentioned, this implies that, although the amount of time dispersion typically increases with the cell size, beyond a certain cell size there is often no reason to increase the cyclic prefix further as the corresponding power loss would have a larger negative impact, compared to the signal corruption due to the residual time dispersion not covered by the cyclic prefix [16].

One situation where a longer cyclic prefix may be needed is in the case of multi-cell transmission using SFN (Single-Frequency Network), as further discussed in Section 3.11.

Thus, to be able to optimize performance to different environments, some OFDM-based systems support multiple cyclic-prefix lengths. The different cyclic-prefix lengths can then be used in different transmission scenarios:

- Shorter cyclic prefix in small-cell environments to minimize the cyclic-prefix overhead.
- Longer cyclic prefix in environments with extreme time dispersion and especially in the case of SFN operation.

3.9 VARIATIONS IN INSTANTANEOUS TRANSMISSION POWER

According to Section 2.3.1, one of the drawbacks of multi-carrier transmission is the corresponding large variations in the instantaneous transmit power, implying a reduced power-amplifier efficiency and higher mobile-terminal power consumption; alternatively, that the power-amplifier output power has to be reduced with a reduced range as a consequence. Being a kind of multi-carrier transmission scheme, OFDM is subject to the same drawback.

However, a large number of different methods have been proposed to reduce the large power peaks of an OFDM signal:

- In the case of *tone reservation* [19], a subset of the OFDM subcarriers are not used for data transmission. Instead, these subcarriers are modulated in such a way that the largest peaks of the overall OFDM signal are suppressed, allowing for a reduced power-amplifier back-off. One drawback of tone reservation is the bandwidth loss due to the fact that a number of subcarriers are not available for actual data transmission. The calculation of what modulation to apply to the reserved tones can also be of relatively high complexity.
- In the case of *prefiltering* or *precoding*, linear processing is applied to the sequence of modulation symbols before OFDM modulation. DFTS-spread OFDM (DFTS-OFDM) as described in the next chapter and which is used for the LTE uplink, can be seen as one kind of prefiltering.

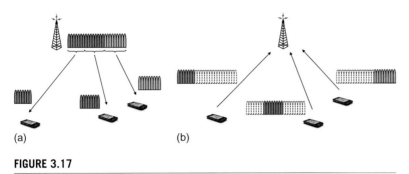

FIGURE 3.17

OFDM as a user-multiplexing/multiple-access scheme: (a) downlink; (b) uplink.

- In the case of *selective scrambling* [20], the coded-bit sequence to be transmitted is scrambled with a number of different scrambling codes. Each scrambled sequence is then OFDM modulated and the signal with the lowest peak power is selected for transmission. After OFDM demodulation at the receiver side, descrambling and subsequent decoding is carried out for all the possible scrambling sequences. Only the decoding carried out for the scrambling code actually used for the transmission will provide a correct decoding result. A drawback of selective scrambling is an increased receiver complexity as multiple decodings need to be carried out in parallel.

Readers are referred to the references above for a more in-depth discussion on different peak-reduction schemes.

3.10 OFDM AS A USER-MULTIPLEXING AND MULTIPLE-ACCESS SCHEME

The discussion has, until now, implicitly assumed that all OFDM subcarriers are transmitted from the same transmitter to a certain receiver, that is:

- downlink transmission of all subcarriers to a *single* terminal;
- uplink transmission of all subcarriers from a *single* terminal.

However, OFDM can also be used as a *user-multiplexing* or *multiple-access scheme*, allowing for simultaneous frequency-separated transmissions to/from multiple terminals (see Figure 3.17).

In the downlink direction, OFDM as a user-multiplexing scheme implies that, in each OFDM symbol interval, different subsets of the overall set of available subcarriers are used for transmission to *different* terminals (see Figure 3.17a).

Similarly, in the uplink direction, OFDM as a *user-multiplexing* or *multiple-access scheme* implies that, in each OFDM symbol interval, different subsets of the overall set of subcarriers are used for data transmission from *different* terminals (see Figure 3.17b). In this case, the term Orthogonal Frequency-Division Multiple Access or OFDMA is also often used.[5]

---

[5]The term OFDMA is sometimes also used to denote the use of OFDM to multiplex multiple users in the downlink, as illustrated in Figure 3.17a.

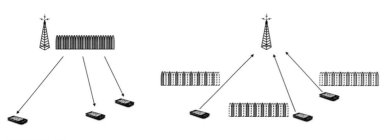

FIGURE 3.18

Distributed user multiplexing.

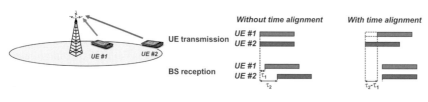

FIGURE 3.19

Uplink transmission-timing control.

Figure 3.17 assumes that *consecutive* subcarriers are used for transmission to/from the same terminal. However, distributing the subcarriers to/from a terminal in the frequency domain is also possible, as illustrated in Figure 3.18. The benefit of such *distributed* user multiplexing or *distributed* multiple access is a possibility for additional frequency diversity as each transmission is spread over a wider bandwidth.

In the case when OFDMA is used as an uplink multiple-access scheme – that is, in the case of frequency multiplexing of OFDM signals from multiple terminals – it is critical that the transmissions from the different terminals arrive approximately time aligned at the base station. More specifically, the transmissions from the different terminals should arrive at the base station with a timing misalignment less than the length of the cyclic prefix to preserve orthogonality between subcarriers received from different terminals and thus avoid inter-user interference.

Due to the differences in distance to the base station for different terminals and the corresponding differences in the propagation time (which may far exceed the length of the cyclic prefix), it is therefore necessary to control the uplink transmission timing of each terminal (see Figure 3.19). Such *transmission-timing control* should adjust the transmit timing of each terminal to ensure that uplink transmissions arrive approximately time aligned at the base station. As the propagation time changes as the terminal is moving within the cell, the transmission-timing control should be an active process, continuously adjusting the exact transmit timing of each terminal.

Furthermore, even in the case of perfect transmission-timing control, there will always be some interference between subcarriers, for example due to frequency errors. Typically this interference is relatively low in the case of reasonable frequency errors, Doppler spread, etc. (see Section 3.8). However, this assumes that the different subcarriers are received with at least approximately the same

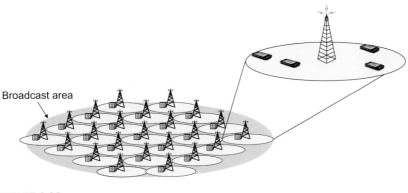

FIGURE 3.20

Broadcast scenario.

power. In the uplink, the propagation distance and thus the path loss of the different mobile-terminal transmissions may differ significantly. If two terminals are transmitting with the same power, the received-signal strengths may thus differ significantly, implying a potentially significant interference from the stronger signal to the weaker signal unless the subcarrier orthogonality is perfectly retained. To avoid this, at least some degree of *uplink transmission-power control* may need to be applied in case of uplink OFDMA, reducing the transmission power of user terminals close to the base station and ensuring that all received signals will be of approximately the same power.

3.11 MULTI-CELL BROADCAST/MULTICAST TRANSMISSION AND OFDM

The provision of broadcast/multicast services in a mobile-communication system implies that the same information is to be *simultaneously* provided to multiple terminals, often dispersed over a large area corresponding to a large number of cells, as shown in Figure 3.20. The broadcast/multicast information may be a TV news clip, information about the local weather conditions, stock-market information, or any other kind of information that, at a given time, may be of interest to a large number of people.

When the same information is to be provided to multiple terminals within a cell it is often beneficial to provide this information as a single "broadcast" radio transmission covering the entire cell and simultaneously being received by all relevant terminals (Figure 3.21a), rather than providing the information by means of individual transmissions to each terminal (unicast transmission; see Figure 3.21b).

As a broadcast transmission according to Figure 3.21 also has to be dimensioned to reach the worst-case terminals, including terminals at the cell border, it will be relative costly in terms of the recourses (base-station transmit power) needed to provide a certain broadcast-service data rate. Alternatively, taking into account the limited signal-to-noise ratio that can be achieved at, for example, the cell edge, the achievable broadcast data rates may be relatively limited, especially in the case of large cells. One way to increase the broadcast data rates would then be to reduce the cell size, thereby increasing the cell-edge receive power. However, this will increase the number of cells to cover a certain area and is thus obviously negative from a cost-of-deployment point of view.

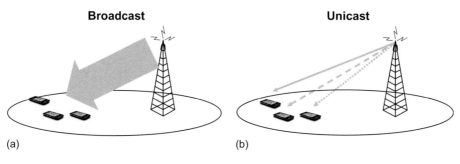

FIGURE 3.21

Broadcast vs. unicast transmission. (a) Broadcast. (b) Unicast.

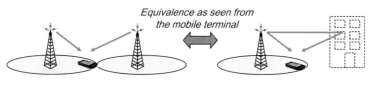

FIGURE 3.22

Equivalence between simulcast transmission and multi-path propagation.

However, as discussed above, the provision of broadcast/multicast services in a mobile-communication network typically implies that identical information is to be provided over a large number of cells. In such a case, the resources (downlink transmit power) needed to provide a certain broadcast data rate can be considerably reduced if terminals at the cell edge can utilize the received power from broadcast transmissions from multiple cells when detecting/decoding the broadcast data.

One way to mitigate this and further improve the provision of broadcast/multicast services in a mobile-communication network is to ensure that the broadcast transmissions from different cells *are truly identical* and *transmitted mutually time aligned*. In this case, the transmissions received from multiple cells will, as seen from the terminal, appear as a single transmission subject to severe multi-path propagation, as illustrated in Figure 3.22. The transmission of identical time-aligned signals from multiple cells, especially in the case of provision of broadcast/multicast services, is sometimes referred to as Single-Frequency Network (SFN) operation [21].

In the case of such identical time-aligned transmissions from multiple cells, the "inter-cell interference" due to transmissions in neighboring cells will, from a terminal point of view, be replaced by signal corruption due to time dispersion. If the broadcast transmission is based on OFDM with a cyclic prefix that covers the main part of this "time dispersion", the achievable broadcast data rates are thus only limited by noise, implying that, especially in smaller cells, very high broadcast data rates can be achieved. Furthermore, the OFDM receiver does not need to explicitly identify the cells to be soft combined. Rather, all transmissions that fall within the cyclic prefix will "automatically" be captured by the receiver.

Wider-Band "Single-Carrier" Transmission

The previous chapters discussed multi-carrier transmission in general (Section 2.3.1) and OFDM transmission in particular (Chapter 3) as means to allow for very high overall transmission bandwidth while still being robust to signal corruption due to radio-channel frequency selectivity.

However, as discussed in the previous chapter, a drawback of OFDM modulation, as well as any kind of multi-carrier transmission, is the large variations in the instantaneous power of the transmitted signal. Such power variations imply a reduced power-amplifier efficiency and higher power-amplifier cost. This is especially critical for the uplink, due to the high importance of low mobile-terminal power consumption and cost.

As discussed in Chapter 3, several methods have been proposed on how to reduce the large power variations of an OFDM signal. However, most of these methods have limitations in terms of to what extent the power variations can be reduced. Furthermore, most of the methods also imply a significant computational complexity and/or a reduced link performance.

Thus, there is an interest to consider also wider-band *single-carrier* transmission as an alternative to multi-carrier transmission, especially for the uplink – that is, for mobile-terminal transmissions. It is then necessary to consider what can be done to handle the corruption to the signal waveform that will occur in most mobile-communication environments due to radio-channel frequency selectivity.

4.1 EQUALIZATION AGAINST RADIO-CHANNEL FREQUENCY SELECTIVITY

Historically, the main method to handle signal corruption due to radio-channel frequency selectivity has been to apply different forms of *equalization* [12] at the receiver side. The aim of equalization is to, by different means, compensate for the channel frequency selectivity and thus, at least to some extent, restore the original signal shape.

4.1.1 Time-Domain Linear Equalization

The most basic approach to equalization is the time-domain linear equalizer, consisting of a linear filter with an impulse response $w(\tau)$ applied to the received signal (see Figure 4.1).

By selecting different filter impulse responses, different receiver/equalizer strategies can be implemented. As an example, in DS-CDMA-based systems a so-called RAKE receiver structure has historically often been used. The RAKE receiver is simply the receiver structure of Figure 4.1 where the filter impulse response has been selected to provide *channel-matched filtering*:

$$w(\tau) = h^*(-\tau). \tag{4.1}$$

4G LTE/LTE-Advanced for Mobile Broadband.

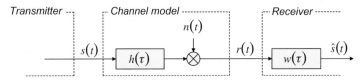

FIGURE 4.1

General time-domain linear equalization.

That is, the filter response has been selected as the complex conjugate of the time-reversed channel impulse response. This is also often referred to as a *Maximum-Ratio Combining* (MRC) filter setting [12].

Selecting the receiver filter according to the MRC criterion – that is, as a channel-matched filter – maximizes the post-filter signal-to-noise ratio (thus the term maximum-ratio combining). However, MRC-based filtering does not provide any compensation for any radio-channel frequency selectivity – that is, no equalization. Thus, MRC-based receiver filtering is appropriate when the received signal is mainly impaired by noise or interference from other transmissions but not when a main part of the overall signal corruption is due to the radio-channel frequency selectivity.

Another alternative is to select the receiver filter to fully compensate for the radio-channel frequency selectivity. This can be achieved by selecting the receiver-filter impulse response to fulfill the relation:

$$h(\tau) \otimes w(\tau) = 1, \tag{4.2}$$

where "$\otimes$" denotes linear convolution. This selecting of the filter setting, also known as *Zero-Forcing* (ZF) equalization [12], provides full compensation for any radio-channel frequency selectivity (complete equalization) and thus full suppression of any related signal corruption. However, zero-forcing equalization may lead to a large, potentially very large, increase in the noise level after equalization and thus to an overall degradation in the link performance. This will be the case especially when the channel has large variations in its frequency response.

A third and, in most cases, better alternative is to select a filter setting that provides a trade-off between signal corruption due to radio-channel frequency selectivity, and noise/interference. This can, for example, be done by selecting the filter to minimize the mean-square error between the equalizer output and the transmitted signal – that is, to minimize:

$$\varepsilon = E\left\{ |\, \hat{s}(t) - s(t)\, |^2 \right\}. \tag{4.3}$$

This is also referred to as a *Minimum Mean-Square Error* (MMSE) equalizer setting [12].

In practice, the linear equalizer has most often been implemented as a *time-discrete FIR filter* [22] with L filter taps applied to the sampled received signal, as illustrated in Figure 4.2. In general, the complexity of such a time-discrete equalizer grows relatively rapidly with the bandwidth of the signal to be equalized:

- A more wideband signal is subject to relatively more radio-channel frequency selectivity or, equivalently, relatively more time dispersion. This implies that the equalizer needs to have a larger

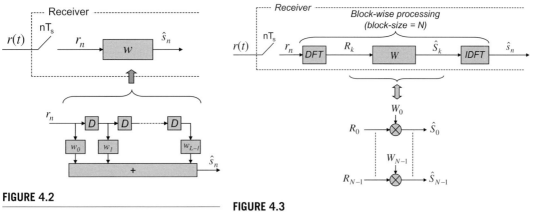

FIGURE 4.2

Linear equalization implemented as a time-discrete FIR filter.

FIGURE 4.3

Frequency-domain linear equalization.

span (larger length L – that is, more filter taps) to be able to properly compensate for the channel frequency selectivity.

- A more wideband signal leads to a correspondingly higher sampling rate for the received signal. Thus, also the receiver-filter processing needs to be carried out with a correspondingly higher rate.

It can be shown [23] that the time-discrete MMSE equalizer setting $\bar{w} = [w_0, w_1, ..., w_{L-1}]^H$ is given by the expression:

$$\bar{w} = R^{-1}\bar{p}. \tag{4.4}$$

In this expression, R is the *channel-output auto-correlation matrix* of size $L \times L$, which depends on the channel impulse response as well as on the noise level, and $\bar{p}$ is the *channel-output/channel-input cross-correlation vector* of size $L \times 1$ that depends on the channel impulse response.

In particular, in the case of a large equalizer span (large L), the time-domain MMSE equalizer may be of relatively high complexity:

- The equalization itself (the actual filtering) may be of relatively high complexity according to the above.
- Calculation of the MMSE equalizer setting, especially the calculation of the inverse of the size $L \times L$ correlation matrix R, may be of relatively high complexity.

4.1.2 Frequency-Domain Equalization

A possible way to reduce the complexity of linear equalization is to carry out the equalization in the frequency domain [24], as illustrated in Figure 4.3. In such *frequency-domain linear equalization*, the equalization is carried out block-wise with block size N. The sampled received signal is first transformed into the frequency domain by means of a size-N DFT. The equalization is then carried out as

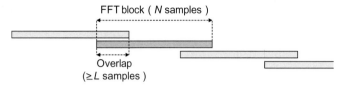

FIGURE 4.4

Overlap-and-discard processing.

frequency-domain filtering, with the frequency-domain filter taps W_0, ..., W_{N-1}, for example, being the DFT of the corresponding time-domain filter taps w_0, ..., w_{L-1} of Figure 4.2. Finally, the equalized frequency-domain signal is transformed back to the time domain by means of a size-N inverse DFT. The block size-N should preferably be selected as $N = 2^n$ for some integer n to allow for computational-efficient radix-2 FFT/IFFT implementation of the DFT/IDFT processing.

For each processing block of size N, the frequency-domain equalization basically consists of:

- A size-N DFT/FFT.
- N complex multiplications (the frequency-domain filter).
- A size-N inverse DFT/FFT.

Especially in the case of channels with extensive frequency selectivity, implying the need for a large span of a time-domain equalizer (large equalizer length L), equalization in the frequency domain according to Figure 4.3 can be of significantly less complexity, compared to time-domain equalization illustrated in Figure 4.2.

However, there are two issues with frequency-domain equalization:

- The time-domain filtering of Figure 4.2 implements a time-discrete *linear convolution*. In contrast, frequency-domain filtering according to Figure 4.3 corresponds to *circular convolution* in the time domain. Assuming a time-domain equalizer of length L, this implies that the first $L - 1$ samples at the output of the frequency-domain equalizer *will not* be identical to the corresponding output of the time-domain equalizer.
- The frequency-domain filter taps W_0, ..., W_{N-1} can be determined by first determining the pulse response of the corresponding time-domain filter and then transforming this filter into the frequency domain by means of a DFT. However, as mentioned above, determining the MMSE time-domain filter may be relatively complex in the case of a large equalizer length L.

One way to address the first issue is to apply an *overlap* in the block-wise processing of the frequency-domain equalizer as outlined in Figure 4.4, where the overlap should be at least $L - 1$ samples. With such an overlap, the first $L - 1$ "incorrect" samples at the output of the frequency-domain equalizer can be *discarded* as the corresponding samples are also (correctly) provided as the last part of the previously received/equalized block. The drawback with this kind of "overlap-and-discard" processing is a computational overhead – that is, somewhat higher receiver complexity.

An alternative approach that addresses both of the above issues is to apply *cyclic-prefix insertion* at the transmitter side (see Figure 4.5). Similar to OFDM, cyclic-prefix insertion in the case of single-carrier transmission implies that a cyclic prefix of length N_{CP} samples is inserted block-wise at the

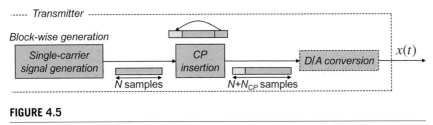

FIGURE 4.5

Cyclic-prefix insertion in single-carrier transmission.

transmitter side. The transmitter-side block size should be the same as the block size N used for the receiver-side frequency-domain equalization.

   With the introduction of a cyclic prefix, the channel will, from a receiver point of view, appear as a circular convolution over a receiver processing block of size N. Thus, there is no need for any receiver overlap-and-discard processing. Furthermore, the frequency-domain filter taps can now be calculated directly from an estimate of the sampled channel frequency response without first determining the time-domain equalizer setting. As an example, in the case of an MMSE equalizer the frequency-domain filter taps can be calculated according to:

$$W_k = \frac{H_k^*}{|H_k|^2 + N_0}, \tag{4.5}$$

where N_0 is the noise power and H_k is the sampled channel frequency response. For large equalizer lengths, this calculation is of much lower complexity compared to the time-domain calculation discussed in the previous section.

   The drawback of cyclic-prefix insertion in the case of single-carrier transmission is the same as for OFDM – that is, it implies an overhead in terms of both power and bandwidth. One method to reduce the relative cyclic-prefix overhead is to increase the block size N of the frequency-domain equalizer. However, for the block-wise equalization to be accurate, the channel needs to be approximately constant over a time span corresponding to the size of the processing block. This constraint provides an upper limit on the block size N that depends on the rate of the channel variations. Note that this is similar to the constraint on the OFDM subcarrier spacing $\Delta f = 1/T_u$ depending on the rate of the channel variations, as discussed in Chapter 3.

4.1.3 Other Equalizer Strategies

The previous sections discussed different approaches to linear equalization as a means to counteract signal corruption of a wideband signal due to radio-channel frequency selectivity. However, there are also other approaches to equalization:

- *Decision-Feedback Equalization* (DFE) [12] implies that previously detected symbols are fed back and used to cancel the contribution of the corresponding transmitted symbols to the overall signal corruption. Such decision feedback is typically used in combination with time-domain linear filtering, where the linear filter transforms the channel response to a shape that is more suitable for the decision-feedback stage. Decision feedback can also very well be used in combination with frequency-domain linear equalization [24].

- *Maximum-Likelihood* (ML) detection, also known as *Maximum-Likelihood Sequence Estimation* (MLSE) [25], is strictly speaking not an equalization scheme but rather a receiver approach where the impact of radio-channel time dispersion is explicitly taken into account in the receiver-side detection process. Fundamentally, an ML detector uses the entire received signal to decide on the most likely transmitted sequence, taking into account the impact of the time dispersion on the received signal. To implement maximum-likelihood detection the *Viterbi algorithm* [26] is often used.[1] However, although maximum-likelihood detection based on the Viterbi algorithm has been extensively used for 2G mobile communication such as GSM, it can be too complex when applied to much wider transmission bandwidths, leading to both much more extensive channel frequency selectivity and much higher sampling rates.

4.2 UPLINK FDMA WITH FLEXIBLE BANDWIDTH ASSIGNMENT

In practice, there are obviously often multiple terminals within a cell and thus multiple terminals that should share the overall uplink radio resource within the cell by means of the *uplink intra-cell multiple-access scheme*.

In mobile communication based on WCDMA and CDMA2000, uplink transmissions within a cell are mutually *non-orthogonal* and the base-station receiver relies on the processing gain due to channel coding and additional direct-sequence spreading to suppress the intra-cell interference. Although non-orthogonal multiple access can, fundamentally, provide higher capacity compared to orthogonal multiple access, in practice the possibility for mutually orthogonal uplink transmissions within a cell is often beneficial from a system performance point of view.

Mutually orthogonal uplink transmissions within a cell can be achieved in the time domain (*Time-Division Multiple Access*, TDMA), as illustrated in Figure 4.6a. TDMA implies that different terminals within in a cell transmit in different non-overlapping time intervals. In each such time interval, the full system bandwidth is then assigned for uplink transmission from a single terminal.

Alternatively, mutually orthogonal uplink transmissions within a cell can be achieved in the frequency domain (*Frequency-Division Multiple Access*, FDMA), as illustrated in Figure 4.6b – that is, by having terminals transmit in different frequency bands.

To be able to provide high-rate packet-data transmission, it should be possible to assign the entire system bandwidth for transmission from a single terminal. At the same time, due to the burstiness of most packet-data services, in many cases terminals will have no uplink data to transmit. Thus, for efficient packet-data access, a TDMA component should always be part of the uplink multiple-access scheme.

However, relying on only TDMA to provide orthogonality between uplink transmissions within a cell could be bandwidth inefficient, especially in case of a very wide overall system bandwidth. As discussed in Chapter 2, a wide bandwidth is needed to support high data rates in a power-efficient way. However, the data rates that can be achieved over a radio link are, in many cases, limited by the

---

[1]Thus, the term *Viterbi equalizer* is sometimes used for the ML detector.

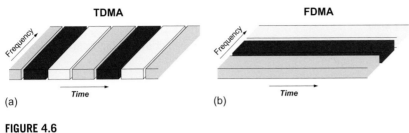

FIGURE 4.6

Orthogonal multiple access: (a) TDMA; (b) FDMA.

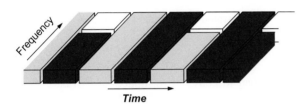

FIGURE 4.7

FDMA with flexible bandwidth assignment.

available signal power (power-limited operation) rather than by the available bandwidth. This is especially the case for the uplink, due to the, in general, more limited mobile-terminal transmit power. Allocating the entire system bandwidth to a single terminal could, in such cases, be highly inefficient in terms of bandwidth utilization. As an example, allocating 20 MHz of transmission bandwidth to a terminal in a scenario where the achievable uplink data rate, due to mobile-terminal transmit-power limitations, is anyway limited to, for example, a few 100 kbit/s would obviously imply a very inefficient usage of the overall available bandwidth. In such cases, a smaller transmission bandwidth should be assigned to the terminal and the remaining part of the overall system bandwidth should be used for uplink transmissions from other terminals. Thus, in addition to TDMA, an uplink transmission scheme should preferably allow for orthogonal user multiplexing also in the frequency domain – that is, FDMA.

At the same time, it should be possible to allocate the entire overall transmission bandwidth to a single terminal when the channel conditions are such that the wide bandwidth can be efficiently utilized – that is, when the achievable data rates are not power limited. Thus, an orthogonal uplink transmission scheme should allow for *FDMA with flexible bandwidth assignment*, as illustrated in Figure 4.7.

Flexible bandwidth assignment is straightforward to achieve with an OFDM-based uplink transmission scheme by dynamically allocating different number of subcarriers to different terminals depending on their instantaneous channel conditions. In the next section, it will be discussed how this can also be achieved in the case of low-PAR "single-carrier" transmission, more specifically by means of so-called *DFT-spread OFDM*.

4.3 **DFT-SPREAD OFDM**

DFT-spread OFDM (DFTS-OFDM) is a transmission scheme that can combine the desired properties discussed in the previous sections, that is:

- Small variations in the instantaneous power of the transmitted signal ("single-carrier" property).
- Possibility for low-complexity high-quality equalization in the frequency domain.
- Possibility for FDMA with flexible bandwidth assignment.

Due to these properties, DFTS-OFDM is, for example, used for uplink data transmission in LTE.

4.3.1 **Basic Principles**

The basic principle of DFTS-OFDM transmission is illustrated in Figure 4.8. One way to interpret DFTS-OFDM is as normal OFDM with a *DFT-based precoding*. Similar to OFDM modulation, DFTS-OFDM relies on block-based signal generation. In case of DFTS-OFDM, a block of M modulation symbols from some modulation alphabet, for example QPSK or 16QAM, is first applied to a size-M DFT. The output of the DFT is then applied to consecutive inputs (subcarriers) of an OFDM modulator where, in practice, the OFDM modulator will be implemented as a size-N *inverse* DFT (IDFT) with $N > M$ and where the unused inputs of the IDFT are set to zero. Typically, the IDFT size N is selected as $N = 2^n$ for some integer n to allow for the IDFT to be implemented by means of computationally efficient radix-2 IFFT processing. Also, similar to normal OFDM, a cyclic prefix is preferably inserted for each transmitted block. As discussed in Section 4.1.2, the presence of a cyclic prefix allows for straightforward low-complexity frequency-domain equalization at the receiver side.

If the DFT size M equals the IDFT size N, the cascaded DFT/IDFT processing would obviously completely cancel each other out. However, if M is smaller than N and the remaining inputs to the IDFT are set to zero, the output of the IDFT will be a signal with "single-carrier" properties – that is, a signal with low power variations, and with a bandwidth that depends on M. More specifically, assuming a sampling rate f_s at the output of the IDFT, the nominal bandwidth of the transmitted signal will be $BW = M/N \cdot f_s$. Thus, by varying the block size M the instantaneous bandwidth of the transmitted signal can be varied, allowing for flexible-bandwidth assignment. Furthermore, by shifting the IDFT inputs to which the DFT outputs are mapped, the transmitted signal can be shifted in the frequency domain, as will be further discussed in Section 4.3.3.

To have a high degree of flexibility in the instantaneous bandwidth, given by the DFT size M, it is typically not possible to ensure that M can be expressed as 2^m for some integer m. However, as long as M can be expressed as a product of relatively small prime numbers, the DFT can still be implemented as relatively low-complexity non-radix-2 FFT processing. As an example, a DFT size $M = 144$ can be implemented by means of a combination of radix-2 and radix-3 FFT processing ($144 = 3^2 \cdot 2^4$).

The main benefit of DFTS-OFDM, compared to normal OFDM, is reduced variations in the instantaneous transmission power, implying the possibility for increased power-amplifier efficiency. This benefit of DFTS-OFDM is shown in Figure 4.9, which illustrates the distribution of the *Peak-to-Average-power Ratio* (PAPR) for DFTS-OFDM and conventional OFDM. The PAPR is defined as the peak power within one DFT block (one OFDM symbol) normalized by the average signal power.

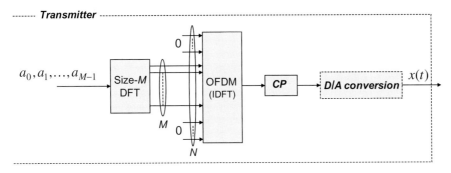

FIGURE 4.8

DFTS-OFDM signal generation.

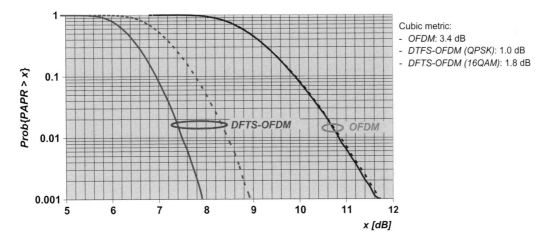

FIGURE 4.9

PAPR distribution for OFDM and DFTS-OFDM respectively. Solid curve: QPSK. Dashed curve: 16QAM.

It should be noted that the PAPR distribution is *not* the same as the distribution of the instantaneous transmission power illustrated in Figure 2.3. Historically, PAPR distributions have often been used to illustrate the power variations of OFDM.

As can be seen in Figure 4.9, the PAPR is significantly lower for DFTS-OFDM, compared to OFDM. In the case of 16QAM modulation, the PAPR of DFTS-OFDM increases somewhat as expected (compare Figure 2.3). On the other hand, in the case of OFDM the PAPR distribution is more or less independent of the modulation scheme. The reason is that, as the transmitted OFDM signal is the sum of a large number of independently modulated subcarriers, the instantaneous power has an approximately exponential distribution, regardless of the modulation scheme applied to the different subcarriers.

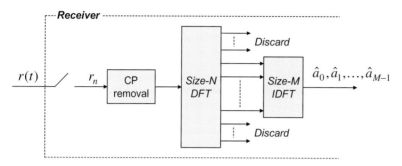

FIGURE 4.10

Basic principle of DFTS-OFDM demodulation.

Although the PAPR distribution can be used to qualitatively illustrate the difference in power variations between different transmission schemes, it is not a very good measure to more accurately quantify the impact of the power variations on, for example, the required power-amplifier back-off.

A better measure of the impact on the required power-amplifier back-off and the corresponding impact on the power-amplifier efficiency is given by the so-called *cubic metric* [27]. The cubic metric is a measure of the amount of additional back-off needed for a certain signal waveform, relative to the back-off needed for some reference waveform.

As can be seen from Figure 4.9, the cubic metric (given to the right of the graph) follows the same trend as the PAPR. However, the differences in cubic metric are somewhat smaller than the corresponding differences in PAPR.

4.3.2 DFTS-OFDM Receiver

The basic principle of demodulation of a DFTS-OFDM signal is illustrated in Figure 4.10. The operations are basically the reverse of those for the DFTS-OFDM signal generation of Figure 4.8 – that is, size-N DFT (FFT) processing, removal of the frequency samples not corresponding to the signal to be received, and size-M inverse DFT processing.

In the ideal case, with no signal corruption on the radio channel, DFTS-OFDM demodulation according to Figure 4.10 will perfectly restore the block of transmitted symbols. However, in the case of a time-dispersive or, equivalently, a frequency-selective radio channel, the DFTS-OFDM signal will be corrupted, with "self-interference" as a consequence. This can be understood in two ways:

1. Being a wideband single-carrier signal, the DFTS-OFDM spread signal is obviously corrupted in the case of a time-dispersive channel.
2. If the channel is frequency selective over the span of the DFT, the inverse DFT at the receiver will not be able to correctly reconstruct the original block of transmitted symbols.

Thus, in the case of DFTS-OFDM, an equalizer is needed to compensate for the radio-channel frequency selectivity. Assuming the basic DFTS-OFDM demodulator structure according to Figure 4.10, frequency-domain equalization as discussed in Section 4.1.2 is especially applicable to DFTS-OFDM transmission (see Figure 4.11).

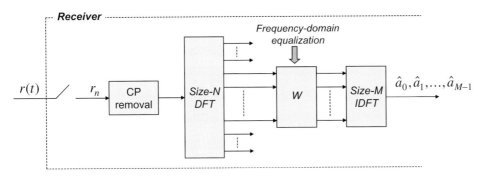

FIGURE 4.11

DFTS-OFDM demodulator with frequency-domain equalization.

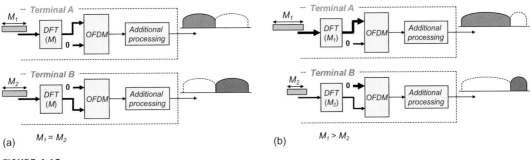

(a) $M_1 = M_2$ (b) $M_1 > M_2$

FIGURE 4.12

Uplink user multiplexing in case of DFTS-OFDM. (a) Equal-bandwidth assignment. (b) Unequal-bandwidth assignment.

4.3.3 User Multiplexing with DFTS-OFDM

As mentioned above, by dynamically adjusting the transmitter DFT size and, consequently, also the size of the block of modulation symbols a_0, a_1, ..., a_{M-1}, the nominal bandwidth of the DFTS-OFDM signal can be dynamically adjusted. Furthermore, by shifting the IDFT inputs to which the DFT outputs are mapped, the exact frequency-domain "position" of the signal to be transmitted can be adjusted. By these means, DFTS-OFDM allows for uplink FDMA with flexible bandwidth assignment, as illustrated in Figure 4.12.

Figure 4.12a illustrates the case of multiplexing the transmissions from two terminals with equal bandwidth assignments – that is, equal DFT sizes M – while Figure 4.12b illustrates the case of differently sized bandwidth assignments.

4.3.4 Distributed DFTS-OFDM

What has been illustrated in Figure 4.8 can more specifically be referred to as *Localized DFTS-OFDM*, referring to the fact that the output of the DFT is mapped to *consecutive* inputs of the OFDM

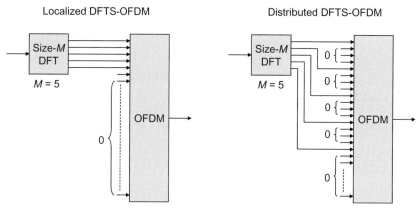

FIGURE 4.13

Localized DFTS-OFDM vs. distributed DFTS-OFDM.

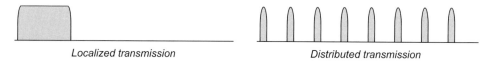

Localized transmission   Distributed transmission

FIGURE 4.14

Spectrum of localized and distributed DFTS-OFDM signals.

Localized transmission   Distributed transmission

FIGURE 4.15

User multiplexing in the case of localized and distributed DFTS-OFDM.

modulator. An alternative is to map the output of the DFT to *equidistant* inputs of the OFDM modulator with zeros inserted in between, as illustrated in Figure 4.13. This can also be referred to as *Distributed DFTS-OFDM*.

Figure 4.14 illustrates the basic structure of the transmitted spectrum for localized and distributed DFTS-OFDM respectively. Although the spectrum of the localized DFTS-OFDM signal clearly indicates a single-carrier transmission, this is not as clearly seen from the spectrum of the distributed DFTS-OFDM signal. However, it can be shown that a distributed DFTS-OFDM signal has similar power variations as localized DFTS-OFDM. Actually, it can be shown that a distributed DFTS-OFDM signal is equivalent to so-called *Interleaved FDMA* (IFDMA) [28]. The benefit of distributed DFTS-OFDM, compared to localized DFTS-OFDM, is the possibility of additional frequency diversity as even a low-rate distributed DFTS-OFDM signal (small DFT size *M*) can be spread over a potentially very large overall transmission bandwidth.

User multiplexing in the frequency domain as well as flexible bandwidth allocation is also possible in the case of distributed DFTS-OFDM. However, in this case, the different users are interleaved in the frequency domain, as illustrated in the right part of Figure 4.15 (thus the alternative term "*Interleaved FDMA*"). As a consequence, distributed DFTS-OFDM is more sensitive to frequency errors and has higher requirements on power control, compared to localized DFTS-OFDM. This is similar to the case of localized OFDM vs. distributed OFDM as discussed in Section 3.10.

Multi-Antenna Techniques

Multi-antenna techniques can be seen as a joint name for a set of techniques with the common theme that they rely on the use of multiple antennas at the receiver and/or the transmitter, in combination with more or less advanced signal processing. Multi-antenna techniques can be used to achieve improved system performance, including improved system capacity (more users per cell) and improved coverage (possibility for larger cells), as well as improved service provisioning – for example, higher per-user data rates. This chapter will provide a general overview of different multi-antenna techniques. How multi-antenna techniques are specifically applied to LTE is discussed in more detail in Chapters 10 and 11.

5.1 MULTI-ANTENNA CONFIGURATIONS

An important characteristic of any multi-antenna configuration is the distance between the different antenna elements, to a large extent due to the relation between the antenna distance and the mutual correlation between the radio-channel fading experienced by the signals at the different antennas.

The antennas in a multi-antenna configuration can be located relatively far from each other, typically implying a relatively low mutual correlation. Alternatively, the antennas can be located relatively close to each other, typically implying a high mutual fading correlation – that is, in essence, that the different antennas experience the same, or at least very similar, instantaneous fading. Whether high or low correlation is desirable depends on what is to be achieved with the multi-antenna configuration (diversity, beam-forming, or spatial multiplexing), as discussed further below.

What actual antenna distance is needed for low (alternatively high) fading correlation depends on the wavelength or, equivalently, the carrier frequency used for the radio communication. However, it also depends on the deployment scenario.

In the case of base-station antennas in typical macro-cell environments (relatively large cells, relatively high base-station antenna positions, etc.), an antenna distance of the order of 10 wavelengths is typically needed to ensure a low mutual fading correlation. At the same time, for a terminal in the same kind of environment, an antenna distance of the order of only half a wavelength (0.5λ) is often sufficient to achieve relatively low mutual correlation [29]. The reason for the difference between the base station and the terminal in this respect is that, in the macro-cell scenario, the multi-path reflections that cause the fading mainly occur in the near-zone around the terminal. Thus, as seen from the terminal, the different paths will typically arrive from a wide angle, implying a low fading correlation already with a relatively small antenna distance. At the same time, as seen from the (macro-cell) base station the different paths will typically arrive within a much smaller angle, implying the need for significantly larger antenna distance to achieve low fading correlation.[1]

On the other hand, in other deployment scenarios, such as micro-cell deployments with base-station antennas below rooftop level and indoor deployments, the environment as seen from the base station is more similar to the environment as seen from the terminal. In such scenarios, a smaller base-station antenna distance is typically sufficient to ensure relatively low mutual correlation between the fading experienced by the different antennas.

The above discussion assumed antennas with the same polarization direction. Another means to achieve low mutual fading correlation is to apply different polarization directions for the different antennas [29]. The antennas can then be located relatively close to each other, implying a compact antenna arrangement, while still experiencing low mutual fading correlation.

5.2 BENEFITS OF MULTI-ANTENNA TECHNIQUES

The availability of multiple antennas at the transmitter and/or the receiver can be utilized in different ways to achieve different aims:

- Multiple antennas at the transmitter and/or the receiver can be used to provide additional diversity against fading on the radio channel. In this case, the channels experienced by the different antennas should have low mutual correlation, implying the need for a sufficiently large inter-antenna distance (*spatial diversity*), or the use of different antenna polarization directions (*polarization diversity*).
- Multiple antennas at the transmitter and/or the receiver can be used to "shape" the overall antenna beam (transmit beam and receive beam respectively) in a certain way – for example, to maximize the overall antenna gain in the direction of the target receiver/transmitter or to suppress specific dominant interfering signals. Such *beam-forming* can be based either on high or low fading correlation between the antennas, as is further discussed in Section 5.4.2.
- The simultaneous availability of multiple antennas at the transmitter and the receiver can be used to create what can be seen as multiple parallel communication "channels" over the radio interface. This provides the possibility for very high bandwidth utilization without a corresponding reduction in power efficiency or, in other words, the possibility for very high data rates within a limited bandwidth without a disproportionately large degradation in terms of coverage. Herein we will refer to this as *spatial multiplexing*. It is often also referred to as MIMO (Multi-Input Multi-Output) antenna processing. Spatial multiplexing is discussed in Section 5.5.

5.3 MULTIPLE RECEIVE ANTENNAS

Perhaps the most straightforward and historically the most commonly used multi-antenna configuration is the use of multiple antennas at the receiver side. This is often referred to as *receive diversity*

---

[1]Although the term "arrive" is used above, the situation is exactly the same in the case of multiple *transmit* antennas. Thus, in the case of multiple base-station transmit antennas in a macro-cell scenario, the antenna distance typically needs to be a number of wavelengths to ensure low fading correlation while, in the case of multiple transmit antennas at the terminal, an antenna distance of a fraction of a wavelength is sufficient.

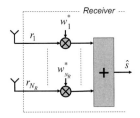

FIGURE 5.1

Linear receive-antenna combination.

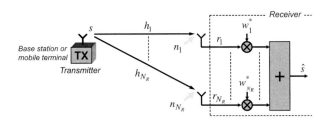

FIGURE 5.2

Channel gains together with linear receive-antenna combination.

or *RX diversity* even if the aim of the multiple receive antennas is not always to achieve additional diversity against radio-channel fading.

Figure 5.1 illustrates the basic principle of linear combination of signals $r_1, \ldots, r_{N_R}$ received at N_R different antennas, with the received signals being multiplied by complex weight factors $w_1^*, \ldots, w_{N_R}^*$ before being added together. In vector notation this *linear receive-antenna combination* can be expressed as:[2]

$$\hat{s} = [w_1^*, \ldots, w_{N_R}^*] \cdot \begin{bmatrix} r_1 \\ \vdots \\ r_{N_R} \end{bmatrix} = \bar{w}^H \cdot \bar{r}. \tag{5.1}$$

What is outlined in Eqn (5.1) and Figure 5.1 is linear receive-antenna combination in general. Different specific antenna-combining approaches then differ in the exact choice of the weight vector $\bar{w}$.

Assuming that the transmitted signal is only subject to non-frequency-selective fading and white noise – that is, there is no radio-channel time dispersion – the signals received at the different antennas in Figure 5.1 can be expressed as:

$$\bar{r} = \begin{pmatrix} r_1 \\ \vdots \\ r_{N_D} \end{pmatrix} = \begin{pmatrix} h_1 \\ \vdots \\ h_{N_D} \end{pmatrix} \cdot s + \begin{pmatrix} n_1 \\ \vdots \\ n_{N_D} \end{pmatrix} = \bar{h} \cdot s + \bar{n}, \tag{5.2}$$

where s is the transmitted signal, the vector $\bar{h}$ consists of the N_R complex channel gains, and the vector $\bar{n}$ consists of the noise impairing the signals received at the different antennas (see also Figure 5.2).

One can easily show that, to maximize the signal-to-noise ratio after linear combination, the weight vector $\bar{w}$ should be selected as [12]:

$$\bar{w}_{\text{MRC}} = \bar{h}. \tag{5.3}$$

---

[2] Note that the weight factors are expressed as complex conjugates of $w_1, \ldots, w_{N_R}$.

This is also known as *Maximum-Ratio Combination* (MRC). The MRC weights fulfill two purposes:

- phase rotate the signals received at the different antennas to compensate for the corresponding channel phases and ensure that the signals are phase aligned when added together (coherent combination);
- weight the signals in proportion to their corresponding channel gains – that is, apply higher weights for stronger received signals.

In the case of mutually uncorrelated antennas – that is, sufficiently large antenna distances or different polarization directions – the channel gains $h_1, \ldots, h_{N_R}$ are uncorrelated and the linear antenna combination provides diversity of order N_R. In terms of receiver-side beam-forming, selecting the antenna weights according to Eqn (5.3) corresponds to a receiver beam with maximum gain N_R in the direction of the target signal. Thus, the use of multiple receive antennas may increase the post-combiner signal-to-noise ratio in proportion to the number of receive antennas.

MRC is an appropriate antenna-combining strategy when the received signal is mainly impaired by noise. However, in many cases of mobile communication the received signal is mainly impaired by interference from other transmitters within the system, rather than by noise. In a situation with a relatively large number of interfering signals of approximately equal strength, maximum-ratio combination is typically still a good choice as, in this case, the overall interference will appear relatively "noise-like" with no specific direction of arrival. However, in situations where there is a single dominating interferer (or, in the general case, a limited number of dominating interferers), as illustrated in Figure 5.3, improved performance can be achieved if, instead of selecting the antenna weights to maximize the received signal-to-noise ratio after antenna combination (MRC), the antenna weights are selected so that the interferer is suppressed. In terms of receiver-side beam-forming this corresponds to a receiver beam with high attenuation in the direction of the interferer, rather than focusing the receiver beam in the direction of the target signal. Applying receive-antenna combination with a target to suppress specific interferers is often referred to as *Interference Rejection Combination* (IRC) [30].

In the case of a single dominating interferer as outlined in Figure 5.3, expression (5.2) can be extended according to:

$$\bar{r} = \begin{pmatrix} r_1 \\ \vdots \\ r_{N_R} \end{pmatrix} = \begin{pmatrix} h_1 \\ \vdots \\ h_{N_R} \end{pmatrix} \cdot s + \begin{pmatrix} h_{I,1} \\ \vdots \\ h_{I,N_R} \end{pmatrix} \cdot s_I + \begin{pmatrix} n_1 \\ \vdots \\ n_{N_R} \end{pmatrix} = \bar{h} \cdot s + \bar{h}_I \cdot s_I + \bar{n}, \tag{5.4}$$

where s_I is the transmitted interferer signal and the vector $\bar{h}_I$ consists of the complex channel gains from the interferer to the N_R receive antennas. By applying expressions (5.1)–(5.4) it is clear that the interfering signal will be completely suppressed if the weight vector $\bar{w}$ is selected to fulfill the expression:

$$\bar{w}^H \cdot \bar{h}_I = 0. \tag{5.5}$$

In the general case, Eqn (5.5) has $N_R - 1$ non-trivial solutions, indicating flexibility in the weight-vector selection. This flexibility can be used to suppress additional dominating interferers. More specifically, in the general case of N_R receive antennas there is a possibility to, at least in theory,

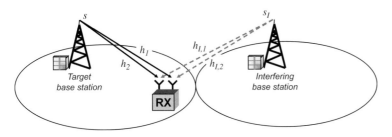

FIGURE 5.3

Downlink scenario with a single dominating interferer (special case of only two receive antennas).

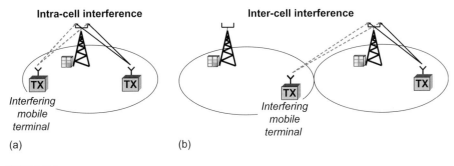

FIGURE 5.4

Receiver scenario with one strong interfering terminal. (a) Intra-cell interference. (b) Inter-cell interference.

completely suppress up to $N_R - 1$ separate interferers. However, such a choice of antenna weights, providing complete suppression of a number of dominating interferers, may lead to a large, potentially very large, increase in the noise level after the antenna combination. This is similar to the potentially large increase in the noise level in the case of a Zero-Forcing equalizer as discussed in Chapter 4.

Thus, similar to the case of linear equalization, a better approach is to select the antenna weight vector $\bar{w}$ to minimize the mean square error:

$$\varepsilon = E\{|\hat{s} - s|^2\}, \tag{5.6}$$

also known as *Minimum Mean Square Error* (MMSE) combination [22,29].

Although Figure 5.3 illustrates a downlink scenario with a dominating interfering base station, IRC can also be applied to the uplink to suppress interference from specific terminals. In this case, the interfering terminal may either be in the same cell as the target terminal (*intra-cell interference*) or in a neighboring cell (*inter-cell interference*) (see Figure 5.4a and b respectively). Suppression of intra-cell interference is relevant in the case of a non-orthogonal uplink – that is, when multiple terminals are transmitting simultaneously using the same time–frequency resource. Uplink intra-cell-interference suppression by means of IRC is sometimes also referred to as *Spatial-Division Multiple Access* (SDMA) [31,32].

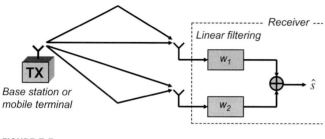

FIGURE 5.5

Two-dimensional space–time linear processing (two receive antennas).

As discussed in Chapter 2, in practice a radio channel is always subject to at least some degree of time dispersion or, equivalently, frequency selectivity, causing corruption to a wideband signal. As discussed in Chapter 4, one method to counteract such signal corruption is to apply linear equalization, either in the time or frequency domain.

It should be clear from the discussion above that linear receive-antenna combination has many similarities to linear equalization:

- Linear time-domain (frequency-domain) filtering/equalization as described in Chapter 4 implies that linear processing is applied to signals received at different time instances (different frequencies) with a target to maximize the post-equalizer SNR (MRC-based equalization), or to suppress signal corruption due to radio-channel frequency selectivity (zero-forcing equalization, MMSE equalization, etc.).
- Linear receive-antenna combination implies that linear processing is applied to signals received at different antennas – that is, processing in the spatial domain – with a target to maximize the post-combiner SNR (MRC-based combination), or to suppress specific interferers (IRC based on e.g. MMSE).

Thus, in the general case of frequency-selective channel and multiple receive antennas, two-dimensional time/space linear processing/filtering can be applied as illustrated in Figure 5.5, where the linear filtering can be seen as a generalization of the antenna weights of Figure 5.1. The filters should be jointly selected to minimize the overall impact of noise, interference, and signal corruption due to radio-channel frequency selectivity.

Alternatively, especially in the case when cyclic-prefix insertion has been applied at the transmitter side, two-dimensional frequency/space linear processing can be applied, as illustrated in Figure 5.6. The frequency-domain weights should then be jointly selected to minimize the overall impact of noise, interference, and signal corruption due to radio-channel frequency selectivity.

The frequency/space processing outlined in Figure 5.6, without the IDFT, is also applicable if receive diversity is to be applied to OFDM transmission. In the case of OFDM transmission, there is no signal corruption due to radio-channel frequency selectivity. Thus, the frequency-domain coefficients of Figure 5.6 can be selected taking into account only noise and interference. In principle,

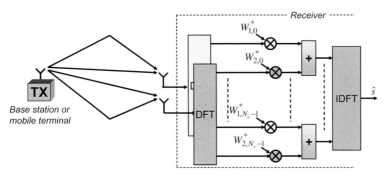

FIGURE 5.6

Two-dimensional space–frequency linear processing (two receive antennas).

this means the antenna-combining schemes discussed above (MRC and IRC) are applied on a *per-subcarrier* basis.

Note that, although Figures 5.5 and 5.6 assume two receive antennas, the corresponding receiver structures can straightforwardly be extended to more than two antennas.

5.4 MULTIPLE TRANSMIT ANTENNAS

As an alternative, or complement, to multiple receive antennas, diversity and beam-forming can also be achieved by applying multiple antennas at the transmitter side. The use of multiple transmit antennas is primarily of interest for the downlink – that is, at the base station. In this case, the use of multiple transmit antennas provides an opportunity for diversity and beam-forming without the need for additional receive antennas and corresponding additional receiver chains at the terminal. On the other hand, due to complexity the use of multiple transmit antennas for the uplink – that is, at the terminal – is less attractive. In this case, it is typically preferred to apply additional receive antennas and corresponding receiver chains at the base station.

5.4.1 Transmit-Antenna Diversity

If no knowledge of the downlink channels of the different transmit antennas is available at the transmitter, multiple transmit antennas cannot provide beam-forming but only diversity. For this to be possible, there should be low mutual correlation between the channels of the different antennas. As discussed in Section 5.1, this can be achieved by means of sufficiently large distance between the antennas, or by the use of different antenna polarization directions. Assuming such antenna configurations, different approaches can be taken to realize the diversity offered by the multiple transmit antennas.

5.4.1.1 *Delay Diversity*

As discussed in Chapter 2, a radio channel subject to time dispersion, with the transmitted signal propagating to the receiver via multiple, independently fading paths with different delays, provides

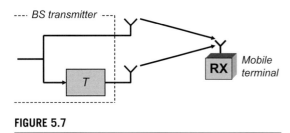

FIGURE 5.7

Two-antenna delay diversity.

the possibility for multi-path diversity or, equivalently, frequency diversity. Thus, multi-path propagation is actually beneficial in terms of radio-link performance, assuming that the amount of multi-path propagation is not too extensive and that the transmission scheme includes tools to counteract signal corruption due to the radio-channel frequency selectivity – for example, by means of OFDM transmission or the use of advanced receiver-side equalization.

If the channel in itself is not time dispersive, the availability of multiple transmit antennas can be used to create *artificial time dispersion* or, equivalently, *artificial frequency selectivity* by transmitting identical signals with different relative delays from the different antennas. In this way, the antenna diversity – that is, the fact that the fading experienced by the different antennas have low mutual correlation – can be transformed into frequency diversity. This kind of *delay diversity* is illustrated in Figure 5.7 for the special case of two transmit antennas. The relative delay T should be selected to ensure a suitable amount of frequency selectivity over the bandwidth of the signal to be transmitted. It should be noted that, although Figure 5.7 assumes two transmit antennas, delay diversity can straightforwardly be extended to more than two transmit antennas with different relative delays for each antenna.

Delay diversity is in essence invisible to the terminal, which will simply see a single radio channel subject to additional time dispersion. Delay diversity can thus straightforwardly be introduced in an existing mobile-communication system without requiring any specific support in a corresponding radio-interface standard. Delay diversity is also applicable to basically any kind of transmission scheme that is designed to handle and benefit from frequency-selective fading.

5.4.1.2 *Cyclic-Delay Diversity*

Cyclic-Delay Diversity (CDD) [33] is similar to delay diversity with the main difference that cyclic-delay diversity operates block-wise and applies *cyclic shifts*, rather than linear delays, to the different antennas (see Figure 5.8). Thus, cyclic-delay diversity is applicable to block-based transmission schemes such as OFDM and DFTS-OFDM.

In the case of OFDM transmission, a cyclic shift of the time-domain signal corresponds to a frequency-dependent phase shift before OFDM modulation, as illustrated in Figure 5.8b. Similar to delay diversity, this will create artificial frequency selectivity as seen by the receiver.

Also similar to delay diversity, CDD can straightforwardly be extended to more than two transmit antennas with different cyclic shifts for each antenna.

5.4.1.3 *Diversity by Means of Space–Time Coding*

Space–time coding is a general term used to indicate multi-antenna transmission schemes where modulation symbols are mapped in the time and spatial (transmit-antenna) domain to capture the

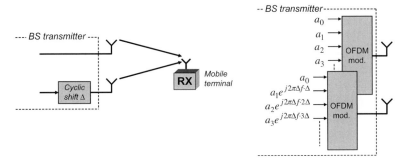

FIGURE 5.8

Two-antenna Cyclic-Delay Diversity (CDD).

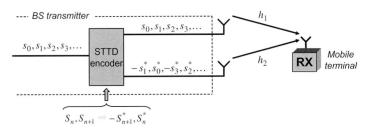

FIGURE 5.9

Space–Time Transmit Diversity (STTD).

diversity offered by the multiple transmit antennas. Two-antenna space–time block coding (STBC), more specifically a scheme referred to as *Space–Time Transmit Diversity* (STTD), has been part of the 3G WCDMA standard already from its first release [34].

As shown in Figure 5.9, STTD operates on pairs of modulation symbols. The modulation symbols are directly transmitted on the first antenna. However, on the second antenna the order of the modulation symbols within a pair is reversed. Furthermore, the modulation symbols are sign-reversed and complex-conjugated, as illustrated in Figure 5.9.

In vector notation, STTD transmission can be expressed as:

$$\bar{r} = \begin{pmatrix} r_{2n} \\ r_{2n+1}^* \end{pmatrix} = \begin{pmatrix} h_1 & -h_2 \\ h_2^* & h_1^* \end{pmatrix} \cdot \begin{pmatrix} s_{2n} \\ s_{2n+1}^* \end{pmatrix} = H \cdot \bar{s}, \tag{5.7}$$

where r_{2n} and r_{2n+1} are the received symbols during the symbol intervals $2n$ and $2n + 1$ respectively.[3] It should be noted that this expression assumes that the channel coefficients h_1 and h_2 are constant over the time corresponding to two consecutive symbol intervals, an assumption that is typically

---

[3] Note that, for convenience, complex conjugates have been applied for the second elements of $\bar{r}$ and $\bar{s}$.

valid. As the matrix H is a scaled unitary matrix, the sent symbols s_{2n} and s_{2n+1} can be recovered from the received symbols r_{2n} and r_{2n+1}, without any interference between the symbols, by applying the matrix $W = H^{-1}$ to the vector $\bar{r}$.

The two-antenna space–time coding of Figure 5.9 can be said to be of rate one, implying that the input symbol rate is the same as the symbol rate at each antenna, corresponding to a bandwidth utilization of 1. Space–time coding can also be extended to more than two antennas. However, in the case of complex-valued modulation, such as QPSK or 16/64QAM, space–time codes of rate one without any inter-symbol interference (*orthogonal space–time codes*) only exist for two antennas [35]. If inter-symbol interference is to be avoided in the case of more than two antennas, space–time codes with rate less than one must be used, corresponding to reduced bandwidth utilization.

5.4.1.4 *Diversity by Means of Space–Frequency Coding*

Space–frequency block coding (SFBC) is similar to space–time block coding, with the difference that the encoding is carried out in the antenna/frequency domains rather than in the antenna/time domains. Thus, space–frequency coding is applicable to OFDM and other "frequency-domain" transmission schemes. The space–frequency equivalence to STTD (which could also be referred to as Space–Frequency Transmit Diversity, SFTD) is illustrated in Figure 5.10. As can be seen, the block of (frequency-domain) modulation symbols $a_0, a_1, a_2, a_3, \ldots$ is directly mapped to OFDM carriers of the first antenna, while the block of symbols $-a_1^*, a_0^*, -a_3^*, a_2^*$ is mapped to the corresponding subcarriers of the second antenna.

Similar to space–time coding, the drawback of space–frequency coding is that there is no straightforward extension to more than two antennas unless a rate reduction is acceptable.

Comparing Figure 5.10 with the right-hand side of Figure 5.8, it can be noted that the difference between SFBC and two-antenna cyclic-delay diversity in essence lies in how the block of frequency-domain modulation symbols are mapped to the second antenna. The benefit of SFBC compared to CDD is that SFBC provides diversity at modulation-symbol level while CDD, in the case of OFDM, must rely on channel coding in combination with frequency-domain interleaving to provide diversity.

5.4.2 Transmitter-Side Beam-Forming

If some knowledge of the downlink channels of the different transmit antennas (more specifically, some knowledge of the relative channel phases) is available at the transmitter side, multiple transmit antennas can, in addition to diversity, also provide beam-forming – that is, the shaping of the overall antenna beam in the direction of a target receiver. In general, such beam-forming can increase the signal strength at the receiver with up to a factor N_T – that is, in proportion to the number of transmit antennas. When discussing transmission schemes relying on multiple transmit antennas to provide beam-forming one can distinguish between the cases of *high* and *low mutual antenna correlation* respectively.

High mutual antenna correlation typically implies an antenna configuration with a small inter-antenna distance, as illustrated in Figure 5.11a. In this case, the channels between the different antennas and a specific receiver are essentially the same, including the same radio-channel fading, except for a direction-dependent phase difference. The overall transmission beam can then be steered in

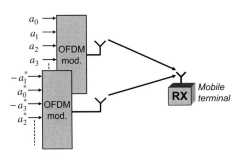

FIGURE 5.10

Space–Frequency Block Coding (SFBC) assuming two transmit antennas.

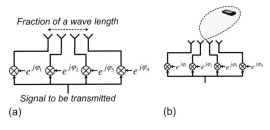

(a)   (b)

FIGURE 5.11

Classical beam-forming with high mutual antennas correlation: (a) antenna configuration; (b) beam-structure.

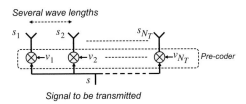

FIGURE 5.12

Pre-coder-based beam-forming in the case of low mutual antenna correlation.

different directions by applying *different phase shifts* to the signals to be transmitted on the different antennas, as illustrated in Figure 5.11b.

This approach to transmitter-side beam-forming, with different phase shifts applied to highly correlated antennas, is sometimes referred to as "classical" beam-forming. Due to the small antenna distance, the overall transmission beam will be relatively wide and any adjustments of the beam direction – in practice, adjustments of the antenna phase shifts – will typically be carried out on a relatively slow basis. The adjustments could, for example, be based on estimates of the direction to the target terminal derived from uplink measurements. Furthermore, due to the assumption of high correlation between the different transmit antennas, classical beam-forming cannot provide any diversity against radio-channel fading but only an increase of the received signal strength.

Low mutual antenna correlation typically implies either a sufficiently large antenna distance, as illustrated in Figure 5.12, or different antenna polarization directions. With low mutual antenna correlation, the basic beam-forming principle is similar to that of Figure 5.11 – that is, the signals to be transmitted on the different antennas are multiplied by different complex weights. However, in contrast to classical beam-forming, the antenna weights should now take general complex values and both the phase and the amplitude of the signals to be transmitted on the different antennas can be adjusted. This reflects the fact that, due to the low mutual antenna correlation, both the phase and the instantaneous gain of the channels of each antenna may differ.

Applying different complex weights to the signals to be transmitted on the different antennas can be expressed, in vector notation, as applying a *precoding vector* $\bar{v}$ to the signal to be transmitted, according to:

$$\bar{s} = \begin{pmatrix} s_1 \\ \vdots \\ s_{N_T} \end{pmatrix} = \begin{pmatrix} v_1 \\ \vdots \\ v_{N_T} \end{pmatrix} \cdot s = \bar{v} \cdot s. \tag{5.8}$$

It should be noted that classical beam-forming according to Figure 5.11 can also be described according to Eqn (5.8) – that is, as *transmit-antenna precoding* – with the constraint that the antenna weights are limited to unit gain and only provide phase shifts to the different transmit antennas.

Assuming that the signals transmitted from the different antennas are only subject to non-frequency-selective fading and white noise – that is, there is no radio-channel time dispersion – it can be shown [36] that, in order to maximize the received signal power, the precoding weights should be selected according to:

$$v_i = \frac{h_i^*}{\sqrt{\sum_{k=1}^{N_T} |h_k|^2}}. \tag{5.9}$$

That is, as the complex conjugate of the corresponding channel coefficient h_i and with a normalization to ensure a fixed overall transmit power. The precoding vector thus:

- phase rotates the transmitted signals to compensate for the instantaneous channel phase and ensure that the received signals are received phase aligned;
- allocates power to the different antennas with, in general, more power being allocated to antennas with good instantaneous channel conditions (high channel gain $|h_i|$);
- ensures an overall unit (or any other constant) transmit power.

A key difference between classical beam-forming according to Figure 5.11, assuming high mutual antenna correlation, and beam-forming according to Figure 5.12, assuming low mutual antenna correlation, is that, in the latter case, there is a need for more detailed channel knowledge, including estimates of the instantaneous channel fading. Updates to the precoding vector are thus typically done on a relatively short time scale to capture the fading variations. As the adjustment of the precoder weights also takes into account the instantaneous fading, including the instantaneous channel gain, fast beam-forming according to Figure 5.12 also provides diversity against radio-channel fading.

Furthermore, at least in the case of communication based on *Frequency-Division Duplex* (FDD), with uplink and downlink communication taking place in different frequency bands, the fading is typically uncorrelated between the downlink and uplink. Thus, in the case of FDD, only the terminal can determine the downlink fading. The terminal would then report an estimate of the downlink channel to the base station by means of uplink signaling. Alternatively, the terminal may, in itself, select a suitable precoding vector from a limited set of possible precoding vectors, the so-called precoder codebook, and report this to the base station.

On the other hand, in the case of *Time-Division Duplex* (TDD), with uplink and downlink communication taking place in the same frequency band but in separate non-overlapping time slots, there

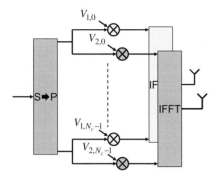

FIGURE 5.13

Per-subcarrier precoding in the case of
OFDM (two transmit antennas).

is typically a high fading correlation between the downlink and uplink. In this case, the base station could, at least in theory, determine the instantaneous downlink fading from measurements on the uplink, thus avoiding the need for any feedback. Note, however, that this assumes that the terminal is continuously transmitting on the uplink.

The above discussion assumed that the channel gain was constant in the frequency domain – that is, there was no radio-channel frequency selectivity. In the case of a frequency-selective channel there is obviously not a single channel coefficient per antenna, based on which the antennas weights can be selected according to Eqn (5.9). However, in the case of OFDM transmission, each subcarrier will typically experience a frequency-non-selective channel. Thus, in the case of OFDM transmission, the precoding of Figure 5.12 can be carried out on a per-subcarrier basis as outlined in Figure 5.13, where the precoding weights of each subcarrier should be selected according to Eqn (5.9).

It should be pointed out that in the case of single-carrier transmission the one-weight-per-antenna approach outlined in Figure 5.12 can be extended to also take into account a time-dispersive/frequency-selective channel [37].

5.5 SPATIAL MULTIPLEXING

The use of multiple antennas at both the transmitter and the receiver can simply be seen as a tool to further improve the signal-to-noise/interference ratio and/or achieve additional diversity against fading, compared to the use of only multiple receive antennas or multiple transmit antennas. However, in the case of multiple antennas at both the transmitter and the receiver there is also the possibility for so-called *spatial multiplexing*, allowing for more efficient utilization of high signal-to-noise/interference ratios and significantly higher data rates over the radio interface.

5.5.1 Basic Principles

It should be clear from the previous sections that multiple antennas at the receiver and the transmitter can be used to improve the receiver signal-to-noise ratio in proportion to the number of antennas by applying beam-forming at the receiver and the transmitter. In the general case of N_T transmit antennas and N_R receive antennas, the receiver signal-to-noise ratio can be made to increase in proportion to the product $N_T \times N_R$. As discussed in Chapter 2, such an increase in the receiver signal-to-noise ratio allows for a corresponding increase in the achievable data rates, assuming that the data rates are power limited rather than bandwidth limited. However, once the bandwidth-limited range of operation is reached, the achievable data rates start to saturate unless the bandwidth is also allowed to increase.

One way to understand this saturation in achievable data rates is to consider the basic expression for the normalized channel capacity:

$$\frac{C}{BW} = \log_2\left(1 + \frac{S}{N}\right), \tag{5.10}$$

where, by means of beam-forming, the signal-to-noise ratio S/N can be made to grow proportionally to $N_T \times N_R$. In general, $\log_2(1 + x)$ is proportional to x for small x, implying that, for low signal-to-noise ratios, the capacity grows approximately proportionally to the signal-to-noise ratio. However, for larger x, $\log_2(1 + x) \approx \log_2(x)$, implying that, for larger signal-to-noise ratios, capacity grows only logarithmically with the signal-to-noise ratio.

However, in the case of multiple antennas at the transmitter *and* the receiver it is, *under certain conditions*, possible to create up to $N_L = \min\{N_T, N_R\}$ parallel "channels" each with N_L times lower signal-to-noise ratio (the signal power is "split" between the channels) – that is, with a channel capacity:

$$\frac{C}{BW} = \log_2\left(1 + \frac{N_R}{N_L} \cdot \frac{S}{N}\right). \tag{5.11}$$

As there are now N_L parallel channels, each with a channel capacity given by Eqn (5.11), the overall channel capacity for such a multi-antenna configuration is thus given by:

$$\begin{aligned}
\frac{C}{BW} &= N_L \cdot \log_2\left(1 + \frac{N_R}{N_L} \cdot \frac{S}{N}\right) \\
&= \min\{N_T, N_R\} \cdot \log_2\left(1 + \frac{N_R}{\min\{N_T, N_R\}} \cdot \frac{S}{N}\right).
\end{aligned} \tag{5.12}$$

Thus, *under certain conditions*, the channel capacity can be made to grow essentially linearly with the number of antennas, avoiding the saturation in the data rates. We will refer to this as *Spatial Multiplexing*. The term MIMO (Multiple-Input/Multiple-Output) antenna processing is also very often used, although the term strictly speaking refers to all cases of multiple transmit antennas and multiple receive antennas, also including the case of combined transmit and receive diversity.[4]

To understand the basic principles how multiple parallel channels can be created in the case of multiple antennas at the transmitter and the receiver, consider a 2×2 antenna configuration – that is, two transmit antennas and two receive antennas – as outlined in Figure 5.14. Furthermore, assume that the transmitted signals are only subject to non-frequency-selective fading and white noise – that is, there is no radio-channel time dispersion.

---

[4]The case of a single transmit antenna and multiple receive antennas is, consequently, often referred to as SIMO (Single-Input/Multiple-Output). Similarly, the case of multiple transmit antennas and a single receiver antenna can be referred to as MISO (Multiple-Input/Single-Output).

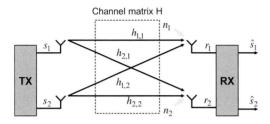

FIGURE 5.14

2×2 antenna configuration.

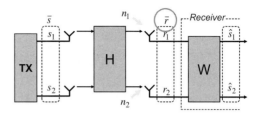

FIGURE 5.15

Linear reception/demodulation of spatially multiplexed signals.

Based on Figure 5.14, the received signals can be expressed as:

$$\bar{r} = \begin{pmatrix} r_1 \\ r_2 \end{pmatrix} = \begin{pmatrix} h_{1,1} & h_{1,2} \\ h_{2,1} & h_{2,2} \end{pmatrix} \cdot \begin{pmatrix} s_1 \\ s_2 \end{pmatrix} + \begin{pmatrix} n_1 \\ n_2 \end{pmatrix} = H \cdot \bar{s} + \bar{n}, \tag{5.13}$$

where H is the 2×2 *channel matrix*. This expression can be seen as a generalization of Eqn (5.2) in Section 5.3 to multiple transmit antennas, with different signals being transmitted from the different antennas.

Assuming no noise and that the channel matrix H is invertible, the vector $\bar{s}$, and thus both signals s_1 and s_2, can be perfectly recovered at the receiver, with no residual interference between the signals, by multiplying the received vector $\bar{r}$ with a matrix $W = H^{-1}$ (see also Figure 5.15):

$$\begin{pmatrix} \hat{s}_1 \\ \hat{s}_2 \end{pmatrix} = W \cdot \bar{r} = \begin{pmatrix} s_1 \\ s_2 \end{pmatrix} + H^{-1} \cdot \bar{n}. \tag{5.14}$$

This is illustrated in Figure 5.15.

Although the vector $\bar{s}$ can be perfectly recovered in the case of no noise, as long as the channel matrix H is invertible, Eqn (5.14) also indicates that the properties of H will determine to what extent the joint demodulation of the two signals will increase the noise level. More specifically, the closer the channel matrix is to being a singular matrix, the larger the increase in the noise level.

One way to interpret the matrix W is to realize that the signals transmitted from the two transmit antennas are two signals causing interference to each other. The two receive antennas can then be used to carry out IRC, in essence completely suppressing the interference from the signal transmitted on the second antenna when detecting the signal transmitted at the first antenna and vice versa. The rows of the receiver matrix W simply implement such IRC.

In the general case, a multiple-antenna configuration will consist of N_T transmit antennas and N_R receive antennas. As discussed above, in such a case, the number of parallel signals that can be spatially multiplexed is, at least in practice, upper limited by $N_L = \min\{N_T, N_R\}$. This can intuitively be understood from the fact that:

- Obviously no more than N_T different signals can be transmitted from N_T transmit antennas, implying a maximum of N_T spatially multiplexed signals.
- With N_R receive antennas, a maximum of $N_R - 1$ interfering signals can be suppressed, implying a maximum of N_R spatially multiplexed signals.

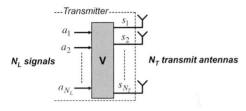

FIGURE 5.16

Pre-coder-based spatial multiplexing.

However, in many cases, the number of spatially multiplexed signals, or the *order of the spatial multiplexing*, will be less than N_L given above:

- In the case of very bad channel conditions (low signal-to-noise ratio) there is no gain of spatial multiplexing as the channel capacity is anyway a linear function of the signal-to-noise ratio. In such a case, the multiple transmit and receive antennas should be used for beam-forming to improve the signal-to-noise ratio, rather than for spatial multiplexing.
- In the more general case, the spatial-multiplexing order should be determined based on the properties of the size $N_R \times N_T$ channel matrix. Any excess antennas should then be used to provide beam-forming. Such combined beam-forming and spatial multiplexing can be achieved by means of *precoder-based* spatial multiplexing, as discussed below.

5.5.2 Precoder-Based Spatial Multiplexing

Linear precoding in the case of spatial multiplexing implies that linear processing by means of a size $N_T \times N_L$ precoding matrix is applied at the transmitter side, as illustrated in Figure 5.16. In line with the discussion above, in the general case N_L is equal or smaller than N_T, implying that N_L signals are spatially multiplexed and transmitted using N_T transmit antennas.

It should be noted that precoder-based spatial multiplexing can be seen as a generalization of precoder-based beam-forming as described in Section 5.4.2, with the precoding vector of size $N_T \times 1$ replaced by a precoding matrix of size $N_T \times N_L$.

The precoding of Figure 5.16 can serve two purposes:

- In the case when the number of signals to be spatially multiplexed equals the number of transmit antennas ($N_L = N_T$), the precoding can be used to "orthogonalize" the parallel transmissions, allowing for improved signal isolation at the receiver side.
- In the case when the number of signals to be spatially multiplexed is less than the number of transmit antennas ($N_L < N_T$), the precoding also provides the mapping of the N_L spatially multiplexed signals to the N_T transmit antennas, including the combination of spatially multiplexing and beam-forming.

To confirm that precoding can improve the isolation between the spatially multiplexed signals, express the channel matrix H as its singular-value decomposition [38]:

$$H = W \cdot \Sigma \cdot V^H,$$ (5.15)

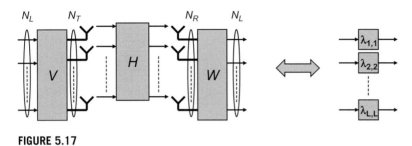

FIGURE 5.17

Orthogonalization of spatially multiplexed signals by means of precoding. $\lambda_{i,i}$ is the ith eigenvalue of the matrix $H^H H$.

where the columns of V and W each form an orthonormal set and Σ is an $N_L \times N_L$ diagonal matrix with the N_L strongest eigenvalues of $H^H H$ as its diagonal elements. By applying the matrix V as precoding matrix at the transmitter side and the matrix W^H at the receiver side, one arrives at an equivalent channel matrix $H' = \Sigma$ (see Figure 5.17). As H' is a diagonal matrix, there is thus no interference between the spatially multiplexed signals at the receiver. At the same time, as both V and W have orthonormal columns, the transmit power as well as the demodulator noise level (assuming spatially white noise) are unchanged.

Clearly, in the case of precoding each received signal will have a certain "quality", depending on the eigenvalues of the channel matrix (see right part of Figure 5.17). This indicates potential benefits of applying dynamic link adaptation in the *spatial domain* – that is, the adaptive selection of the coding rates and/or modulation schemes for each signal to be transmitted.

As the precoding matrix will never perfectly match the channel matrix in practice, there will always be some residual interference between the spatially multiplexed signals. This interference can be taken care of by means of additional receiver-size linear processing according to Figure 5.15 or nonlinear processing as discussed in Section 5.5.3 below.

To determine the precoding matrix V, knowledge about the channel matrix H is obviously needed. Similar to precoder-based beam-forming, a common approach is to have the receiver estimate the channel and decide on a suitable precoding matrix from a set of available precoding matrices (the precoder *codebook*). The receiver then feeds back information about the selected precoding matrix to the transmitter.

5.5.3 Nonlinear Receiver Processing

The previous sections discussed the use of linear receiver processing to jointly recover spatially multiplexed signals. However, improved demodulation performance can be achieved if nonlinear receiver processing can be applied in the case of spatial multiplexing.

The "optimal" receiver approach for spatially multiplexed signals is to apply *Maximum-Likelihood* (ML) detection [25]. However, in many cases ML detection is too complex to use. Thus, several different proposals have been made for reduced complexity, almost ML, schemes (see, for example, [39]).

Another nonlinear approach to the demodulation of spatially multiplexed signals is to apply *Successive Interference Cancellation* (SIC) [40]. Successive Interference Cancellation is based on an assumption that the spatially multiplexed signals are separately coded before the spatial multiplexing.

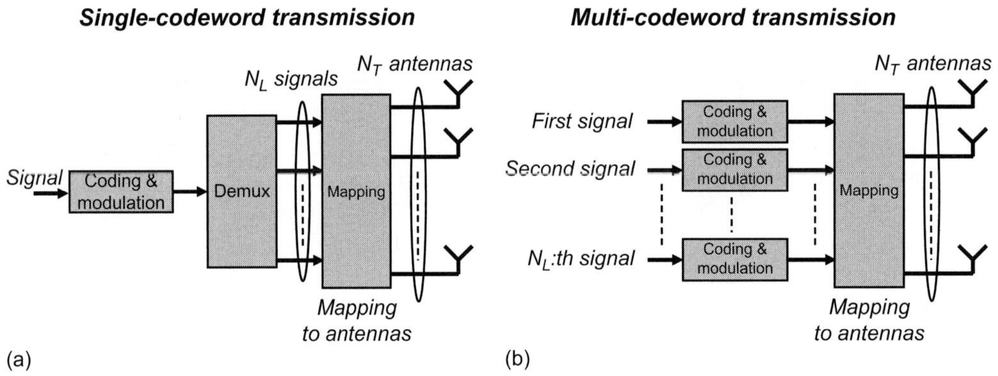

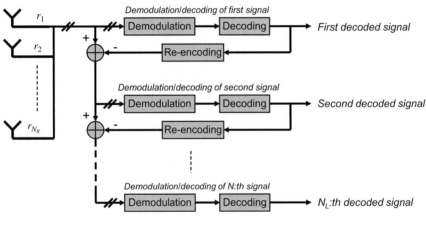

FIGURE 5.18

Single-codeword transmission (a) vs. multi-codeword transmission (b).

FIGURE 5.19

Demodulation/decoding of spatially multiplexed signals based on Successive
Interference Cancellation.

This is often referred to as *Multi-Codeword* transmission, in contrast to *Single-Codeword* transmission where the spatially multiplexed signals are assumed to be jointly coded (Figure 5.18). It should be understood that, also in the case of multi-codeword transmission, the data may originate from the same source but is then demultiplexed into different signals to be spatially multiplexed before channel coding.

As shown in Figure 5.19, in the case of successive interference cancellation the receiver first demodulates and decodes one of the spatially multiplexed signals. The corresponding decoded data is then, if correctly decoded, re-encoded and subtracted from the received signals. A second spatially multiplexed signal can then be demodulated and decoded without, at least in the ideal case, any interference from the first signal – that is, with an improved signal-to-interference ratio. The decoded data of the second signal is then, if correctly decoded, re-encoded and subtracted from the received signal before decoding of a third signal. These iterations continue until all spatially multiplexed signals have been demodulated and decoded.

Clearly, in the case of Successive Interference Cancellation, the first signals to be decoded are subject to higher interference level, compared to later decoded signals. To work properly, there should thus be a differentiation in the robustness of the different signals with, at least in principle, the first signal to be decoded being more robust than the second signal, the second signal being more robust than the third signal, etc. Assuming multi-codeword transmission according to Figure 5.18b, this can be achieved by applying different modulation schemes and coding rates to the different signals with, typically, lower-order modulation and lower coding rate, implying a lower data rate, for the first signals to be decoded. This is often referred to as *Per-Antenna Rate Control* (PARC) [41].

Scheduling, Link Adaptation, and Hybrid ARQ

One key characteristic of mobile radio communication is the typically rapid and significant variations in the instantaneous channel conditions. There are several reasons for these variations. Frequency-selective fading will result in rapid and random variations in the channel attenuation. Shadow fading and distance-dependent path loss will also significantly affect the average received signal strength. Finally, the interference at the receiver due to transmissions in other cells and by other terminals will also impact the interference level. Hence, to summarize, there will be rapid, and to some extent random, variations in the experienced quality of each radio link in a cell, variations that must be taken into account and preferably exploited.

In this chapter, some of the techniques for handling variations in the instantaneous radio-link quality will be discussed. *Channel-dependent scheduling* in a mobile-communication system deals with the question of how to share, between different users (different terminals), the radio resource(s) available in the system to achieve as efficient resource utilization as possible. Typically, this implies minimizing the amount of resources needed per user and thus allowing for as many users as possible in the system, while still satisfying whatever quality-of-service requirements that may exist. Closely related to scheduling is *link adaptation*, which deals with how to set the transmission parameters of a radio link to handle variations of the radio-link quality.

Both channel-dependent scheduling and link adaptation try to exploit the channel variations through appropriate processing *prior* to transmission of the data. However, due to the random nature of the variations in the radio-link quality, perfect adaptation to the instantaneous radio-link quality is never possible. *Hybrid ARQ,* which requests retransmission of erroneously received data packets, is therefore useful. This can be seen as a mechanism for handling variations in the instantaneous radio-link quality *after* transmission and nicely complements channel-dependent scheduling and link adaptation. Hybrid ARQ also serves the purpose of handling random errors due to, for example, noise in the receiver.

6.1 LINK ADAPTATION: POWER AND RATE CONTROL

Historically, *dynamic transmit-power control* has been used in CDMA-based mobile-communication systems such as WCDMA and CDMA2000 to compensate for variations in the instantaneous channel conditions.[1] As the name suggests, dynamic power control dynamically adjusts the radio-link transmit

---

[1] Power control is also used in GSM, but operating on a much slower basis than in CDMA-based systems. In CDMA-based systems, power control is also essential to handle the near–far problem and the associated performance impact from non-orthogonal uplink transmissions.

power to compensate for variations and differences in the instantaneous channel conditions. The aim of these adjustments is to maintain a (near) constant E_b/N_0 at the receiver to successfully transmit data without the error probability becoming too high. In principle, transmit-power control increases the power at the transmitter when the radio link experiences poor radio conditions (and vice versa). Thus, the transmit power is in essence inversely proportional to the channel quality, as illustrated in Figure 6.1a. This results in a basically constant data rate, regardless of the channel variations. For services such as circuit-switched voice, this is a desirable property. Transmit-power control can be seen as one type of link adaptation – that is, the adjustment of transmission parameters, in this case the transmit power – to adapt to differences and variations in the instantaneous channel conditions to maintain the received E_b/N_0 at a desired level.

However, in many cases of mobile communication, especially in the case of packet-data traffic, there is not a strong need to provide a certain constant data rate over a radio link. Rather, from a user perspective, the data rate provided over the radio interface should simply be as "high as possible". Actually, even in case of typical "constant-rate" services such as voice and video, (short-term)

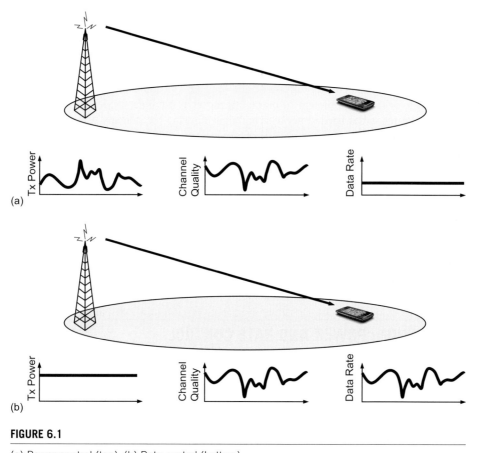

FIGURE 6.1

(a) Power control (top). (b) Rate control (bottom).

variations in the data rate are often not an issue, as long as the average data rate remains constant, assuming averaging over some relatively short time interval. In such cases – that is, when a constant data rate is not required – an alternative to transmit-power control is link adaptation by means of *dynamic rate control*. Rate control does not aim at keeping the instantaneous radio-link data rate constant, regardless of the instantaneous channel conditions. Instead, with rate control, the data rate is dynamically adjusted to compensate for the varying channel conditions. In situations with advantageous channel conditions, the data rate is increased, and vice versa. Thus, rate control maintains the $E_b/N_0 \sim P/R$ at the desired level, not by adjusting the transmission power P, but rather by adjusting the data rate R. This is illustrated in Figure 6.1b.

It can be shown that rate control is more efficient than power control [42,43]. Rate control in principle implies that the power amplifier is always transmitting at full power and therefore efficiently utilized. Power control, on the other hand, results in the power amplifier in most situations not being efficiently utilized as the transmission power is less than its maximum.

In practice, the radio-link data rate is controlled by adjusting the modulation scheme and/or the channel coding rate. In the case of advantageous radio-link conditions, the E_b/N_0 at the receiver is high and the main limitation of the data rate is the bandwidth of the radio link. Hence, in such situations higher-order modulation, for example 16QAM or 64QAM, together with a high code rate, is appropriate as discussed in Chapter 2. Similarly, in the case of poor radio-link conditions, QPSK and low-rate coding is used. For this reason, link adaptation by means of rate control is sometimes also referred to as *Adaptive Modulation and Coding* (AMC).

6.2 CHANNEL-DEPENDENT SCHEDULING

Scheduling controls the allocation of the shared resources among users at each time instant. It is closely related to link adaptation, and often scheduling and link adaptation is seen as one joint function. The scheduling principles, as well as which resources are shared between users, differ depending on the radio-interface characteristics – for example, whether uplink or downlink is considered and whether different users' transmissions are mutually orthogonal or not.

6.2.1 Downlink Scheduling

In the downlink, transmissions to different terminals within a cell are typically mutually orthogonal, implying that, at least in theory, there is no interference between the transmissions (no intra-cell interference). Downlink intra-cell orthogonality can be achieved in the time domain (*Time-Division Multiplexing*, TDM), in the frequency domain (*Frequency-Domain Multiplexing*, FDM), or in the code domain (*Code-Domain Multiplexing*, CDM). In addition, the spatial domain can also be used to separate users, at least in a quasi-orthogonal way, through different antenna arrangements. This is sometimes referred to as *Spatial-Division Multiplexing* (SDM), although in most cases it is used in combination with one or several of the above multiplexing strategies and is not discussed further in this chapter.

For packet data, where the traffic is often very bursty, it can be shown that TDM is preferable from a theoretical point of view [44,45] and is therefore typically the main component in the downlink [46,47]. However, as discussed in Chapter 3, the TDM component is often combined with sharing of the radio resource also in the frequency domain (FDM) or in the code domain (CDM). For example, in the case of HSPA (see Chapter 19), downlink multiplexing is a combination of TDM and CDM. On the other hand, in the case of LTE, downlink multiplexing is a combination of TDM and FDM. The reasons for sharing the resources not only in the time domain will be elaborated upon later in this section.

When transmissions to multiple users occur in parallel, either by using FDM or CDM, there is also an instantaneous sharing of the total available cell transmit power. In other words, not only are the time/frequency/code resources shared resources, but also the power resource in the base station. In contrast, in the case of sharing only in the time domain there is, by definition, only a single transmission at a time and thus no instantaneous sharing of the total available cell transmit power.

For the purpose of discussion, assume initially a TDM-based downlink with a single user being scheduled at a time. In this case, the utilization of the radio resources is maximized if, at each time instant, all resources are assigned to the user with the best instantaneous channel condition:

- In the case of link adaptation based on power control, this implies that the lowest possible transmit power can be used for a given data rate and thus minimizes the interference to transmissions in other cells for a given link utilization.
- In the case of link adaptation based on rate control, this implies that the highest data rate is achieved for a given transmit power or, in other words, for a given interference to other cells, the highest link utilization is achieved.

However, if applied to the downlink, transmit-power control in combination with TDM scheduling implies that the total available cell transmit power will, in most cases, not be fully utilized. Thus, rate control is generally preferred [42,44,47,48].

The strategy outlined above is an example of channel-dependent scheduling, where the scheduler takes the instantaneous radio-link conditions into account. Scheduling the user with the instantaneously best radio-link conditions is often referred to as *max-C/I* (or *maximum rate*) scheduling. Since the radio conditions for the different radio links within a cell typically vary independently, at each point in time there is almost always a radio link whose channel quality is near its peak (see Figure 6.2). Thus, the channel eventually used for transmission will typically have a high quality and, with rate control, a correspondingly high data rate can be used. This translates into a high system capacity. The gain obtained by transmitting to users with favorable radio-link conditions is commonly known as multi-user diversity; the multi-user diversity gains are larger, the larger the channel variations and the larger the number of users in a cell. Hence, in contrast to the traditional view that fast fading – that is, rapid variations in the radio-link quality – is an undesirable effect that has to be combated, the possibility of channel-dependent scheduling implies that *fading is in fact potentially beneficial and should be exploited*.

Mathematically, the max-C/I (maximum rate) scheduler can be expressed as scheduling user k given by:

$$k = \arg\max_i R_i,$$

where R_i is the instantaneous data rate for user i. Although, from a system capacity perspective, max-C/I scheduling is beneficial, this scheduling principle will not be fair in all situations. If all terminals are, on average, experiencing similar channel conditions and large variations in the instantaneous channel conditions are only due to, for example, fast multi-path fading, all users will experience the

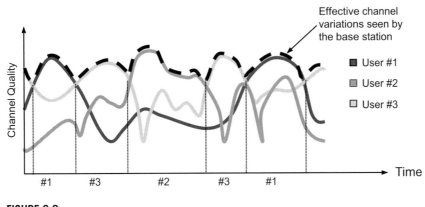

FIGURE 6.2

Channel-dependent scheduling.

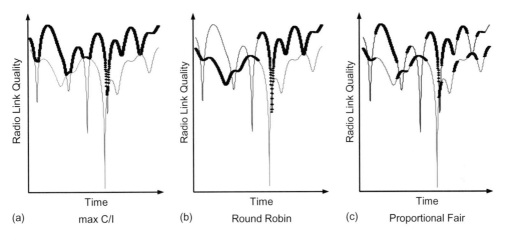

FIGURE 6.3

Examples of three different scheduling behaviors for two users with different average channel quality: (a) max-C/I; (b) round robin; (c) proportional fair. The selected user is shown with bold lines.

same average data rate. Any variations in the instantaneous data rate are rapid and often not even noticeable by the user. However, in practice different terminals will also experience differences in the (short-term) average channel conditions – for example, due to differences in the distance and shadow fading between the base station and the terminal. In this case, the channel conditions experienced by one terminal may, for a relatively long time, be worse than the channel conditions experienced by other terminals. A pure max-C/I-scheduling strategy may then, in essence, "starve" the terminal with the bad channel conditions, and the terminal with bad channel conditions will never be scheduled. This is illustrated in Figure 6.3a, where a max-C/I scheduler is used to schedule between two different users with different average channel quality. The same user is scheduled virtually all the time. Although resulting in the highest system capacity, this situation is often not acceptable from a quality-of-service point of view.

An alternative to the max-C/I scheduling strategy is so-called *round-robin* scheduling, illustrated in Figure 6.3b. This scheduling strategy lets users take turns in using the shared resources, without taking the instantaneous channel conditions into account. Round-robin scheduling can be seen as fair scheduling in the sense that the same amount of radio resources (the same amount of time) is given to each communication link. However, round-robin scheduling is not fair in the sense of providing the same service quality to all communication links. In that case more radio resources (more time) must be given to communication links with bad channel conditions. Furthermore, as round-robin scheduling does not take the instantaneous channel conditions into account in the scheduling process, it will lead to lower overall system performance but more equal service quality between different communication links, compared to max-C/I scheduling.

Thus, what is needed is a scheduling strategy that is able to utilize the fast channel variations to improve the overall cell throughput while still ensuring the same average user throughput for all users or at least a certain minimum user throughput for all users. When discussing and comparing different scheduling algorithms it is important to distinguish between different types of variations in the service quality:

- Fast variations in the service quality corresponding to, for example, fast multi-path fading and fast variations in the interference level. For many packet-data applications, relatively large short-term variations in service quality are often acceptable or not even noticeable to the user.
- More long-term differences in the service quality between different communication links corresponding to, for example, differences in the distance to the cell site and shadow fading. In many cases there is a need to limit such long-term differences in service quality.

A practical scheduler should thus operate somewhere in between the max-C/I scheduler and the round-robin scheduler – that is, try to utilize fast variations in channel conditions as much as possible while still satisfying some degree of fairness between users.

One example of such a scheduler is the proportional-fair scheduler [49–51], illustrated in Figure 6.3c. In this strategy, the shared resources are assigned to the user with the *relatively* best radio-link conditions – that is, at each time instant, user k is selected for transmission according to:

$$k = \arg\max_i \frac{R_i}{\overline{R_i}},$$

where R_i is the instantaneous data rate for user i and $\overline{R_i}$ is the average data rate for user i. The average is calculated over a certain averaging period T_{PF}. To ensure efficient usage of the short-term channel variations and, at the same time, limit the long-term differences in service quality to an acceptable level, the time constant T_{PF} should be set longer than the time constant for the short-term variations. At the same time, T_{PF} should be sufficiently short so that quality variations within the interval T_{PF} are not strongly noticed by a user. Typically, T_{PF} can be set to be of the order of one second.

In the above discussion, it was assumed that all the radio resources in the downlink were assigned to a single user at a time – that is, scheduling was done purely in the time domain using TDM between users. However, in several situations, TDM is complemented by CDM or FDM. In principle, there are two reasons for not relying solely on TDM in the downlink:

- In the case of insufficient payload – that is, the amount of data to transfer to a user is not sufficiently large to utilize the full channel capacity, and a fraction of resources could be assigned to another user, either through FDM or CDM.
- In the case where channel variations in the frequency domain are exploited through FDM, as discussed further below.

The scheduling strategies in these cases can be seen as generalizations of the schemes discussed for the TDM-only cases above. For example, to handle small payloads, a greedy filling approach can be used, where the scheduled user is selected according to max-C/I (or any other scheduling scheme). Once this user has been assigned resources matching the amount of data awaiting transmission, the second best user according to the scheduling strategy is selected and assigned (a fraction of) the residual resources and so on.

Finally, it should also be noted that the scheduling algorithm typically is a base-station-implementation issue and nothing that is normally specified in any standard. What need to be specified in a standard to support channel-dependent scheduling are channel-quality measurements/reports and the signaling needed for dynamic resource allocation.

6.2.2 Uplink Scheduling

The previous section discussed scheduling from a downlink perspective. However, scheduling is equally applicable to uplink transmissions and to a large extent the same principles can be considered, although there are some differences between the two.

Fundamentally, the uplink power resource is *distributed* among the users, while in the downlink the power resource is *centralized* within the base station. Furthermore, the maximum uplink transmission power of a single terminal is typically significantly lower than the output power of a base station. This has a significant impact on the scheduling strategy. Unlike the downlink, where pure TDMA can often be used, uplink scheduling typically has to rely on sharing in the frequency and/or code domain in addition to the time domain, as a single terminal may not have sufficient power for efficiently utilizing the link capacity.

Similar to the downlink case, channel-dependent scheduling is also beneficial in the uplink case. However, the characteristics of the underlying radio interface, most notably whether the uplink relies on orthogonal or non-orthogonal multiple access and the type of link adaptation scheme used, also have a significant impact on the uplink scheduling strategy.

In the case of a non-orthogonal multiple-access scheme such as CDMA, power control is typically essential for proper operation. As discussed earlier in this chapter, the purpose of power control is to control the received E_b/N_0 such that the received information can be recovered. However, in a non-orthogonal multiple-access setting, power control also serves the purpose of controlling the amount of interference affecting *other* users. This can be expressed as the *maximum tolerable interference level* at the base station in a shared resource. Even if, from a single user's perspective, it would be beneficial to transmit at full power to maximize the data rate, this may not be acceptable from an interference perspective as other terminals in this case may not be able to successfully transfer any data. Thus, with non-orthogonal multiple access, scheduling a terminal when the channel conditions are favorable may not directly translate into a higher data rate as the interference generated to other simultaneously transmitting terminals in the cell must be taken into account. Stated differently, the *received* power (and thus the data rate) is, thanks to power control, in principle constant, regardless of the channel conditions at the time of transmission, while the *transmitted* power depends on the channel conditions at the time of transmission. Hence, even though channel-dependent scheduling in this example does not give a direct gain in terms of a higher data rate from the terminal, channel-dependent scheduling will still provide a gain for the system in terms of lower intra-cell interference.

The above discussion on non-orthogonal multiple access was simplified in the sense that no bounds on the terminals' transmission power were assumed. In practice, the transmission power of a terminal is upper bounded, due both to implementation and regulatory reasons, and scheduling a terminal for transmission in favorable channel conditions decreases the probability that the terminal has insufficient power to utilize the channel capacity.

In the case of an orthogonal multiple-access scheme, intra-cell power control is fundamentally not necessary and the benefits of channel-dependent scheduling become more similar to the downlink case. In principle, from an intra-cell perspective, a terminal can transmit at full power and the scheduler assigns a suitable part of the orthogonal resources (in practice a suitable part of the overall bandwidth) to the terminal for transmission. The remaining orthogonal resources can be assigned to other users. However, implementation constraints, for example leakage between the received signals or limited dynamic range in the receiver circuitry, may pose restrictions on the maximum tolerable power difference between the signals from simultaneously transmitting terminals. As a consequence, a certain degree of power control may be necessary, making the situation somewhat similar to the non-orthogonal case.

The discussion on non-orthogonal and orthogonal multiple access mainly considered intra-cell multiple access. However, in many practical systems universal frequency reuse between cells is applied. In this case, the inter-cell multiple access is non-orthogonal, regardless of the intra-cell multiple access, which sets limits on the allowable transmission power from a terminal.

Regardless of whether orthogonal or non-orthogonal multiple access is used, the same basic scheduling principles as for the downlink can be used. A max-C/I scheduler would assign all the uplink resources to the terminal with the best uplink channel conditions. Neglecting any power limitations in the terminal, this would result in the highest capacity (in an isolated cell) [44].

In the case of a non-orthogonal multiple-access scheme, *greedy filling* is one possible scheduling strategy [48]. With greedy filling, the terminal with the best radio conditions is assigned as high a data rate as possible. If the interference level at the receiver is smaller than the maximum tolerable level, the terminal with the second best channel conditions is allowed to transmit as well, continuing with more and more terminals until the maximum tolerable interference level at the receiver is reached. This strategy maximizes the air interface utilization but is achieved at the cost of potentially large differences in data rates between users. In the extreme case, a user at the cell border with poor channel conditions may not be allowed to transmit at all.

Strategies between greedy filling and max-C/I can also be envisioned – for example, different proportional-fair strategies. This can be achieved by including a weighting factor for each user, proportional to the ratio between the instantaneous and average data rates, into the greedy filling algorithm.

The schedulers above all assume knowledge of the instantaneous radio-link conditions, knowledge that can be hard to obtain in the uplink scenario, as discussed in Section 7.2.4. In situations when no information about the uplink radio-link quality is available at the scheduler, *round-robin* scheduling can be used. Similar to the downlink, round-robin implies terminals taking turns in transmitting, thus creating a TDMA-like operation with inter-user orthogonality in the time domain. Although the round-robin scheduler is simple, it is far from the optimal scheduling strategy.

However, as already discussed in Chapter 4, the transmission power in a terminal is limited and therefore additional sharing of the uplink resources in the frequency and/or code domain is required. This also impacts the scheduling decisions. For example, terminals far from the base station typically operate in the power-limited region, in contrast to terminals close to the base stations, which often are in the bandwidth-limited region (for a discussion on power-limited vs. bandwidth-limited operation, see Chapter 2). Thus, for a terminal far from the base station, increasing the bandwidth will not result in an increased data rate and it is better to only assign a small amount of the bandwidth to this terminal and assign the remaining bandwidth to other terminals. On the other hand, for terminals close to the base station, an increase in the assigned bandwidth will provide a higher data rate.

6.2.3 Link Adaptation and Channel-Dependent Scheduling in the Frequency Domain

In the previous section, TDM-based scheduling was assumed and it was explained how, in this case, channel variations in the time domain could be utilized to improve system performance by applying channel-dependent scheduling, especially in combination with dynamic rate control. However, if the scheduler has access to the frequency domain, for example through the use of OFDM transmission, scheduling and link adaptation can also take place in the frequency domain.

Link adaptation in the frequency domain implies that, based on knowledge about the instantaneous channel conditions also in the frequency domain – that is, knowledge about the attenuation as well as the noise/interference level of, in the extreme case, every OFDM subcarrier – the power and/or the data rate of each OFDM carrier can be individually adjusted for optimal utilization.

Similarly, channel-dependent scheduling in the frequency domain implies that, based on knowledge about the instantaneous channel conditions also in the frequency domain, different subcarriers are used for transmission to or from different terminals. The scheduling gains from exploiting variations in the frequency domain are similar to those obtained from time-domain variations. Obviously, in situations where the channel quality varies significantly with the frequency while the channel quality only varies slowly with time, channel-dependent scheduling in the frequency domain can enhance system capacity. An example of such a situation is a wideband indoor system with low mobility, where the quality only varies slowly with time.

6.2.4 Acquiring on Channel-State Information

To select a suitable data rate, in practice a suitable modulation scheme and channel-coding rate, the transmitter needs information about the radio-link channel conditions. Such information is also required for the purpose of channel-dependent scheduling. In the case of a system based on frequency-division duplex (FDD), only the receiver can accurately estimate the radio-link channel conditions.

For the downlink, most systems provide a downlink signal of a predetermined structure, known as the downlink pilot or the downlink reference signal. This reference signal is transmitted from the base station with constant power and can be used by the terminal to estimate the instantaneous downlink channel conditions. Information about the instantaneous downlink conditions can then be reported to the base station.

Basically, what is relevant for the transmitter is an estimate reflecting the channel conditions *at the time of transmission*. Hence, in principle, the terminal could apply a predictor, trying to predict the future channel conditions and report this predicted value to the base station. However, as this would require specification of prediction algorithms and how they would operate when the terminal is moving at different speeds, most practical systems simply report the measured channel conditions to the base station. This can be seen as a very simple predictor, basically assuming the conditions in the near future will be similar to the current conditions. Thus, the more rapid the time-domain channel variations, the less efficient the link adaptation.

As there inevitably will be a delay between the point in time when the terminal measured the channel conditions and the application of the reported value in the transmitter, channel-dependent scheduling and link adaptation typically operates at its best at low terminal mobility. If the terminal starts to move at high speed, the measurement reports will be outdated once arriving at the base station. In such cases, it is often preferable to perform link adaptation on the long-term average channel quality and rely on hybrid ARQ with soft combining for the rapid adaptation.

For the uplink, estimation of the uplink channel conditions is not as straightforward, as there is typically no reference signal transmitted with constant power from each terminal. Means to estimate uplink channel conditions for LTE are discussed in Chapter 11.

In the case of a system with time-division duplex (TDD), where uplink and downlink communication are time multiplexed within the same frequency band, the instantaneous uplink signal quality attenuation could be estimated from downlink measurements of the terminal, due to the reciprocity of the multi-path fading in the case of TDD. However, it should then be noted that this may not provide full knowledge of the downlink channel conditions. As an example, the interference situations at the terminal and the base station are different also in the case of TDD.

6.2.5 Traffic Behavior and Scheduling

It should be noted that there is little difference between different scheduling algorithms at low system load – that is, when only one or, in some cases, a few users have data waiting for transmission at the base station at each scheduling instant. The differences between different scheduling algorithms are primarily visible at high load. However, not only the load, but also the traffic behavior affects the overall scheduling performance.

As discussed above, channel-dependent scheduling tries to exploit short-term variations in radio quality. Generally, a certain degree of long-term fairness in service quality is desirable, which should be accounted for in the scheduler design. However, since system throughput decreases the more fairness is enforced, a trade-off between fairness and system throughput is necessary. In this trade-off, it is important to take traffic characteristics into account as they have a significant influence on the trade-off between system throughput and service quality.

To illustrate this, consider three different downlink schedulers:

1. *Round-robin (RR) scheduler,* where channel conditions are not taken into account.
2. *Proportional-fair (PF) scheduler,* where short-term channel variations are exploited while maintaining the long-term average user data rate.
3. *Max-C/I scheduler,* where the user with the best instantaneous channel quality in absolute terms is scheduled.

For a full buffer scenario when there is always data available at the base station for all terminals in the cell, a max-C/I scheduler will result in no, or very low, user throughput for users at the cell edge with a low average channel quality. The reason is the fundamental strategy of the max-C/I

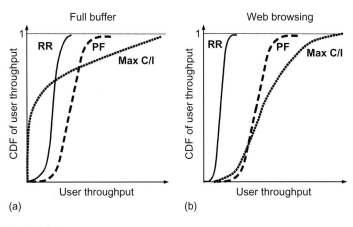

FIGURE 6.4

Illustration of the principle behavior of different scheduling strategies:
(a) for full buffers; (b) for web-browsing traffic model.

scheduler – all resources are allocated for transmission to the terminal whose channel conditions support the highest data rate. Only in the rare, not to say unlikely, case of a cell-edge user having better conditions than a cell-center user, for example due to a deep fading dip for the cell-center user, will the cell-edge user be scheduled. A proportional-fair scheduler, on the other hand, will ensure some degree of fairness by selecting the user supporting the highest data rate relative to its average data rate. Hence, users tend to be scheduled on their fading peaks, regardless of the absolute quality. Thus, users on the cell edge will also be scheduled, thereby resulting in some degree of fairness between users.

For a scenario with bursty packet data, the situation is different. In this case, the users' buffers will be finite and in many cases also empty. For example, a web page has a certain size and, after transmitting the page, there is no more data to be transmitted to the terminal in question until the users requests a new page by clicking on a link. In this case, a max-C/I scheduler can still provide a certain degree of fairness. Once the buffer for the user with the highest C/I has been emptied, another user with non-empty buffers will have the highest C/I and be scheduled and so on. This is the reason for the difference between full buffer and web-browsing traffic illustrated in Figure 6.4. The proportional-fair scheduler has similar performance in both scenarios.

Clearly, the degree of fairness introduced by the traffic properties depends heavily on the actual traffic; a design made with certain assumptions may be less desirable in an actual network where the traffic pattern may be different from the assumptions made during the design. Therefore, relying solely on the traffic properties for fairness is not a good strategy, but the discussion above emphasizes the need to design the scheduler not only for the full buffer case.

6.3 ADVANCED RETRANSMISSION SCHEMES

Transmissions over wireless channels are subject to errors, for example due to variations in the received signal quality. To some degree, such variations can be counteracted through link adaptation, as discussed above. However, receiver noise and unpredictable interference variations cannot be counteracted. Therefore, virtually all wireless communications systems employ some form of *Forward Error Correction* (FEC), tracing its roots to the pioneering work of Claude Shannon in 1948 [11]. There is a rich literature in the area of error-correction coding (see, for example, [52,53] and references therein), and a detailed description is beyond the scope of this book. In short, the basic principle beyond forward error-correction coding is to introduce redundancy in the transmitted signal. This is achieved by adding *parity bits* to the information bits prior to transmission (alternatively, the transmission could consists of parity bits alone, depending on the coding scheme used). The parity bits are computed from the information bits using a method given by the coding structure used. Thus, the number of bits transmitted over the channel is larger than the number of original information bits and a certain amount of *redundancy* has been introduced in the transmitted signal.

Another approach to handle transmissions errors is to use *Automatic Repeat Request* (ARQ). In an ARQ scheme, the receiver uses an error-detection code, typically a Cyclic Redundancy Check (CRC), to detect if the received packet is in error or not. If no error is detected in the received data packet, the received data is declared error-free and the transmitter is notified by sending a positive acknowledgement (ACK). On the other hand, if an error is detected, the receiver discards the received data and notifies the transmitter via a return channel by sending a negative acknowledgement (NAK). In response to an NAK, the transmitter retransmits the same information.

Virtually all modern communication systems, including LTE, employ a combination of forward error-correction coding and ARQ, known as *hybrid ARQ*. Hybrid ARQ uses forward error-correction codes to correct a subset of all errors and relies on error detection to detect uncorrectable errors. Erroneously received packets are discarded and the receiver requests retransmissions of corrupted packets. Thus, it is a combination of FEC and ARQ as described above. Hybrid ARQ was first proposed in [54] and numerous publications on hybrid ARQ have appeared since (see [52] and references therein). Most practical hybrid ARQ schemes are built around a CRC code for error detection and convolutional or Turbo codes for error correction, but in principle any error-detection and error-correction code can be used.

6.4 HYBRID ARQ WITH SOFT COMBINING

The hybrid-ARQ operation described above discards erroneously received packets and requests retransmission. However, despite it not being possible to decode the packet, the received signal still contains information, which is lost by discarding erroneously received packets. This shortcoming is addressed by *hybrid ARQ with soft combining*. In hybrid ARQ with soft combining, the erroneously received packet is stored in a buffer memory and later combined with the retransmission to obtain a single, combined packet that is more reliable than its constituents. Decoding of the error-correction code operates on the combined signal. If the decoding fails (typically a CRC code is used to detect this event), a retransmission is requested.

Retransmission in any hybrid-ARQ scheme must, by definition, represent the same set of information bits as the original transmission. However, the set of coded bits transmitted in each retransmission may be selected differently as long as they represent the same set of information bits. Hybrid ARQ with soft combining is therefore usually categorized into *Chase combining* and *incremental redundancy*, depending on whether the retransmitted bits are required to be identical to the original transmission or not.

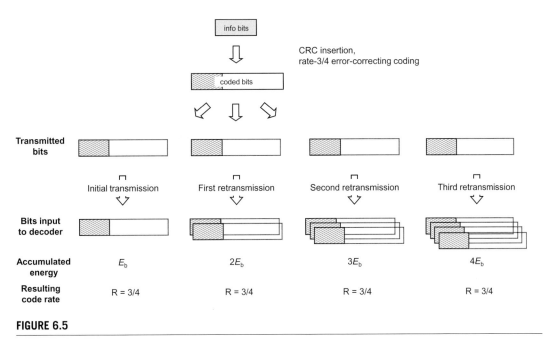

FIGURE 6.5

Example of Chase combining.

Chase combining, where the retransmissions consist of the same set of coded bits as the original transmission, was first proposed in [55]. After each retransmission, the receiver uses maximum-ratio combination to combine each received channel bit with any previous transmissions of the same bit, and the combined signal is fed to the decoder. As each retransmission is an identical copy of the original transmission, retransmissions with Chase combining can be seen as additional repetition coding. Therefore, as no new redundancy is transmitted, Chase combining does not give any additional coding gain but only increases the accumulated received E_b/N_0 for each retransmission (Figure 6.5).

Several variants of Chase combining exist. For example, only a subset of the bits transmitted in the original transmission might be retransmitted, so-called partial Chase combining. Furthermore, although combination is often done after demodulation but before channel decoding, combination can also be carried out at the modulation symbol level before demodulation, as long as the modulation scheme is unchanged between transmission and retransmission.

With *Incremental Redundancy* (IR), each retransmission does not have to be identical to the original transmission. Instead, *multiple sets* of coded bits are generated, each representing the same set of information bits [56,57]. Whenever a retransmission is required, the retransmission typically uses a different set of coded bits than the previous transmission. The receiver combines the retransmission with previous transmission attempts of the same packet. As the retransmission may contain additional parity bits not included in the previous transmission attempts, the resulting code rate is generally lowered by a retransmission. Furthermore, each retransmission does not necessarily have to consist of the same number of coded bits as the original and, in general, the modulation scheme can also be different for different

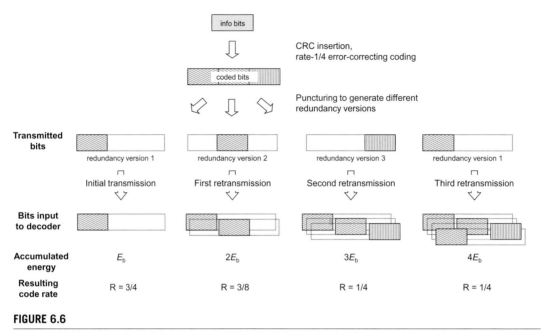

FIGURE 6.6

Example of incremental redundancy.

retransmissions. Hence, incremental redundancy can be seen as a generalization of Chase combining or, stated differently, Chase combining is a special case of incremental redundancy.

Typically, incremental redundancy is based on a low-rate code and the different redundancy versions are generated by puncturing the output of the encoder. In the first transmission only a limited number of coded bits are transmitted, effectively leading to a high-rate code. In the retransmissions, additional coded bits are transmitted. As an example, assume a basic rate-1/4 code. In the first transmission, only every third coded bit is transmitted, effectively giving a rate-3/4 code, as illustrated in Figure 6.6. In the case of a decoding error and a subsequent request for a retransmission, additional bits are transmitted, effectively leading to a rate-3/8 code. After a second retransmission the code rate is 1/4. In the case of more than two retransmissions, transmitted coded bits would already be repeated. In addition to a gain in accumulated received E_b/N_0, incremental redundancy also results in a coding gain for each retransmission. The gain with IR compared to Chase is larger for high initial code rates, while at lower initial coding rates Chase combining is almost as good as IR [58]. Furthermore, as shown in [59], the performance gain of incremental redundancy compared to Chase combining can also depend on the relative power difference between the transmission attempts.

With incremental redundancy, the code used for the first transmission should provide good performance not only when used alone, but also when used in combination with the code for the second transmission. The same holds for subsequent retransmissions. Thus, as the different redundancy versions typically are generated through puncturing of a low-rate mother code, the puncturing patterns should be defined such that all the code bits used by a high-rate code should also be part of any lower-rate codes. In other words, the resulting code rate R_i after transmission attempt i, consisting of the coded bits from redundancy versions RV_k, $k = 1, \ldots, i$, should have similar performance as a good code designed directly for rate R_i. Examples of this for convolutional codes are the so-called rate-compatible convolutional codes [60].

In the discussion so far, it has been assumed that the receiver has received all the previously transmitted redundancy versions. If all redundancy versions provide the same amount of information about the data packet, the order of the redundancy versions is not critical. However, for some code structures, not all redundancy versions are of equal importance. One example here is Turbo codes, where the systematic bits are of higher importance than the parity bits. Hence, the initial transmission should at least include all the systematic bits and some parity bits. In the retransmission(s), parity bits not in the initial transmission can be included. However, if the initial transmission was received with poor quality or not at all, a retransmission with only parity bits is not appropriate as a retransmission of (some of) the systematic bits provides better performance. Incremental redundancy with Turbo codes can therefore benefit from multiple levels of feedback, for example by using two different negative acknowledgements – NAK to request additional parity bits and LOST to request a retransmission of the systematic bits. In general, the problem of determining the amount of systematic and parity bits in a retransmission based on the signal quality of previous transmission attempts is non-trivial.

Hybrid ARQ with soft combining, regardless of whether Chase or incremental redundancy is used, leads to an implicit reduction of the data rate by means of retransmissions and can thus be seen as *implicit* link adaptation. However, in contrast to link adaptation based on explicit estimates of the instantaneous channel conditions, hybrid ARQ with soft combining implicitly adjusts the coding rate based on the result of the decoding. In terms of overall throughput this kind of implicit link adaptation can be superior to explicit link adaptation, as additional redundancy is only added *when needed* – that is, when previous higher-rate transmissions were not possible to decode correctly. Furthermore, as it does not try to predict any channel variations, it works equally well, regardless of the speed at which the terminal is moving. Since implicit link adaptation can provide a gain in system throughput, a valid question is why explicit link adaptation is necessary at all. One major reason for having explicit link adaptation is the reduced delay. Although relying on implicit link adaptation alone is sufficient from a system throughput perspective, the end-user service quality may not be acceptable from a delay perspective.

LTE Radio Access: An Overview

The work on LTE was initiated in late 2004 with the overall aim of providing a new radio-access technology focusing on packet-switched data only. The first phase of the 3GPP work on LTE was to define a set of performance and capability targets for LTE [10]. This included targets on peak data rates, user/system throughput, spectral efficiency, and control/user-plane latency. In addition, requirements were also set on spectrum flexibility, as well as on interaction/compatibility with other 3GPP radio-access technologies (GSM, WCDMA/HSPA, and TD-SCDMA).

Once the targets were set, 3GPP studies on the feasibility of different technical solutions considered for LTE were followed by development of the detailed specifications. The first release of the LTE specifications, release 8, was completed in the spring of 2008 and commercial network operation began in late 2009. Release 8 has then been followed by additional LTE releases, introducing additional functionality and capabilities in different areas, as illustrated in Figure 7.1.

In parallel to the development of LTE, there has also been an evolution of the overall 3GPP network architecture, termed *System Architecture Evolution* (SAE), including both the radio-access network and the core network. Requirements were also set on the architecture evolution, leading to a new flat radio-access-network architecture with a single type of node, the *eNodeB*[1], as well as a new core-network architecture. An excellent description of the LTE-associated core-network architecture, the *Evolved Packet Core* (EPC), can be found in [9].

The remaining part of this chapter provides an overview of LTE up to and including release 10. The most important technologies used by LTE release 8 – including transmission schemes, scheduling, multi-antenna support, and spectrum flexibility – are covered, as well as the additional features introduced in LTE releases 9 and 10. The chapter can either be read on its own to get a high-level overview of LTE, or as an introduction to the subsequent chapters.

The following chapters, Chapters 8–18, then contain a detailed description of the LTE radio-access technology. Chapter 8 provides an overview of the LTE protocol structure, including RLC, MAC, and the physical layer, explaining the logical and physical channels, and the related data flow. The time–frequency structure on which LTE is based is covered in Chapter 9, followed by a detailed description of the physical layer for downlink and uplink transmission in Chapters 10 and 11 respectively. Chapter 12 contains a description of the retransmission mechanisms used in LTE, followed by a discussion on power control, scheduling, and interference management in Chapter 13. Access procedures, necessary for a terminal to connect to the network, are the topic of Chapter 14. Chapter 15 covers the multicast/broadcast functionality of LTE and Chapter 16 describes relaying operation. Chapter 17 addresses how radio-frequency (RF) requirements are defined in LTE, taking into account the spectrum flexibility. Finally, Chapter 18 contains an assessment of the system performance of LTE.

---

[1] eNodeB is a 3GPP term that can roughly be seen as being equivalent to a base station, see further in Chapter 8.

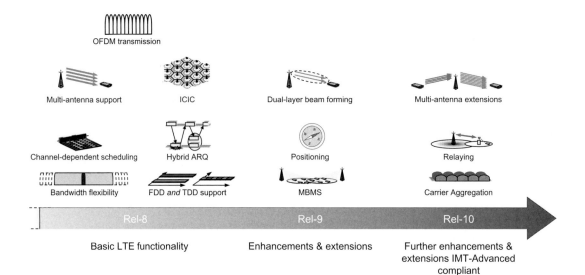

FIGURE 7.1

LTE and its evolution.

7.1 BASIC PRINCIPLES

Building upon the basic technologies described in the previous chapters, the main principles behind LTE will be described in the following.

7.1.1 Transmission Scheme

The LTE downlink transmission scheme is based on conventional OFDM. As discussed in Chapter 3, OFDM is an attractive transmission scheme for several reasons. Due to the relatively long OFDM symbol time in combination with a cyclic prefix, OFDM provides a high degree of robustness against channel frequency selectivity. Although signal corruption due to a frequency-selective channel can, in principle, be handled by equalization at the receiver side, the complexity of such equalization starts to become unattractively high for implementation in a terminal at larger bandwidths and especially in combination with advanced multi-antenna transmission schemes such as spatial multiplexing, thereby making OFDM an attractive choice for LTE for which a wide bandwidth and support for advanced multi-antenna transmission were key requirements.

OFDM also provides some additional benefits relevant for LTE:

- OFDM provides access to the frequency domain, thereby enabling an additional degree of free-dom to the channel-dependent scheduler compared to time-domain-only scheduling used in major 3G systems.
- Flexible transmission bandwidth to support operation in spectrum allocations of different size is straightforward with OFDM, at least from a baseband perspective, by varying the number of OFDM subcarriers used for transmission. Note, however, that support of a flexible transmission

bandwidth also requires flexible RF filtering, etc., for which the exact transmission scheme is irrelevant. Nevertheless, maintaining the same baseband-processing structure, regardless of the bandwidth, eases terminal development and implementation.

- Broadcast/multicast transmission, where the same information is transmitted from multiple base stations, is straightforward with OFDM as already described in Chapter 3 and elaborated upon from an LTE perspective in Chapter 15.

Also, the LTE uplink is based on OFDM transmission. However, different means are taken to reduce the *cubic metric* of the uplink transmission, thereby enabling higher terminal power-amplifier efficiency. More specifically, for uplink data transmission, the OFDM modulator is preceded by a DFT precoder, leading to *DFT-spread OFDM* (DFTS-OFDM) as described in Chapter 4. Often, the term DFTS-OFDM is used to describe the LTE uplink transmission scheme in general. However, it should be understood that DFTS-OFDM is only applicable to uplink data transmission. Other means are used to achieve a low cubic metric for other types of uplink transmissions. Thus, the LTE uplink transmission scheme should be described as OFDM with different techniques, including DFT precoding for data transmission, being used to reduce the cubic metric of the transmitted signal.

The use of DFTS-OFDM on the LTE uplink allows for orthogonal separation of uplink transmissions also in the frequency domain. Orthogonal separation is in many cases beneficial as it avoids interference between uplink transmissions from different terminals within the cell (*intra-cell interference*). As discussed in Chapter 4, allocating a very large instantaneous bandwidth for transmission from a single terminal is not an efficient strategy in situations where the data rate is mainly limited by the available terminal transmit power rather than the bandwidth. In such situations, a terminal can instead be allocated only a part of the total available bandwidth and other terminals within the cell can be scheduled to transmit in parallel on the remaining part of the spectrum. In other words, the LTE uplink transmission scheme allows for both *time division* (TDMA) and *frequency division* (FDMA) between users.

7.1.2 Channel-Dependent Scheduling and Rate Adaptation

At the core of the LTE transmission scheme is the use of *shared-channel transmission* with the overall time–frequency resource dynamically shared between users. The use of shared-channel transmission is well matched to the rapidly varying resource requirements posed by packet-data communication and also enables several of the other key technologies on which LTE is based.

The scheduler controls, for each time instant, to which users the different parts of the shared resource should be assigned. The scheduler also determines the data rate to be used for each transmission. Thus, *rate adaptation* can be seen as a part of the scheduling functionality. The scheduler is thus a key element and to a large extent determines the overall system performance, especially in a highly loaded network. Both downlink and uplink transmissions are subject to tight scheduling in LTE. From Chapter 6 it is well known that a substantial gain in system capacity can be achieved if the channel conditions are taken into account in the scheduling decision, so-called *channel-dependent scheduling*. Due to the use of OFDM in both the downlink and uplink transmission directions, the scheduler has access to both the time and frequency domains. In other words, the scheduler can, for each time instant and frequency region, select the user with the best channel conditions, as illustrated in Figure 7.2.

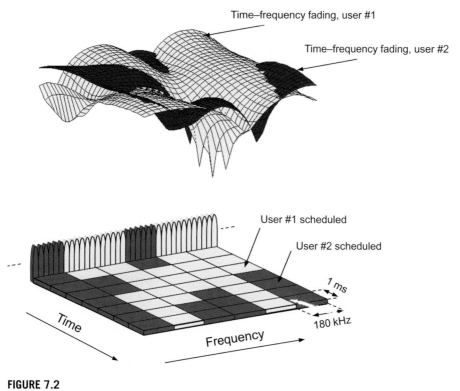

FIGURE 7.2

Downlink channel-dependent scheduling in the time and frequency domains.

The possibility of channel-dependent scheduling in the frequency domain is particularly useful at low terminal speeds – in other words, when the channel is varying slowly in time. As discussed in Chapter 6, channel-dependent scheduling relies on channel-quality variations between users to obtain a gain in system capacity. For delay-sensitive services, a time-domain-only scheduler may, due to the delay constraints, be forced to schedule a particular user, despite the channel quality not being at its peak. In such situations, exploiting channel-quality variations also in the frequency domain will help to improve the overall performance of the system. For LTE, scheduling decisions can be taken as often as once every 1 ms and the granularity in the frequency domain is 180 kHz. This allows for relatively rapid channel variations in both the time and frequency domains to be tracked and utilized by the scheduler.

To support downlink scheduling, a terminal may provide the network with *channel-state* reports indicating the instantaneous downlink channel quality in both the time and frequency domains. The channel state is typically obtained by measuring on *reference signals* transmitted in the downlink. Based on the channel-state reports, also referred to as *channel-state information* (CSI), the downlink scheduler can assign resources for downlink transmission to different terminals, taking the channel quality into account in the scheduling decision. In principle, a scheduled terminal can be assigned an arbitrary combination of 180 kHz wide *resource blocks* in each 1 ms scheduling interval.

As already mentioned, the LTE uplink is based on orthogonal separation of different uplink transmissions and it is the task of the uplink scheduler to assign resources in both the time and frequency domains to different terminals. Scheduling decisions, taken once per 1 ms, control what set of terminals are allowed to transmit within a cell during a given time interval and, for each terminal, on what frequency resources the transmission is to take place and what transmission parameters, including the data rate, to use.

Channel conditions can also be taken into account in the uplink scheduling process, similar to the downlink scheduling. However, as will be discussed in more detail in subsequent chapters, obtaining information about the uplink channel conditions may not be feasible or desirable in all situations. Therefore, different means to obtain *uplink diversity* are important as a complement in situations where uplink channel-dependent scheduling is not suitable.

7.1.3 Inter-Cell Interference Coordination

LTE is designed to operate with a one-cell frequency reuse, implying that the same time–frequency resources can be used in neighboring cells. In particular, the basic control channels are designed to operate properly also with the relatively low signal-to-interference ratio that may be experienced in a reuse-one deployment.

From an overall system-efficiency point-of-view, having access to the entire available spectrum in each cell and operating with one-cell reuse is always beneficial. However, it may also lead to relatively large variations in the signal-to-interference ratio, and thus also in the achievable data rates, over the cell area with potentially only relatively low data rates being available at the cell border. Thus, system performance, and especially the cell-edge user quality, can be further enhanced by allowing for some coordination in the scheduling between cells. The basic aim of such *inter-cell interference coordination* (ICIC) is to, if possible, avoid scheduling transmissions to/from terminals at the cell border simultaneously in neighboring cells, thereby avoiding the worst-case interference situations.

To support such interference coordination, the LTE specification includes several messages that can be communicated between eNodeBs using the so-called *X2 interface*, see Chapter 8. These messages provide information about the interference situation and/or scheduling strategies of the eNodeB issuing the message and can be used by an eNodeB receiving the message as input to its scheduling process. LTE interference coordination is discussed in more detail in Chapter 13.

An even more complicated interference situation may occur in so-called heterogeneous network deployments consisting of overlapping cell layers with large differences in the cell output power. This will be briefly discussed in Section 7.3.4 as part of the discussion on LTE release-10 features. The interference handling in heterogeneous network deployments will then be discussed in more detail in Chapter 13.

7.1.4 Hybrid ARQ with Soft Combining

Fast hybrid ARQ with soft combining is used in LTE to allow the terminal to rapidly request retransmissions of erroneously received transport blocks and to provide a tool for implicit rate adaptation. Retransmissions can be rapidly requested after each packet transmission, thereby minimizing the impact on end-user performance from erroneously received packets. Incremental redundancy is used as the soft combining strategy and the receiver buffers the soft bits to be able to perform soft combining between transmission attempts.

7.1.5 Multi-Antenna Transmission

Already from its first release, LTE included support for different multi-antenna transmission techniques as an integral part of the radio-interface specifications. In many respects, the use of multiple antennas is the key technology to reach many of the aggressive LTE performance targets. As discussed in Chapter 5, multiple antennas can be used in different ways for different purposes:

- Multiple receive antennas can be used for receive diversity. For uplink transmissions, this has been used in many cellular systems for several years. However, as dual receive antennas are the baseline for all LTE terminals, the downlink performance is also improved. The simplest way of using multiple receive antennas is classical receive diversity to collect additional energy and suppress fading, but additional gains can be achieved in interference-limited scenarios if the antennas are used not only to provide diversity, but also to suppress interference, as discussed in Chapter 5.
- Multiple transmit antennas at the base station can be used for transmit diversity and different types of beam-forming. The main goal of beam-forming is to improve the received SINR and, eventually, improve system capacity and coverage.
- *Spatial multiplexing*, sometimes referred to as MIMO, using multiple antennas at both the transmitter and receiver is supported by LTE. Spatial multiplexing results in an increased data rate, channel conditions permitting, in bandwidth-limited scenarios by creating several parallel "channels", as described in Chapter 5. Alternatively, by combining the spatial properties with the appropriate interference-suppressing receiver processing, multiple terminals can transmit on the same time–frequency resource in order to improve the overall cell capacity. This is sometimes referred to as *multi-user MIMO*.

In general, the different multi-antenna techniques are beneficial in different scenarios. As an example, at relatively low SINR, such as at high load or at the cell edge, spatial multiplexing provides relatively limited benefits. Instead, in such scenarios multiple antennas at the transmitter side should be used to raise the SINR by means of beam-forming. On the other hand, in scenarios where there already is a relatively high SINR, for example in small cells, raising the signal quality further provides relatively minor gains as the achievable data rates are then mainly bandwidth limited rather than SINR limited. In such scenarios, spatial multiplexing should instead be used to fully exploit the good channel conditions. The multi-antenna scheme used is under control of the base station, which therefore can select a suitable scheme for each transmission.

Up to four layers can be spatially multiplexed in release 8. Later releases further enhance the multi-antenna support, as described later.

7.1.6 Spectrum Flexibility

A high degree of spectrum flexibility is one of the main characteristics of the LTE radio-access technology. The aim of this spectrum flexibility is to allow for the deployment of LTE radio access in difference frequency bands with different characteristics, including different duplex arrangements and different sizes of the available spectrum. Chapter 17 outlines further details of how spectrum flexibility is achieved in LTE.

7.1.6.1 Flexibility in Duplex Arrangement

One important part of the LTE requirements in terms of spectrum flexibility is the possibility to deploy LTE-based radio access in both paired *and* unpaired spectrum. Therefore, LTE supports both

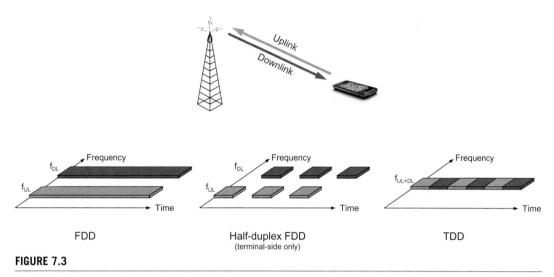

FIGURE 7.3

Frequency- and time-division duplex.

frequency- and time-division-based duplex arrangements. *Frequency-Division Duplex* (FDD), as illustrated on the left in Figure 7.3, implies that downlink and uplink transmission take place in different, sufficiently separated, frequency bands. *Time-Division Duplex* (TDD), as illustrated on the right in Figure 7.3, implies that downlink and uplink transmission take place in different, non-overlapping time slots. Thus, TDD can operate in unpaired spectrum, whereas FDD requires paired spectrum. The required flexibility and resulting requirements to support LTE operation in different paired and unpaired frequency arrangements are further discussed in Chapter 17.

Operation in both paired and unpaired spectrum has been supported by 3GPP radio-access technologies even before the introduction of LTE by means of FDD-based WCDMA/HSPA in combination with TDD-based TD-SCDMA radio. However, this was then achieved by means of, at least in the details, relatively different radio-access technologies leading to additional effort and complexity when developing and implementing dual-mode terminals capable of both FDD and TDD operation. LTE, on the other hand, supports both FDD and TDD *within a single radio-access technology*, leading to a minimum of deviation between FDD and TDD for LTE-based radio access. As a consequence of this, the overview of the LTE radio access provided in the following chapters is, to a large extent, valid for both FDD and TDD. In the case of differences between FDD and TDD, these differences will be explicitly indicated. Furthermore, the TDD mode, also known as TD-LTE, is designed with coexistence between TD-LTE and TD-SCDMA in mind to simplify a gradual migration from TD-SCDMA to TD-LTE.

LTE also supports *half-duplex* FDD at the terminal (illustrated in the middle of Figure 7.3). In half-duplex FDD, transmission and reception *at a specific terminal* are separated in both frequency and time. The base station still uses full-duplex FDD as it simultaneously may schedule *different* terminals in uplink and downlink; this is similar to, for example, GSM operation. The main benefit with half-duplex FDD is the reduced terminal complexity as no duplex filter is needed in the terminal. This is especially beneficial in the case of multi-band terminals which otherwise would need multiple sets of duplex filters.

7.1.6.2 *Bandwidth Flexibility*

An important characteristic of LTE is the possibility for different transmission bandwidths on both downlink and uplink. The main reason for this is that the amount of spectrum available for LTE deployment may vary significantly between different frequency bands and also depending on the exact situation of the operator. Furthermore, the possibility of operating in different spectrum allocations gives the possibility for gradual migration of spectrum from other radio-access technologies to LTE.

LTE supports operation in a wide range of spectrum allocations, achieved by a flexible transmission bandwidth being part of the LTE specifications. To efficiently support very high data rates when spectrum is available, a wide transmission bandwidth is necessary, as discussed in Chapter 2. However, a sufficiently large amount of spectrum may not always be available, either due to the band of operation or due to a gradual migration from another radio-access technology, in which case LTE can be operated with a more narrow transmission bandwidth. Obviously, in such cases, the maximum achievable data rates will be reduced accordingly. As discussed below, the spectrum flexibility is further improved in later releases of LTE.

The LTE physical-layer specifications [61–64] are bandwidth agnostic and do not make any particular assumption on the supported transmission bandwidths beyond a minimum value. As will be seen in the following, the basic radio-access specification, including the physical-layer and protocol specifications, allows for any transmission bandwidth ranging from roughly 1 MHz up to around 20 MHz. At the same time, at an initial stage, radio-frequency requirements are only specified for a limited subset of transmission bandwidths, corresponding to what is predicted to be relevant spectrum-allocation sizes and relevant migration scenarios. Thus, in practice the LTE radio-access technology supports a limited set of transmission bandwidths, but additional transmission bandwidths can easily be introduced by updating only the RF specifications.

7.2 LTE RELEASE 9

After completing the first release of LTE, work continued in 3GPP with introducing additional functionality in the second release of the LTE specifications, release 9. The main enhancements seen in release 9, completed in late 2009, were support for multicast transmission, support for network-assisted positioning services, and enhancements to beam-forming in the downlink.

7.2.1 Multicast and Broadcast Support

Multi-cell broadcast implies transmission of the same information from multiple cells, as described in Chapter 3. By exploiting this at the terminal, effectively using signal power from multiple cell sites at the detection, a substantial improvement in coverage (or higher broadcast data rates) can be achieved. By transmitting not only identical signals from multiple cell sites (with identical coding and modulation), but also synchronizing the transmission timing between the cells, the signal at the terminal will appear exactly as a signal transmitted from a single cell site and subject to multi-path propagation. Due to the OFDM robustness to multi-path propagation, such multi-cell transmission, in 3GPP also referred to as *Multicast/Broadcast Single-Frequency Network* (MBSFN) transmission, will then not only improve the received signal strength, but also eliminate the inter-cell interference as described in

Chapter 3. Thus, with OFDM, multi-cell broadcast/multicast throughput may eventually be limited by noise only and can then, in the case of small cells, reach extremely high values.

It should be noted that the use of MBSFN transmission for multi-cell broadcast/multicast assumes the use of tight synchronization and time alignment of the signals transmitted from different cell sites.

7.2.2 Positioning

Positioning, as the name implies, refers to functionality in the radio-access network to determine the location of individual terminals. Determining the position of a terminal can, in principle, be done by including a GPS receiver in the terminal. Although this is a quite common feature, not all terminals include the necessary GSP receiver and there may also be cases when the GPS service is not available. LTE release 9 therefore introduces positioning support inherent in the radio-access network. By measuring on special reference signals transmitted regularly from different cell sites, the location of the terminal can be determined.

7.2.3 Dual-Layer Beam-Forming

Release 9 enhances the support for combining spatial multiplexing with beam-forming. Although the combination of beam-forming and spatial multiplexing was already possible in release 8, this was then restricted to so-called *codebook-based precoding* (see Chapter 10). In release 9, spatial multiplexing can be combined with so-called *non-codebook-based precoding*, thereby improving the flexibility in deploying various multi-antenna schemes. The enhancements can also be seen as the basis for a further improvement in the area in LTE release 10.

7.3 LTE RELEASE 10 AND IMT-ADVANCED

As described in Chapter 1, *IMT-Advanced* is the term used by the ITU for radio-access technologies beyond IMT-2000. As a first step in defining IMT-Advanced, the ITU defined a set of requirements that any IMT-Advanced compliant technology should fulfill [65]. Examples of these requirements are support for at least 40 MHz bandwidth, peak spectral efficiencies of 15 bit/s/Hz in downlink and 6.75 bit/s/Hz in uplink (corresponding to peak rates of at least 600 and 270 Mbit/s respectively), and control and user plane latency of less than 100 and 10 ms respectively.

One of the main targets of LTE release 10 was to ensure that the LTE radio-access technology would be fully compliant with the IMT-Advanced requirements, thus the name *LTE-Advanced* is often used for LTE release 10. However, in addition to the ITU requirements, 3GPP also defined its own targets and requirements [66] for LTE release 10 (LTE-Advanced). These targets/requirements extended the ITU requirements both in terms of being more aggressive as well as including additional requirements. One important requirement was *backwards compatibility*. Essentially this means that an earlier-release LTE terminal should always be able to access a carrier supporting LTE release-10 functionality, although obviously not being able to utilize all the release-10 features of that carrier.

LTE release 10 was completed in late 2010 and enhances LTE spectrum flexibility through *carrier aggregation*, further extends multi-antenna transmission, introduces support for *relaying*, and provides improvements in the area of inter-cell interference coordination in *heterogeneous network deployments*.

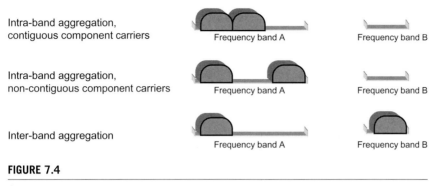

Intra-band aggregation, contiguous component carriers — Frequency band A — Frequency band B

Intra-band aggregation, non-contiguous component carriers — Frequency band A — Frequency band B

Inter-band aggregation — Frequency band A — Frequency band B

FIGURE 7.4

Carrier aggregation.

7.3.1 Carrier Aggregation

As mentioned earlier, the first release of LTE already provided extensive support for deployment in spectrum allocations of various characteristics, with bandwidths ranging from roughly 1 MHz up to 20 MHz in both paired and unpaired bands. In LTE release 10 the transmission bandwidth can be further extended by means of so-called *carrier aggregation* (CA), where multiple *component carriers* are aggregated and jointly used for transmission to/from a single terminal. Up to five component carriers, possibly each of different bandwidth, can be aggregated, allowing for transmission bandwidths up to 100 MHz. Backwards compatibility is catered for as each component carrier uses the release-8 structure. Hence, to a release-8/9 terminal each component carrier will appear as an LTE release-8 carrier, while a carrier-aggregation-capable terminal can exploit the total aggregated bandwidth, enabling higher data rates. In the general case, a different number of component carriers can be aggregated for the downlink and uplink.

Component carriers do not have to be contiguous in frequency, which enables exploitation of *fragmented spectrum*; operators with a fragmented spectrum can provide high-data-rate services based on the availability of a wide overall bandwidth even though they do not posses a single wideband spectrum allocation. From a baseband perspective, there is no difference between the cases in Figure 7.4 and they are all supported by LTE release 10. However, the RF-implementation complexity is vastly different with the first case being the least complex. Thus, although spectrum aggregation is supported by the basic specifications, the actual release-10 RF requirements will be strongly constrained, including specification of only a limited number of aggregation scenarios and including support of inter-band aggregation only for the most advanced terminals, but excluding non-contiguous intra-band aggregation.

7.3.2 Extended Multi-Antenna Transmission

In release 10, downlink spatial multiplexing is expanded to support up to eight transmission layers. Along with this, and as discussed in more detail in Chapter 10 an enhanced *reference-signal structure* is introduced to improve the support of various beam-forming solutions. This can be seen as an extension of the release-9 dual-layer beam-forming to support up to eight antenna ports and eight corresponding layers. Together with the support for carrier aggregation, this enables downlink data rates up to 3 Gbit/s or 30 bit/s/Hz.

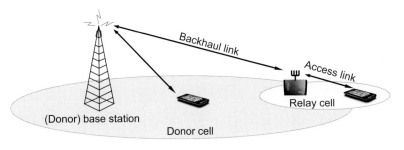

FIGURE 7.5

Example of relaying.

Uplink spatial multiplexing of up to four layers is also part of release 10. It consists of a code-book-based scheme under control of the base station, which means that the structure can also be used for uplink transmit-side beam-forming. Together with the possibility for uplink carrier aggregations, this allows for uplink data rates up to 1.5 Gbit/s or 15 bit/s/Hz.

7.3.3 Relaying

Relaying implies that the terminal communicates with the network via a *relay node* that is *wirelessly connected* to a *donor cell* using the LTE radio-interface technology (Figure 7.5). From a terminal point of view, the relay node will appear as an ordinary cell. This has the important advantage of simplifying the terminal implementation and making the relay node backwards compatible – that is, also accessible to LTE release-8/9 terminals. In essence, the relay is a low-power base station wirelessly connected to the remaining part of the network. One of the attractive features of a relay is the LTE-based wireless backhaul, as this could provide a simple way of improving coverage, for example in indoor environments, by simply placing relays at the problematic locations. At a later stage, if motivated by the traffic situation, the wireless donor-relay link could be replaced by an optical fiber in order to use the precious radio resources in the donor cell for terminal communication instead of serving the relay.

7.3.4 Heterogeneous Deployments

Heterogeneous deployments refer to deployments with a mixture of cells with different downlink transmission power, operating on (partially) the same set of frequencies and with overlapping geographical coverage (Figure 7.6). A typical example is a pico cell placed within the coverage area of a macro cell. Although such deployments were already supported in release 8, release 10 introduced improved inter-cell interference handling focusing on scenarios with large power differences between overlapping cells.

7.4 TERMINAL CAPABILITIES

To support different scenarios, which may call for different terminal capabilities in terms of data rates, as well as to allow for market differentiation in terms of low- and high-end terminals with a

FIGURE 7.6

Example of heterogeneous deployment with a pico cell inside a macro cell.

Table 7.1 UE Categories

	Category							
	Release 8/9/10					Release 10 only		
	1	2	3	4	5	6	7	8
Downlink peak rate (Mbit/s)	10	50	100	150	300	300	300	3000
Uplink peak rate (Mbit/s)	5	25	50	50	75	50	150	1500
Maximum downlink modulation	64QAM							
Maximum uplink modulation	16QAM				64QAM	16QAM	64QAM	
Max. number of layers for downlink spatial multiplexing	1	2			4	Signaled separately		

corresponding difference in price, not all terminals support all capabilities. Furthermore, terminals from an earlier release of the standard will not support features introduced in later versions of LTE. For example, a release-8 terminal will obviously not support carrier aggregation as this feature was introduced in release 10. Therefore, as part of the connection setup, the terminal indicates not only which release of LTE it supports, but also its capabilities within the release.

In principle, the different parameters could be specified separately, but to limit the number of combinations and avoid a parameter combination that does not make sense, a set of physical-layer capabilities are lumped together to form a UE category (UE, *User Equipment*, is the term used in 3GPP to denote a terminal). In total five different UE categories have been specified for LTE release 8/9, ranging from the low-end category 1 not supporting spatial multiplexing to the high-end category 5 supporting the full set of features in the release-8/9 physical layer specifications. The categories are summarized in Table 7.1 (in simplified form; for the full set of details, see [67]). Note that, regardless of the category, a terminal is always capable of receiving transmissions from up to four antenna ports. This is necessary as the system information can be transmitted on up to four antenna ports.

In LTE release 10, features such as carrier aggregation and uplink spatial multiplexing are introduced, which calls for additional capability signaling compared to release 8/9, either in the form of additional UE categories or as separate capabilities. Defining new categories for each foreseen combination of the maximum number of component carriers and maximum degree of spatial multiplexing could be done in principle, although the number of categories might become very large and which categories a terminal support may be frequency-band dependent. Therefore, in release 10, three additional UE categories were defined as seen in Table 7.1, and the maximum number of component carriers and degree of spatial multiplexing supported, both in uplink and downlink, are signaled separately from the category number. A release-10 terminal may therefore declare itself as, for example, category 4 but capable of uplink spatial multiplexing. Hence, categories 1–5 may have a slightly different meaning for a release-8/9 and a release-10 terminal, depending on the value of the separately declared capabilities. Furthermore, in order to be able to operate in release-8/9 networks, a release-10 UE has to be able to declare both release-8/9 and release-10 categories.

In addition to the capabilities mentioned in the different UE categories, there are some capabilities specified outside the categories. The duplexing schemes supported is one such example, the support of UE-specific reference signals for FDD in release 8 is another. Whether the terminal supports other radio-access technologies, for example GSM and WCDMA, is also declared separately.

Radio-Interface Architecture

This chapter contains a brief overview of the overall architecture of an LTE radio-access network and the associated core network, followed by descriptions of the radio-access network user-plane and control-plane protocols.

8.1 OVERALL SYSTEM ARCHITECTURE

In parallel to the work on the LTE radio-access technology in 3GPP, the overall system architecture of both the *Radio-Access Network* (RAN) and the *Core Network* (CN) was revisited, including the split of functionality between the two network parts. This work was known as the *System Architecture Evolution* (SAE) and resulted in a flat RAN architecture, as well as a new core network architecture referred to as the *Evolved Packet Core* (EPC). Together, the LTE RAN and the EPC can be referred to as the *Evolved Packet System* (EPS).[1]

The RAN is responsible for all radio-related functionality of the overall network including, for example, scheduling, radio-resource handling, retransmission protocols, coding and various multi-antenna schemes. These functions will be discussed in detail in the subsequent chapters.

The EPC is responsible for functions not related to the radio interface but needed for providing a complete mobile-broadband network. This includes, for example, authentication, charging functionality, and setup of end-to-end connections. Handling these functions separately, instead of integrating them into the RAN, is beneficial as it allows for several radio-access technologies to be served by the same core network.

Although this book focuses on the LTE RAN, a brief overview of the EPC, as well as how it connects to the RAN, is useful. For an excellent in-depth discussion of EPC, the reader is referred to [9].

8.1.1 Core Network

The EPC is a radical evolution from the GSM/GPRS core network used for GSM and WCDMA/HSPA. EPC supports access to the *packet-switched domain* only, with no access to *the circuit-switched domain*. It consists of several different types of nodes, some of which are briefly described below and illustrated in Figure 8.1.

---

[1] UTRAN, the WCDMA/HSPA radio-access network, is also part of the EPS.

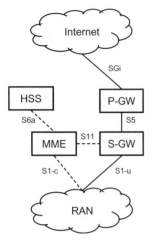

FIGURE 8.1

Core-network (EPC) architecture.

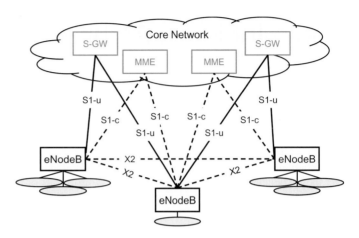

FIGURE 8.2

Radio-access-network interfaces.

The *Mobility Management Entity* (MME) is the control-plane node of the EPC. Its responsibilities include connection/release of bearers to a terminal, handling of IDLE to ACTIVE transitions, and handling of security keys. The functionality operating between the EPC and the terminal is sometimes referred to as the *Non-Access Stratum* (NAS), to separate it from the *Access Stratum* (AS) which handles functionality operating between the terminal and the radio-access network.

The *Serving Gateway* (S-GW) is the user-plane node connecting the EPC to the LTE RAN. The S-GW acts as a mobility anchor when terminals move between eNodeBs (see next section), as well as a mobility anchor for other 3GPP technologies (GSM/GPRS and HSPA). Collection of information and statistics necessary for charging is also handled by the S-GW.

The *Packet Data Network Gateway* (PDN Gateway, P-GW) connects the EPC to the internet. Allocation of the IP address for a specific terminal is handled by the P-GW, as well as quality-of-service enforcement according to the policy controlled by the PCRF (see below). The P-GW is also the mobility anchor for non-3GPP radio-access technologies, such as CDMA2000, connected to the EPC.

In addition, the EPC also contains other types of nodes such as *Policy and Charging Rules Function* (PCRF) responsible for quality-of-service (QoS) handling and charging, and the *Home Subscriber Service* (HSS) node, a database containing subscriber information. There are also some additional nodes present as regards network support of *Multimedia Broadcast Multicast Services* (MBMS) (see Chapter 15 for a more detailed description of MBMS, including the related architecture aspects).

It should be noted that the nodes discussed above are *logical* nodes. In an actual physical implementation, several of them may very well be combined. For example, the MME, P-GW, and S-GW could very well be combined into a single physical node.

8.1.2 **Radio-Access Network**

The LTE radio-access network uses a flat architecture with a single type of node[2] – the *eNodeB*. The eNodeB is responsible for all radio-related functions in one or several cells. It is important to note that an eNodeB is a *logical* node and not a physical implementation. One common implementation of an eNodeB is a three-sector site, where a base station is handling transmissions in three cells, although other implementations can be found as well, such as one baseband processing unit to which a number of remote radio heads are connected. One example of the latter is a large number of indoor cells, or several cells along a highway, belonging to the same eNodeB. Thus, a base station is a *possible* implementation of, but not *the same* as, an eNodeB.

As can be seen in Figure 8.2, the eNodeB is connected to the EPC by means of the *S1 interface*, more specifically to the S-GW by means of the *S1 user-plane part*, S1-u, and to the MME by means of the *S1 control-plane part*, S1-c. One eNodeB can be connected to multiple MMEs/S-GWs for the purpose of load sharing and redundancy.

The *X2 interface*, connecting eNodeBs to each other, is mainly used to support active-mode mobility. This interface may also be used for multi-cell *Radio Resource Management* (RRM) functions such as *Inter-Cell Interference Coordination* (ICIC) discussed in Chapter 13. The X2 interface is also used to support lossless mobility between neighboring cells by means of packet forwarding.

8.2 **RADIO PROTOCOL ARCHITECTURE**

With the overall network architecture in mind, the RAN protocol architecture for the user as well as the control planes can be discussed. Figure 8.3 illustrates the RAN protocol architecture (the MME is, as discussed in the previous section, not part of the RAN but is included in the figure for completeness). As seen in the figure, many of the protocol entities are common to the user and control planes. Therefore, although this section mainly describes the protocol architecture from a user-plane perspective, the description is in many respects also applicable to the control plane. Control-plane-specific aspects are discussed in Section 8.3.

The LTE radio-access network provides one or more *Radio Bearers* to which IP packets are mapped according to their Quality-of-Service requirements. A general overview of the LTE (user-plane) protocol architecture for the downlink is illustrated in Figure 8.4. As will become clear in the subsequent discussion, not all the entities illustrated in Figure 8.4 are applicable in all situations. For example, neither MAC scheduling nor hybrid ARQ with soft combining is used for broadcast of the basic system information. The LTE protocol structure related to uplink transmissions is similar to the downlink structure in Figure 8.4, although there are some differences with respect to, for example, transport-format selection.

The different protocol entities of the radio-access network are summarized below and described in more detail in the following sections.

- *Packet Data Convergence Protocol* (PDCP) performs IP header compression to reduce the number of bits to transmit over the radio interface. The header-compression mechanism is based

---

[2]The introduction of MBMS (see Chapter 15) in release 9 and relaying (see Chapter 16) in release 10 brings additional node types to the RAN.

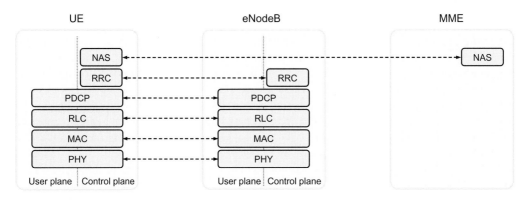

FIGURE 8.3

Overall RAN protocol architecture.

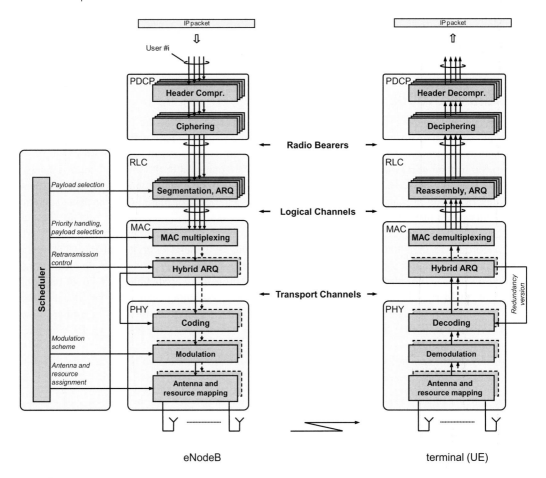

FIGURE 8.4

LTE protocol architecture (downlink).

on Robust Header Compression (ROHC) [68], a standardized header-compression algorithm also used for several mobile-communication technologies. PDCP is also responsible for ciphering and, for the control plane, integrity protection of the transmitted data, as well as in-sequence delivery and duplicate removal for handover. At the receiver side, the PDCP protocol performs the corresponding deciphering and decompression operations. There is one PDCP entity per radio bearer configured for a terminal.

- *Radio-Link Control* (RLC) is responsible for segmentation/concatenation, retransmission handling, duplicate detection, and in-sequence delivery to higher layers. The RLC provides services to the PDCP in the form of *radio bearers*. There is one RLC entity per radio bearer configured for a terminal.
- *Medium-Access Control* (MAC) handles multiplexing of logical channels, hybrid-ARQ retransmissions, and uplink and downlink scheduling. The scheduling functionality is located in the eNodeB for both uplink and downlink. The hybrid-ARQ protocol part is present in both the transmitting and receiving ends of the MAC protocol. The MAC provides services to the RLC in the form of *logical channels*.
- *Physical Layer* (PHY) handles coding/decoding, modulation/demodulation, multi-antenna mapping, and other typical physical-layer functions. The physical layer offers services to the MAC layer in the form of *transport channels*.

To summarize the flow of downlink data through all the protocol layers, an example illustration for a case with three IP packets, two on one radio bearer and one on another radio bearer, is given in Figure 8.5. The data flow in the case of uplink transmission is similar. The PDCP performs (optional) IP-header compression, followed by ciphering. A PDCP header is added, carrying information required for deciphering in the terminal. The output from the PDCP is forwarded to the RLC.

The RLC protocol performs concatenation and/or segmentation of the PDCP SDUs[3] and adds an RLC header. The header is used for in-sequence delivery (per logical channel) in the terminal and for identification of RLC PDUs in the case of retransmissions. The RLC PDUs are forwarded to the MAC layer, which multiplexes a number of RLC PDUs and attaches a MAC header to form a transport block. The transport-block size depends on the instantaneous data rate selected by the link-adaptation mechanism. Thus, the link adaptation affects both the MAC and RLC processing. Finally, the physical layer attaches a CRC to the transport block for error-detection purposes, performs coding and modulation, and transmits the resulting signal, possibly using multiple transmit antennas.

The remainder of the chapter contains an overview of the RLC, MAC, and physical layers. A more detailed description of the LTE physical-layer processing is given in Chapters 10 (downlink) and 11 (uplink), followed by descriptions of some specific uplink and downlink radio-interface functions and procedures in the subsequent chapters.

8.2.1 **Radio-Link Control**

The RLC protocol is responsible for segmentation/concatenation of (header-compressed) IP packets, also known as RLC SDUs, from the PDCP into suitably sized RLC PDUs. It also handles

---

[3]In general, the data entity from/to a higher protocol layer is known as a Service Data Unit (SDU) and the corresponding entity to/from a lower protocol layer entity is called a Protocol Data Unit (PDU).

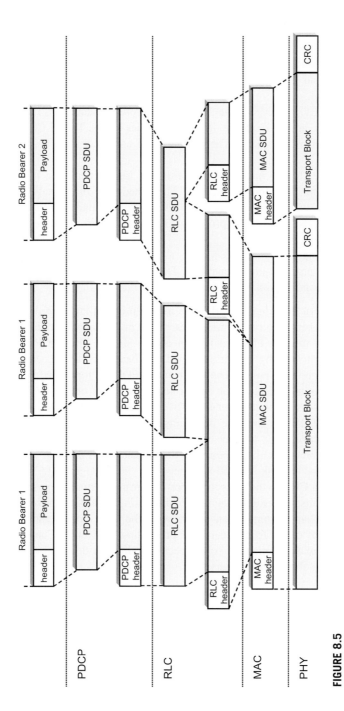

FIGURE 8.5

Example of LTE data flow.

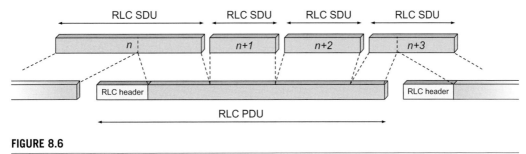

FIGURE 8.6

RLC segmentation and concatenation.

retransmission of erroneously received PDUs, as well as removal of duplicated PDUs. Finally, the RLC ensures in-sequence delivery of SDUs to upper layers. Depending on the type of service, the RLC can be configured in different modes to perform some or all of these functions.

Segmentation and concatenation, one of the main RLC functions, is illustrated in Figure 8.6. Depending on the scheduler decision, a certain amount of data is selected for transmission from the RLC SDU buffer and the SDUs are segmented/concatenated to create the RLC PDU. Thus, for LTE the RLC PDU size varies *dynamically*. For high data rates, a large PDU size results in a smaller relative overhead, while for low data rates, a small PDU size is required as the payload would otherwise be too large. Hence, as the LTE data rates may range from a few kbit/s up to 3 Gbit/s, dynamic PDU sizes are motivated for LTE in contrast to earlier mobile-communication technologies, which typically use a fixed PDU size. Since the RLC, scheduler, and rate adaptation mechanisms are all located in the eNodeB, dynamic PDU sizes are easily supported for LTE. In each RLC PDU, a header is included, containing, among other things, a sequence number used for in-sequence delivery and by the retransmission mechanism.

The RLC retransmission mechanism is also responsible for providing error-free delivery of data to higher layers. To accomplish this, a retransmission protocol operates between the RLC entities in the receiver and transmitter. By monitoring the sequence numbers of the incoming PDUs, the receiving RLC can identify missing PDUs. Status reports are then fed back to the transmitting RLC entity, requesting retransmission of missing PDUs. Based on the received status report, the RLC entity at the transmitter can take the appropriate action and retransmit the missing PDUs if needed.

Although the RLC is capable of handling transmission errors due to noise, unpredictable channel variations, etc., error-free delivery is in most cases handled by the MAC-based hybrid-ARQ protocol. The use of a retransmission mechanism in the RLC may therefore seem superfluous at first. However, as will be discussed in Section 8.2.2.3, this is not the case and the use of both RLC- and MAC-based retransmission mechanisms is in fact well motivated by the differences in the feedback signaling.

The details of RLC are further described in Chapter 12.

8.2.2 Medium-Access Control

The MAC layer handles logical-channel multiplexing, hybrid-ARQ retransmissions, and uplink and downlink scheduling. It is also responsible for multiplexing/demultiplexing data across multiple component carriers when carrier aggregation is used.

8.2.2.1 *Logical Channels and Transport Channels*

The MAC provides services to the RLC in the form of *logical channels*. A logical channel is defined by the *type* of information it carries and is generally classified as a *control channel*, used for transmission of control and configuration information necessary for operating an LTE system, or as a *traffic channel*, used for the user data. The set of logical-channel types specified for LTE includes:

- *The Broadcast Control Channel* (BCCH), used for transmission of *system information* from the network to all terminals in a cell. Prior to accessing the system, a terminal needs to acquire the system information to find out how the system is configured and, in general, how to behave properly within a cell.
- The *Paging Control Channel* (PCCH), used for paging of terminals whose location on a cell level is not known to the network. The paging message therefore needs to be transmitted in multiple cells.
- The *Common Control Channel* (CCCH), used for transmission of control information in conjunction with random access.
- The *Dedicated Control Channel* (DCCH), used for transmission of control information to/from a terminal. This channel is used for individual configuration of terminals such as different handover messages.
- The *Multicast Control Channel* (MCCH), used for transmission of control information required for reception of the MTCH (see below).
- The *Dedicated Traffic Channel* (DTCH), used for transmission of user data to/from a terminal. This is the logical channel type used for transmission of all uplink and non-MBSFN downlink user data.
- The *Multicast Traffic Channel* (MTCH), used for downlink transmission of MBMS services.

From the physical layer, the MAC layer uses services in the form of *transport channels*. A transport channel is defined by *how* and *with what characteristics* the information is transmitted over the radio interface. Data on a transport channel is organized into *transport blocks*. In each *Transmission Time Interval* (TTI), at most one transport block of dynamic size is transmitted over the radio interface to/from a terminal in the absence of spatial multiplexing. In the case of spatial multiplexing (MIMO), there can be up to two transport blocks per TTI.

Associated with each transport block is a *Transport Format* (TF), specifying *how* the transport block is to be transmitted over the radio interface. The transport format includes information about the transport-block size, the modulation-and-coding scheme, and the antenna mapping. By varying the transport format, the MAC layer can thus realize different data rates. Rate control is therefore also known as *transport-format selection*.

The following transport-channel types are defined for LTE:

- The *Broadcast Channel* (BCH) has a fixed transport format, provided by the specifications. It is used for transmission of parts of the BCCH system information, more specifically the so-called *Master Information Block* (MIB), as described in Chapter 14.
- The *Paging Channel* (PCH) is used for transmission of paging information from the PCCH logical channel. The PCH supports *discontinuous reception* (DRX) to allow the terminal to save battery power by waking up to receive the PCH only at predefined time instants. The LTE paging mechanism is also described in Chapter 14.

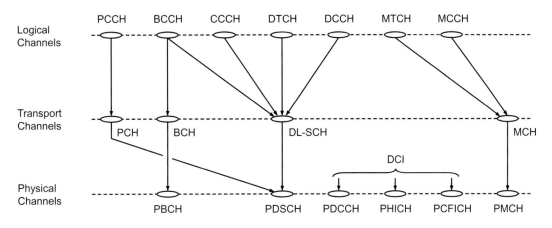

FIGURE 8.7

Downlink channel mapping.

- The *Downlink Shared Channel* (DL-SCH) is the main transport channel used for transmission of downlink data in LTE. It supports key LTE features such as dynamic rate adaptation and channel-dependent scheduling in the time and frequency domains, hybrid ARQ with soft combining, and spatial multiplexing. It also supports DRX to reduce terminal power consumption while still providing an always-on experience. The DL-SCH is also used for transmission of the parts of the BCCH system information not mapped to the BCH. There can be multiple DL-SCHs in a cell, one per terminal[4] scheduled in this TTI, and, in some subframes, one DL-SCH carrying system information.
- The *Multicast Channel* (MCH) is used to support MBMS. It is characterized by a semi-static transport format and semi-static scheduling. In the case of multi-cell transmission using MBSFN, the scheduling and transport format configuration is coordinated among the transmission points involved in the MBSFN transmission. MBSFN transmission is described in Chapter 15.
- The Uplink Shared Channel (UL-SCH) is the uplink counterpart to the DL-SCH – that is, the uplink transport channel used for transmission of uplink data.

In addition, the *Random-Access Channel* (RACH) is also defined as a transport channel, although it does not carry transport blocks.

Part of the MAC functionality is multiplexing of different logical channels and mapping of the logical channels to the appropriate transport channels. The supported mappings between logical-channel types and transport-channel types are given in Figure 8.7 for the downlink and Figure 8.8 for the uplink. The figures clearly indicate how DL-SCH and UL-SCH are the main downlink and uplink transport channels respectively. In the figures, the corresponding physical channels, described further below, are also included and the mapping between transport channels and physical channels is illustrated.

To support priority handling, multiple logical channels, where each logical channel has its own RLC entity, can be multiplexed into one transport channel by the MAC layer. At the receiver, the

---

[4]For carrier aggregation, a UE may receive multiple DL-SCHs, one per component carrier.

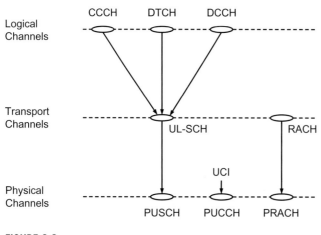

FIGURE 8.8

Uplink channel mapping.

MAC layer handles the corresponding demultiplexing and forwards the RLC PDUs to their respective RLC entity for in-sequence delivery and the other functions handled by the RLC. To support the demultiplexing at the receiver, a MAC header, shown in Figure 8.9, is used. To each RLC PDU, there is an associated sub-header in the MAC header. The sub-header contains the identity of the logical channel (LCID) from which the RLC PDU originated and the length of the PDU in bytes. There is also a flag indicating whether this is the last sub-header or not. One or several RLC PDUs, together with the MAC header and, if necessary, padding to meet the scheduled transport-block size, form one transport block which is forwarded to the physical layer.

In addition to multiplexing of different logical channels, the MAC layer can also insert the so-called *MAC control elements* into the transport blocks to be transmitted over the transport channels. A MAC control element is used for inband control signaling – for example, timing-advance commands and random-access response, as described in Sections 11.5 and 14.3 respectively. Control elements are identified with reserved values in the LCID field, where the LCID value indicates the type of control information. Furthermore, the length field in the sub-header is removed for control elements with a fixed length.

The MAC multiplexing functionality is also responsible for handling of multiple component carriers in the case of carrier aggregation. The basic principle for carrier aggregation is independent processing of the component carriers in the physical layer, including control signaling, scheduling and hybrid-ARQ retransmissions, while carrier aggregation is invisible to RLC and PDCP. Carrier aggregation is therefore mainly seen in the MAC layer, as illustrated in Figure 8.10, where logical channels, including any MAC control elements, are multiplexed to form one (two in the case of spatial multiplexing) transport block(s) per component carrier with each component carrier having its own hybrid-ARQ entity.

8.2.2.2 *Scheduling*

One of the basic principles of LTE radio access is shared-channel transmission – that is, time–frequency resources are dynamically shared between users. The *scheduler* is part of the MAC layer

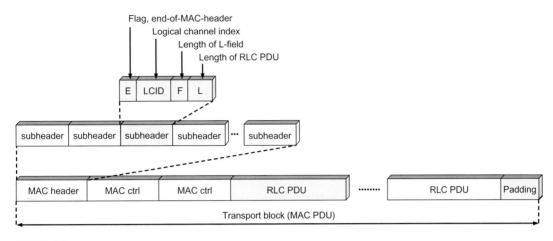

FIGURE 8.9

MAC header and SDU multiplexing.

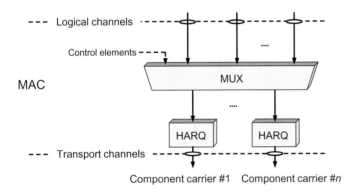

FIGURE 8.10

MAC operation in the case of multiple component carriers.

(although often better viewed as a separate entity as illustrated in Figure 8.4) and controls the assignment of uplink and downlink resources in terms of so-called *resource-block pairs*. Resource blocks correspond to a time–frequency unit of 1 ms times 180 kHz, as described in more detail in Chapter 9.

The basic operation of the scheduler is so-called *dynamic* scheduling, where the eNodeB in each 1 ms interval takes a scheduling decision and sends scheduling information to the selected set of terminals. However, there is also a possibility for semi-persistent scheduling where a semi-static scheduling pattern is signaled in advance to reduce the control-signaling overhead. Coordination of scheduling decisions across multiple cells residing in different eNodeBs is supported using signaling over the X2 interface.

Uplink and downlink scheduling are separated in LTE, and uplink and downlink scheduling decisions can be taken independently of each other (within the limits set by the uplink/downlink split in the case of half-duplex FDD operation).

The downlink scheduler is responsible for (dynamically) controlling which terminal(s) to transmit to and, for each of these terminals, the set of resource blocks upon which the terminal's DL-SCH should be transmitted. Transport-format selection (selection of transport-block size, modulation scheme, and antenna mapping) and logical-channel multiplexing for downlink transmissions are controlled by the eNodeB, as illustrated in the left part of Figure 8.11. As a consequence of the scheduler controlling the data rate, the RLC segmentation and MAC multiplexing will also be affected by the scheduling decision. The outputs from the downlink scheduler can be seen in Figure 8.4.

The uplink scheduler serves a similar purpose, namely to (dynamically) control which terminals are to transmit on their respective UL-SCH and on which uplink time–frequency resources (including component carrier). Despite the fact that the eNodeB scheduler determines the transport format for the terminal, it is important to point out that the uplink scheduling decision is taken *per terminal* and not per radio bearer. Thus, although the eNodeB scheduler controls the payload of a scheduled terminal, the terminal is still responsible for selecting *from which radio bearer(s)* the data is taken. The terminal autonomously handles logical-channel multiplexing according to rules, the parameters of which can be configured by the eNodeB. This is illustrated in the right part of Figure 8.11, where the eNodeB scheduler controls the transport format and the terminal controls the logical-channel multiplexing.

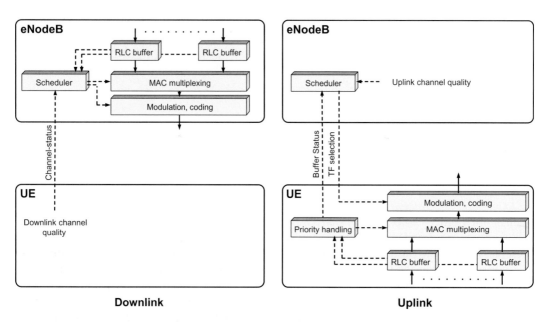

FIGURE 8.11

Transport-format selection in downlink and uplink.

Although the scheduling strategy is implementation specific and not specified by 3GPP, the over-all goal of most schedulers is to take advantage of the channel variations between terminals and pref-erably schedule transmissions to a terminal on resources with advantageous channel conditions. A benefit of the use of OFDM in LTE is the possibility to exploit channel variations in both time and frequency domains through channel-dependent scheduling. This was mentioned earlier in Chapter 7 and illustrated in Figure 7.2. For the larger bandwidths supported by LTE, where a significant amount of frequency-selective fading will often be experienced, the possibility for the scheduler to also exploit frequency-domain channel variations becomes increasingly important compared to exploiting time-domain variations only. The possibility to also exploit frequency-domain variations is beneficial especially at low speeds, where the variations in the time domain are relatively slow compared to the delay requirements set by many services.

Downlink channel-dependent scheduling is supported through *channel-state reports*, transmitted by the UE and reflecting the instantaneous channel quality in the time and frequency domains, as well as information necessary to determine the appropriate antenna processing in the case of spa-tial multiplexing. In the uplink, the channel-state information necessary for uplink channel-dependent scheduling can be based on a *sounding reference signal* transmitted from each terminal for which the eNodeB wants to estimate the uplink channel quality. To aid the uplink scheduler in its decisions, the terminal can transmit buffer-status information to the eNodeB using a MAC message. Obviously, this information can only be transmitted if the terminal has been given a valid scheduling grant. For situ-ations when this is not the case, an indicator that the terminal needs uplink resources is provided as part of the uplink L1/L2 control-signaling structure (see Chapter 11).

Interference coordination, which tries to control the inter-cell interference on a slow basis, as mentioned in Chapter 7, is also part of the scheduler. As the scheduling strategy is not mandated by the specifications, the interference-coordination scheme (if used) is vendor specific and may range from simple higher-order reuse deployments to more advanced schemes. The mechanisms supporting inter-cell interference coordination are discussed in Chapter 13.

8.2.2.3 *Hybrid ARQ with Soft Combining*

Hybrid ARQ with soft combining provides robustness against transmission errors. As hybrid-ARQ retransmissions are fast, many services allow for one or multiple retransmissions, thereby forming an implicit (closed loop) rate-control mechanism. The hybrid-ARQ protocol is part of the MAC layer, while the actual soft combining is handled by the physical layer.[5]

Obviously, hybrid ARQ is not applicable for all types of traffic. For example, broadcast transmis-sions, where the same information is intended for multiple terminals, typically do not rely on hybrid ARQ. Hence, hybrid ARQ is only supported for the DL-SCH and the UL-SCH, although its usage is optional.

The LTE hybrid-ARQ protocol uses multiple parallel stop-and-wait processes. Upon reception of a transport block, the receiver makes an attempt to decode the transport block and informs the trans-mitter about the outcome of the decoding operation through a single acknowledgement bit indicating whether the decoding was successful or if a retransmission of the transport block is required. Further details on transmission of hybrid-ARQ acknowledgements are found in Chapters 10–12. Clearly, the receiver must know to which hybrid-ARQ process a received acknowledgement is associated. This is

---

[5]The soft combining is done before or as part of the channel decoding, which clearly is a physical-layer functionality.

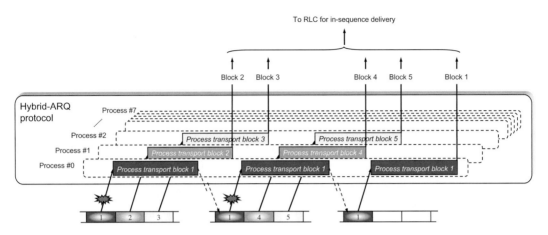

FIGURE 8.12

Multiple parallel hybrid-ARQ processes.

solved by using the timing of the acknowledgement for association with a certain hybrid-ARQ process. Note that, in the case of TDD operation, the time relation between the reception of data in a certain hybrid-ARQ process and the transmission of the acknowledgement is also affected by the uplink/downlink allocation.

The use of multiple parallel hybrid-ARQ processes, illustrated in Figure 8.12, for each user can result in data being delivered from the hybrid-ARQ mechanism out of sequence. For example, transport block 5 in the figure was successfully decoded before transport block 1 which required retransmissions. In-sequence delivery of data is therefore ensured by the RLC layer. The same reordering mechanism in the RLC is also used to handle in-sequence delivery across multiple component carriers in the case of carrier aggregation. As mentioned above, the hybrid-ARQ retransmissions are handled independently per component carrier, which may result in out-of-sequence delivery in a similar way as within a component carrier.

Downlink retransmissions may occur at any time after the initial transmission – that is, the protocol is asynchronous – and an explicit hybrid-ARQ process number is used to indicate which process is being addressed. In an asynchronous hybrid-ARQ protocol, the retransmissions are in principle scheduled similarly to the initial transmissions. Uplink retransmissions, on the other hand, are based on a synchronous protocol, the retransmission occurs at a predefined time after the initial transmission and the process number can be implicitly derived. In a synchronous protocol the time instant for the retransmissions is fixed once the initial transmission has been scheduled, which must be accounted for in the scheduling operation. However, note that the scheduler knows from the hybrid-ARQ entity in the eNodeB whether a terminal will perform a retransmission or not.

The hybrid-ARQ mechanism will rapidly correct transmission errors due to noise or unpredictable channel variations. As discussed above, the RLC is also capable of requesting retransmissions, which at first sight may seem unnecessary. However, the reason for having two retransmission mechanisms on top of each other can be seen in the feedback signaling – hybrid ARQ provides fast retransmissions but due to errors in the feedback the residual error rate is typically too high for, for example, good TCP performance, while RLC ensures (almost) error-free data delivery but slower retransmissions than the hybrid-ARQ protocol. Hence, the combination of hybrid ARQ and RLC provides an attractive combination of small round-trip time and reliable data delivery. Furthermore, as the RLC and hybrid ARQ are located in the same node, tight interaction between the two is possible, as discussed in Chapter 12.

8.2.3 Physical Layer

The physical layer is responsible for coding, physical-layer hybrid-ARQ processing, modulation, multiantenna processing, and mapping of the signal to the appropriate physical time–frequency resources. It also handles mapping of transport channels to physical channels, as shown in Figures 8.7 and 8.8.

As mentioned in the introduction, the physical layer provides services to the MAC layer in the form of transport channels. Data transmission in downlink and uplink use the DL-SCH and UL-SCH transport-channel types respectively. There is at most one or, in the case of spatial multiplexing, two transport blocks per TTI on a DL-SCH or UL-SCH. In the case of carrier aggregation, there is one DL-SCH (or UL-SCH) per component carrier.

A *physical channel* corresponds to the set of time–frequency resources used for transmission of a particular transport channel and each transport channel is mapped to a corresponding physical channel, as shown in Figures 8.7 and 8.8. In addition to the physical channels with a corresponding transport channel, there are also physical channels without a corresponding transport channel. These channels, known as L1/L2 control channels, are used for *downlink control information* (DCI), providing the terminal with the necessary information for proper reception and decoding of the downlink data transmission, and *uplink control information* (UCI) used for providing the scheduler and the hybrid-ARQ protocol with information about the situation at the terminal.

The physical-channel types defined in LTE include the following:

- The *Physical Downlink Shared Channel* (PDSCH) is the main physical channel used for unicast data transmission, but also for transmission of paging information.
- The *Physical Broadcast Channel* (PBCH) carries part of the system information, required by the terminal in order to access the network.
- The *Physical Multicast Channel* (PMCH) is used for MBSFN operation.
- The *Physical Downlink Control Channel* (PDCCH) is used for downlink control information, mainly scheduling decisions, required for reception of PDSCH, and for scheduling grants enabling transmission on the PUSCH.
- The *Physical Hybrid-ARQ Indicator Channel* (PHICH) carries the hybrid-ARQ acknowledgement to indicate to the terminal whether a transport block should be retransmitted or not.
- The *Physical Control Format Indicator Channel* (PCFICH) is a channel providing the terminals with information necessary to decode the set of PDCCHs. There is only one PCFICH per component carrier.
- The *Physical Uplink Shared Channel* (PUSCH) is the uplink counterpart to the PDSCH. There is at most one PUSCH per uplink component carrier per terminal.
- The *Physical Uplink Control Channel* (PUCCH) is used by the terminal to send hybrid-ARQ acknowledgements, indicating to the eNodeB whether the downlink transport block(s) was successfully received or not, to send channel-state reports aiding downlink channel-dependent scheduling, and for requesting resources to transmit uplink data upon. There is at most one PUCCH per terminal.

- The *Physical Random-Access Channel* (PRACH) is used for random access, as described in Chapter 14.

Note that some of the physical channels, more specifically the channels used for downlink control information (PCFICH, PDCCH, PHICH) and uplink control information (PUCCH), do not have a corresponding transport channel.

The remaining downlink transport channels are based on the same general physical-layer processing as the DL-SCH, although with some restrictions in the set of features used. This is especially true for PCH and MCH transport channels. For the broadcast of system information on the BCH, a terminal must be able to receive this information channel as one of the first steps prior to accessing the system. Consequently, the transmission format must be known to the terminals a priori and there is no dynamic control of any of the transmission parameters from the MAC layer in this case. The BCH is also mapped to the physical resource (the OFDM time–frequency grid) in a different way, as described in more detail in Chapter 14.

For transmission of paging messages on the PCH, dynamic adaptation of the transmission parameters can, to some extent, be used. In general, the processing in this case is similar to the generic DL-SCH processing. The MAC can control modulation, the amount of resources, and the antenna mapping. However, as an uplink has not yet been established when a terminal is paged, hybrid ARQ cannot be used as there is no possibility for the terminal to transmit a hybrid-ARQ acknowledgement.

The MCH is used for MBMS transmissions, typically with single-frequency network operation, as described in Chapter 3, by transmitting from multiple cells on the same resources with the same format at the same time. Hence, the scheduling of MCH transmissions must be coordinated between the cells involved and dynamic selection of transmission parameters by the MAC is not possible.

8.3 CONTROL-PLANE PROTOCOLS

The control-plane protocols are, among other things, responsible for connection setup, mobility, and security. Control messages transmitted from the network to the terminals can originate either from the MME, located in the core network, or from the *Radio Resource Control* (RRC), located in the eNodeB.

NAS control-plane functionality, handled by the MME, includes EPS bearer management, authentication, security, and different idle-mode procedures such as paging. It is also responsible for assigning an IP address to a terminal. For a detailed discussion about the NAS control-plane functionality, see [9].

The RRC is located in the eNodeB and is responsible for handling the RAN-related procedures, including:

- Broadcast of system information necessary for the terminal to be able to communicate with a cell. Acquisition of system information is described in Chapter 14.
- Transmission of paging messages originating from the MME to notify the terminal about incoming connection requests. Paging, discussed further in Chapter 14, is used in the RRC_IDLE state (described further below) when the terminal is not connected to a particular cell. Indication of system-information update is another use of the paging mechanism, as is public warning systems.

- Connection management, including setting up bearers and mobility within LTE. This includes establishing an RRC context – that is, configuring the parameters necessary for communication between the terminal and the radio-access network.
- Mobility functions such as cell (re)selection.
- Measurement configuration and reporting.
- Handling of UE capabilities; when connection is established the terminal will announce its capabilities as all terminals are not capable of supporting all the functionality described in the LTE specifications, as briefly discussed in Chapter 7.

RRC messages are transmitted to the terminal using *signaling radio bearers* (SRBs), using the same set of protocol layers (PDCP, RLC, MAC and PHY) as described in Section 8.2. The SRB is mapped to the common control channel (CCCH) during establishment of connection and, once a connection is established, to the dedicated control channel (DCCH). Control-plane and user-plane data can be multiplexed in the MAC layer and transmitted to the terminal in the same TTI. The aforementioned MAC control elements can also be used for control of radio resources in some specific cases where low latency is more important than ciphering, integrity protection, and reliable transfer.

8.3.1 **State Machine**

In LTE, a terminal can be in two different states, as illustrated in Figure 8.13, RRC_CONNECTED and RRC_IDLE.

In RRC_CONNECTED, there is an RRC context established – that is, the parameters necessary for communication between the terminal and the radio-access network are known to both entities. The cell to which the terminal is belongs is known and an identity of the terminal, the *Cell Radio-Network Temporary Identifier* (C-RNTI), used for signaling purposes between the terminal and the network, has been configured. RRC_CONNECTED is intended for data transfer to/from the terminal, but *discontinuous reception* (DRX) can be configured in order to reduce terminal power consumption (DRX is described in further detail in Chapter 13). Since there is an RRC context established in the eNodeB in RRC_CONNECTED, leaving DRX and starting to receive/transmit data is relatively fast as no connection with its associated signaling is needed.

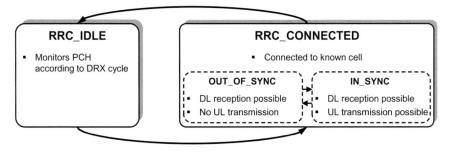

FIGURE 8.13

LTE states.

Although expressed differently in the specifications, RRC_CONNECTED can be thought of as having two substates, IN_SYNC and OUT_OF_SYNC, depending on whether the uplink is synchronized to the network or not. Since LTE uses an orthogonal FDMA/TDMA-based uplink, it is necessary to synchronize the uplink transmission from different terminals such that they arrive at the receiver at (approximately) the same time. The procedure for obtaining and maintaining uplink synchronization is described in Chapter 11, but in short the receiver measures the arrival time of the transmissions from each actively transmitting terminal and sends timing-correction commands in the downlink. As long as the uplink is synchronized, uplink transmission of user data and L1/L2 control signaling is possible. If no uplink transmission has taken place within a configurable time window, timing alignment is obviously not possible and the uplink is declared to be non-synchronized. In this case, the terminal needs to perform a random-access procedure to restore uplink synchronization prior to transmission of uplink data or control information.

In RRC_IDLE, there is no RRC context in the radio-access network and the terminal does not belong to a specific cell. No data transfer may take place as the terminal sleeps most of the time in order to reduce battery consumption. Uplink synchronization is not maintained and hence the only uplink transmission activity that may take place is random access, discussed in Chapter 14, to move to RRC_CONNECTED. When moving to RRC_CONNECTED the RRC context needs to be established in both the radio-access network and the terminal. Compared to leaving DRX this takes a somewhat longer time. In the downlink, terminals in RRC_IDLE periodically wake up in order to receive paging messages, if any, from the network, as described in Chapter 14.

Physical Transmission Resources

In Chapter 8, the overall LTE architecture was discussed, including an overview of the different protocol layers. Prior to discussing the detailed LTE downlink and uplink transmission schemes, a description of the basic time- and frequency-domain structure of LTE will be provided in this chapter.

9.1 OVERALL TIME–FREQUENCY STRUCTURE

OFDM is the basic transmission scheme for both the downlink and uplink transmission directions in LTE although, for the uplink, specific means are taken to reduce the cubic metric of the transmitted signal, thereby allowing for improved efficiency for the terminal transmitter power amplifier. Thus, for data transmission, DFT precoding is applied before OFDM modulation, leading to *DFT-spread OFDM* or *DFTS-OFDM*, as described in Chapter 4. It should be noted though that DFTS-OFDM is only applied to uplink *data* transmission – that is, for the transmission of the UL-SCH transport channel. As will be seen in Chapter 11, for other uplink transmissions, such as the transmission of *L1/L2 control signaling* and different types of *reference-signal transmissions*, other means are taken to limit the cubic metric of the transmitted signal.

The LTE OFDM subcarrier spacing equals 15 kHz for both downlink and uplink. As discussed in Chapter 3, the selection of the subcarrier spacing in an OFDM-based system needs to carefully balance overhead from the cyclic prefix against sensitivity to Doppler spread/shift and other types of frequency errors and inaccuracies. The choice of 15 kHz for the LTE subcarrier spacing was found to offer a good balance between these two constraints.

Assuming an FFT-based transmitter/receiver implementation, 15 kHz subcarrier spacing corresponds to a sampling rate $f_s = 15\,000 \cdot N_{\text{FFT}}$, where N_{FFT} is the FFT size. It is important to understand though that the LTE specifications do not in any way mandate the use of FFT-based transmitter/receiver implementations and even less so a particular FFT size or sampling rate. Nevertheless, FFT-based implementations of OFDM are common practice and an FFT size of 2048, with a corresponding sampling rate of 30.72 MHz, is suitable for the wider LTE carrier bandwidths, such as bandwidths of the order of 15 MHz and above. However, for smaller carrier bandwidths, a smaller FFT size and a correspondingly lower sampling rate can very well be used. The sampling rate above illustrates another factor influencing the choice of the LTE subcarrier spacing, namely a desire to simplify implementation of dual-mode LTE/HSPA terminals. Assuming a power-of-two FFT size and a subcarrier spacing of 15 kHz, the sampling rate $\Delta f \cdot N_{\text{FFT}}$ will be a multiple or sub-multiple of the HSPA chip rate of 3.84 Mchip/s.

In addition to the 15 kHz subcarrier spacing, a *reduced subcarrier spacing* of 7.5 kHz, with a corresponding OFDM symbol time that is twice as long, is also defined for LTE. The introduction of the reduced subcarrier spacing specifically targeted MBSFN-based multicast/broadcast transmissions (see Chapter 15). However, currently the 7.5 kHz subcarrier numerology is only partly implemented in the LTE specifications. Thus, at least for LTE up to and including release 10, only the 15 kHz subcarrier spacing is fully supported. The remaining discussions within this and the following chapters will assume the 15 kHz subcarrier spacing unless explicitly stated otherwise.

In the time domain, LTE transmissions are organized into (*radio*) *frames* of length 10 ms, each of which is divided into ten equally sized *subframes* of length 1 ms, as illustrated in Figure 9.1. Each subframe consists of two equally sized *slots* of length $T_{slot} = 0.5$ ms, with each slot consisting of a number of OFDM symbols including cyclic prefix.[1]

To provide consistent and exact timing definitions, different time intervals within the LTE specifications are defined as multiples of a basic time unit $T_s = 1/(15\,000 \cdot 2048)$. The basic time unit T_s can thus be seen as the sampling time of an FFT-based transmitter/receiver implementation with an FFT size equal to 2048. The time intervals outlined in Figure 9.1 can thus also be expressed as $T_{frame} = 307\,200 \cdot T_s$, $T_{subframe} = 30\,720 \cdot T_s$, and $T_{slot} = 15\,360 \cdot T_s$ for the frame, subframe, and slot durations respectively.

On a higher level, each frame is identified by a *System Frame Number* (SFN). The SFN is used to control different transmission cycles that may have a period longer than one frame, such as paging sleep-mode cycles and periods for channel-status reporting. The SFN period equals 1024, thus the SFN repeats itself after 1024 frames or roughly 10 seconds.

The 15 kHz LTE subcarrier spacing corresponds to a useful symbol time $T_u = 2048 \cdot T_s$ or approximately 66.7 μs. The overall OFDM symbol time is then the sum of the useful symbol time and the cyclic-prefix length T_{CP}. As illustrated in Figure 9.1, LTE defines two cyclic-prefix lengths, the *normal* cyclic prefix and an *extended* cyclic prefix, corresponding to seven and six OFDM symbols per slot respectively. The exact cyclic-prefix lengths, expressed in the basic time unit T_s, are given in Figure 9.1. It can be noted that, in the case of the normal cyclic prefix, the cyclic-prefix length for the first OFDM symbol of a slot is somewhat larger compared to the remaining OFDM symbols. The reason for this is simply to fill the entire 0.5 ms slot, as the number of basic time units T_s per slot (15 360) is not divisible by seven.

The reasons for defining two cyclic-prefix lengths for LTE are twofold:

- A longer cyclic prefix, although less efficient from a cyclic-prefix-overhead point of view, may be beneficial in specific environments with extensive delay spread, for example in very large cells. It is important to have in mind, though, that a longer cyclic prefix is not necessarily beneficial in the case of large cells, even if the delay spread is very extensive in such cases. If, in large cells, link performance is limited by noise rather than by signal corruption due to residual time dispersion not covered by the cyclic prefix, the additional robustness to radio-channel time dispersion, due to the use of a longer cyclic prefix, may not justify the corresponding additional energy overhead of a longer cyclic prefix.

---

[1] This is valid for the "normal" downlink subframes. As described in Section 9.5.2, the *special subframe* present in the case of TDD operation is not divided into two slots but rather into three fields. However, the number of OFDM symbols in the special subframe is the same as for the normal subframes illustrated in Figure 9.1.

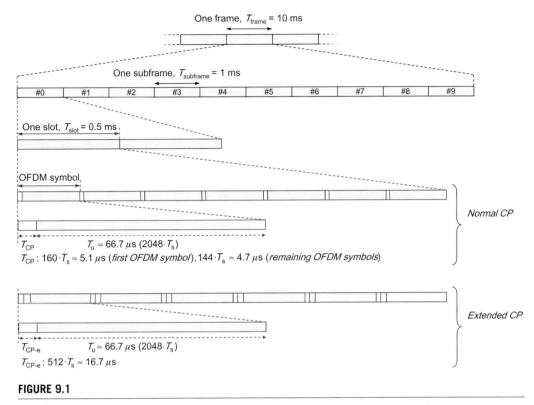

FIGURE 9.1

LTE time-domain structure.

- As already discussed in Chapter 3, in the case of MBSFN-based multicast/broadcast transmission, the cyclic prefix should not only cover the main part of the actual channel time dispersion, but also the timing difference between the transmissions received from the cells involved in the MBSFN transmission. In the case of MBSFN operation, the extended cyclic prefix is therefore used.

It should be noted that different cyclic-prefix lengths may be used for different subframes within a frame. As an example, to be discussed further in Chapter 15, MBSFN-based multicast/broadcast transmission is always confined to a limited set of subframes, in which case the use of the extended cyclic prefix, with its associated additional cyclic-prefix overhead, may only be applied to these subframes.[2]

A *resource element*, consisting of one subcarrier during one OFDM symbol, is the smallest physical resource in LTE. Furthermore, as illustrated in Figure 9.2, resource elements are grouped into *resource blocks*, where each resource block consists of 12 consecutive subcarriers in the frequency domain and one 0.5 ms slot in the time domain. Each resource block thus consists of $7 \cdot 12 = 84$

---

[2]The extended cyclic prefix is then actually applied only to the so-called MBSFN part of the MBSFN subframes (see Section 9.2).

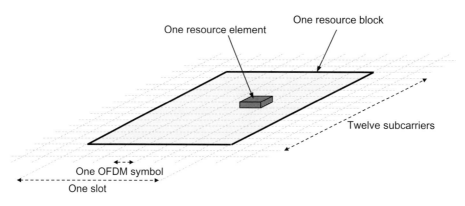

One resource element

One resource block

Twelve subcarriers

One OFDM symbol

One slot

FIGURE 9.2

The LTE physical time–frequency resource.

resource elements in the case of a normal cyclic prefix and $6 \bullet 12 = 72$ resource elements in the case of an extended cyclic prefix.

Although resource blocks are defined over one slot, the basic time-domain unit for dynamic scheduling in LTE is one subframe, consisting of two consecutive slots. The reason for defining the resource blocks over one slot is that *distributed downlink transmission*, as described in Chapter 10 and *uplink frequency hopping* (described in Chapter 11) are defined on a slot or resource-block basis. The minimum scheduling unit, consisting of two time-consecutive resource blocks within one sub-frame (one resource block per slot), can be referred to as a *resource-block pair*.

The LTE physical-layer specification allows for a carrier to consist of any number of resource blocks in the frequency domain, ranging from a minimum of six resource blocks up to a maximum of 110 resource blocks. This corresponds to an overall transmission bandwidth ranging from roughly 1 MHz up to in the order of 20 MHz with very fine granularity and thus allows for a very high degree of LTE bandwidth flexibility, at least from a physical-layer-specification point of view. However, as mentioned in Chapter 7, LTE radio-frequency requirements are, at least initially, only specified for a limited set of transmission bandwidths, corresponding to a limited set of possible values for the number of resource blocks within a carrier. Also note that, in LTE release 10, the total bandwidth of the transmitted signal can be as large as 100 MHz by aggregating multiple carriers, as mentioned in Chapter 7 and further described in Section 9.3.

The resource-block definition above applies to both the downlink and uplink transmission directions. However, there is a minor difference between the downlink and uplink in terms of where the carrier center frequency is located in relation to the subcarriers.

In the downlink (upper part of Figure 9.3), there is an unused *DC-subcarrier* that coincides with the carrier center frequency. The reason why the DC-subcarrier is not used for downlink transmission is that it may be subject to disproportionately high interference due, for example, to local-oscillator leakage.

On the other hand, in the uplink (lower part of Figure 9.3), no unused DC-subcarrier is defined and the center frequency of an uplink carrier is located *between* two uplink subcarriers. The presence of an unused DC-carrier in the center of the spectrum would have prevented the assignment of the entire cell bandwidth to a single terminal and still retain the assumption of mapping to consecutive

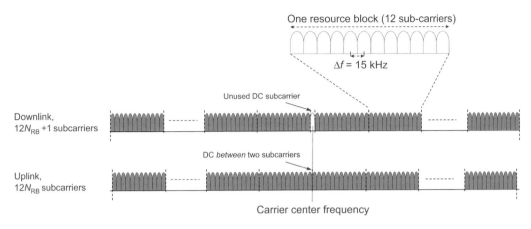

FIGURE 9.3

Frequency-domain structure for LTE.

inputs of the OFDM modulator, something that is needed to retain the low-cubic-metric property of the DFTS-OFDM modulation used for uplink data transmission.

9.2 NORMAL SUBFRAMES AND MBSFN SUBFRAMES

In LTE, each downlink subframe (and the DwPTS in the case of TDD; see Section 9.5.2 for a discussion of the TDD frame structure) is normally divided into a *control region*, consisting of the first few OFDM symbols, and a *data region*, consisting of the remaining part of the subframe. The detailed usage of the resource elements in the two regions will be discussed in detail in Chapter 10; at this stage all we need to know is that the control region carries L1/L2 signaling necessary to control uplink and downlink data transmissions.

Additionally, already from the first release of LTE, so-called *MBSFN subframes* have also been defined. The original intention with MBSFN subframes was, as indicated by the name, to support MBSFN transmission, as described in Chapter 15. However, MBSFN subframes have also been found to be useful in other contexts, for example as part of relaying functionality, as discussed in Chapter 16. Hence, MBSFN subframes are therefore better seen as a generic tool and not related only to MBSFN transmission.

An MBSFN subframe, illustrated in Figure 9.4, consists of a control region of length one or two OFDM symbols, which is in essence identical to its counterpart in a normal subframe, followed by an *MBSFN region* whose contents depend on the usage of the MBSFN subframe. The reason for keeping the control region also in MBSFN subframes is, for example, to transmit control signaling necessary for uplink transmissions. All terminals, from LTE release 8 and onwards, are capable of receiving the control region of an MBSFN subframe. This is the reason why MBSFN subframes have been found useful as a generic tool to introduce, in a backwards-compatible way, new types of signaling and transmission not part of an earlier release of the LTE radio-access specification. Terminals not capable of receiving transmissions in the MBSFN region will simply ignore those transmissions.

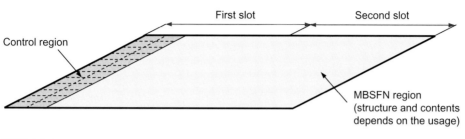

FIGURE 9.4

Resource-block structure for MBSFN subframes, assuming normal cyclic prefix for the control region and extended cyclic prefix for the MBSFN region.

Information about the set of subframes that are configured as MBSFN subframes in a cell is provided as part of the system information. In principle, an arbitrary pattern of MBSFN subframes can be configured with the pattern repeating after 40 ms.[3] However, as information necessary to operate the system (to be more specific, synchronization signals, system information and paging, all of which will be discussed in detail in later chapters) needs to be transmitted in order for terminals to find and connect to a cell, subframes where such information is provided cannot be configured as MBSFN subframes. Therefore, subframes 0, 4, 5, and 9 for FDD and subframes 0, 1, 5, and 6 for TDD cannot be configured as MBSFN subframes, leaving the remaining six subframes as candidates for MBSFN subframes.

9.3 CARRIER AGGREGATION

The possibility for *carrier aggregation* was introduced in LTE release 10. In the case of carrier aggregation, multiple LTE carriers, each with a bandwidth up to 20 MHz, can be transmitted in parallel to/from the same terminal, thereby allowing for an overall wider bandwidth and correspondingly higher per-link data rates. In the context of carrier aggregation, each carrier is referred to as a *component carrier*[4] as, from an RF point-of-view, the entire set of aggregated carriers can be seen as a single (RF) carrier.

Up to five component carriers, possibly of different bandwidths up to 20 MHz, can be aggregated allowing for overall transmission bandwidths up to 100 MHz. A terminal capable of carrier aggregation may receive or transmit simultaneously on multiple component carriers. Each component carrier can also be accessed by an LTE terminal from earlier releases – that is, component carriers are *backwards compatible*. Thus, in most respects and unless otherwise mentioned, the physical-layer description in the following chapters applies to each component carrier separately in the case of carrier aggregation.

---

[3]The reason for this value is that both the 10 ms frame length and the 8 ms hybrid-ARQ round-trip time are factors in the 40 ms, which is important for relaying operation (see also Chapter 16).

[4]In the specifications, the term *cell* is used instead of component carrier, but as the term cell is something of a misnomer in the uplink case, the term component carrier is used herein.

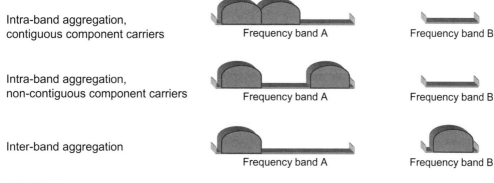

Intra-band aggregation,
contiguous component carriers

Frequency band A   Frequency band B

Intra-band aggregation,
non-contiguous component carriers

Frequency band A   Frequency band B

Inter-band aggregation

Frequency band A   Frequency band B

FIGURE 9.5

Different types of carrier aggregation.

It should be noted that aggregated component carriers do not need to be contiguous in the frequency domain. Rather, with respect to the frequency location of the different component carriers, three different cases can be identified (see also Figure 9.5):

- Intra-band aggregation with frequency-contiguous component carriers
- Intra-band aggregation with non-contiguous component carriers
- Inter-band aggregation with non-contiguous component carriers.

The possibility to aggregate non-adjacent component carriers allows for exploitation of a fragmented spectrum; operators with a fragmented spectrum can provide high-data-rate services based on the availability of a wide overall bandwidth even though they do not posses a single wideband spectrum allocation. Except from an RF point of view there is no difference between the three different cases outlined in Figure 9.5 and they are all supported by the basic LTE release-10 specification. However, the complexity of RF implementation is vastly different, with the first case being the least complex. Thus, although spectrum aggregation is supported by the physical-layer and protocol specifications, the actual implementation will be strongly constrained, including specification of only a limited number of aggregation scenarios and aggregation over a dispersed spectrum only being supported by the most advanced terminals.

A terminal capable of carrier aggregation has one downlink *primary* component carrier and an associated uplink primary component carrier. In addition, it may have one or several *secondary* component carriers in each direction. Different terminals may have different carriers as their primary component carrier – that is, the configuration of the primary component carrier is terminal specific. The association between the downlink primary carrier and the corresponding uplink primary carrier is cell specific and signaled as part of the system information. This is similar to the case without carrier aggregation, although in the latter case the association is trivial. The reason for such an association is, for example, to determine to which uplink component carrier a certain scheduling grant transmitted on the downlink relates without having to explicitly signal the component-carrier number.

All idle mode procedures apply to the primary component carrier only or, expressed differently, carrier aggregation with additional secondary carriers configured only applies to terminals in the RRC_CONNECTED state. Upon connection to the network, the terminal performs the related procedures such

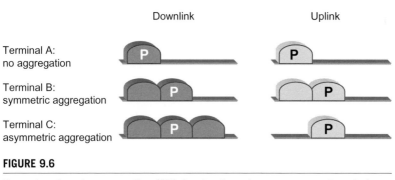

FIGURE 9.6

Examples of carrier aggregation ("P" denotes the primary component carrier).

as cell search and random access (see Chapter 14 for a detailed description of these procedures) following the same steps as in the absence of carrier aggregation. Once the communication between the network and the terminal is established, additional secondary component carriers can be configured.

The fact that carrier aggregation is terminal specific – that is, different terminals may be configured to use different sets of component carriers – is useful not only from a network perspective to balance the load across component carriers, but also to handle different capabilities between terminals. Some terminals may be able to transmit/receive on multiple component carriers, while other terminals may do so on only a single carrier. This is an obvious consequence of being able to serve terminals from earlier releases at the same time as a carrier-aggregation-capable release-10 terminal, but it also allows for different capabilities in terms of carrier aggregation for different terminals as well as a differentiation between downlink and uplink carrier-aggregation capability. For example, a terminal may be capable of two component carriers in the downlink but of only a single component carrier – that is, no carrier aggregation – in the uplink, as is the case for terminal C in Figure 9.6. Note also that the primary component-carrier configuration can differ between terminals. Asymmetric carrier aggregation can also be useful to handle different spectrum allocations, for example if an operator has more spectrum available for downlink transmissions than uplink transmissions.

In release 10, only downlink-heavy asymmetries are supported – that is, the number of uplink component carriers configured for a terminal is always equal to or smaller than the number of configured downlink component carriers. Uplink-heavy asymmetries are less likely to be of practical interest and would also complicate the overall control signaling structure, as in such a case multiple uplink component carriers would need to be associated with the same downlink component carrier.

Carrier aggregation is supported for both FDD and TDD, although all component carriers need to have the same duplex scheme. Furthermore, in the case of TDD, the uplink–downlink configuration should be the same across component carriers. The special subframe configuration can be different for the different components carriers though, as long as the resulting downlink–uplink switch time is sufficiently large.

9.4 FREQUENCY-DOMAIN LOCATION OF LTE CARRIERS

In principle, an LTE carrier could be positioned anywhere within the spectrum and, actually, the basic LTE physical-layer specification does not say anything about the exact frequency location of an LTE

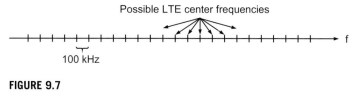

Possible LTE center frequencies

100 kHz

FIGURE 9.7

LTE carrier raster.

carrier, including the frequency band. However, in practice, there is obviously a need for restrictions on where an LTE carrier can be positioned in the frequency domain.

- In the end, an LTE terminal must be implemented and RF-wise such a terminal can obviously only support certain frequency bands. The frequency bands for which LTE is specified to operate are discussed in Chapter 17.
- After being activated, an LTE terminal has to search for a network-transmitted carrier within the frequency bands supported by the terminal. In order for that carrier search not to take an unreasonably long time, there is a need to limit the set of frequencies to be searched.

For this reason it is assumed that, within each supported frequency band, LTE carriers may exist on a 100 kHz *carrier raster* or *carrier grid* – that is, the carrier center frequency can be expressed as $m \cdot 100$ kHz, where m is an integer (see Figure 9.7).

In the case of carrier aggregation, multiple carriers can be transmitted to/from the same terminal. In order for the different component carriers to be accessible also by earlier-release terminals, each component carrier should fall on the 100 kHz carrier grid. However, in the case of carrier aggregation, there is an additional constraint that the carrier spacing between adjacent component carriers should be a multiple of the 15 kHz subcarrier spacing to allow transmission/reception with a single FFT.[5] Thus, in the case of carrier aggregation, the carrier spacing between the different component carriers should be a multiple of 300 kHz, the smallest carrier spacing being a multiple of both 100 kHz (the raster grid) and 15 kHz (the subcarrier spacing). A consequence of this is that there will always be a small number of unused subcarriers between two component carriers, even when they are located as close as possible to each other, as illustrated in Figure 9.8.

9.5 DUPLEX SCHEMES

Spectrum flexibility is one of the key features of LTE. In addition to the flexibility in transmission bandwidth, LTE also supports operation in both paired and unpaired spectrum by supporting both FDD- and TDD-based duplex operation with the time–frequency structures illustrated in Figure 9.9. Although the time-domain structure is, in most respects, the same for FDD and TDD, there are some differences, most notably the presence of a *special subframe* in the case of TDD. The special subframe is used to provide the necessary guard time for downlink–uplink switching, as discussed below.

[5]This is obviously only relevant for component carriers that are contiguous in the frequency domain. Furthermore, in the case of independent frequency errors between component carriers, separate FFTs may be needed at the receiver.

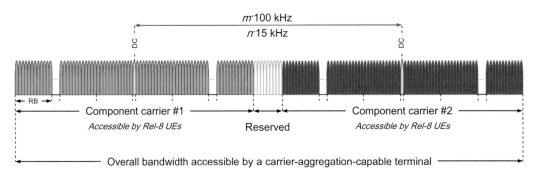

FIGURE 9.8

LTE carrier raster and carrier aggregation.

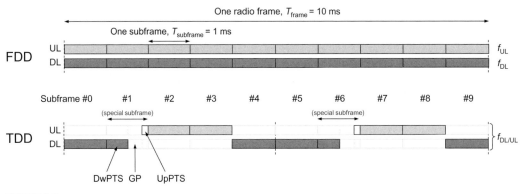

FIGURE 9.9

Uplink/downlink time–frequency structure for FDD and TDD.

9.5.1 Frequency-Division Duplex (FDD)

In the case of FDD operation (upper part of Figure 9.9), there are two carrier frequencies, one for uplink transmission (f_{UL}) and one for downlink transmission (f_{DL}). During each frame, there are thus ten uplink subframes and ten downlink subframes, and uplink and downlink transmission can occur simultaneously within a cell. Isolation between downlink and uplink transmissions is achieved by transmission/reception filters, known as duplex filters, and a sufficiently large *duplex separation* in the frequency domain.

Even if uplink and downlink transmission can occur simultaneously within a cell in the case of FDD operation, a terminal may be capable of *full-duplex* operation or only *half-duplex* operation for a certain frequency band, depending on whether or not it is capable of simultaneous transmission/reception. In the case of full-duplex capability, transmission and reception may also occur simultaneously at a terminal, whereas a terminal capable of only half-duplex operation cannot transmit and receive simultaneously. As mentioned in Chapter 7, supporting only half-duplex operation allows for simplified terminal implementation due to relaxed duplex-filter requirements. This applies especially

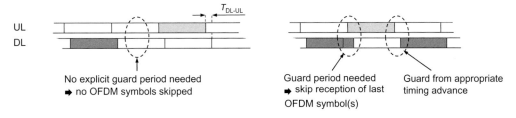

FIGURE 9.10

Guard time at the terminal for half-duplex FDD.

for certain frequency bands with a narrow duplex gap. Hence, full duplex support is *frequency-band dependent* such that a terminal may support only half-duplex operation in certain frequency bands while being capable of full-duplex operation in the remaining supported bands. It should be noted that full/half-duplex capability is a property of the *terminal*; the base station is operating in full duplex irrespective of the terminal capabilities. Hence, as the relevant transmission structures and timing relations are identical between full-duplex and half-duplex FDD, a single cell may simultaneously support a mixture of full-duplex and half-duplex FDD terminals. Half-duplex operation has an impact on the sustained data rates that can be provided to/from a single mobile terminal as it cannot transmit in all uplink subframes, but the cell capacity is hardly affected as typically it is possible to schedule different terminals in uplink and downlink in a given subframe.

Since a half-duplex terminal is not capable of simultaneous transmission and reception, the scheduling decisions must take this into account and half-duplex operation can be seen as a scheduling restriction, as will be discussed in more detail in Chapter 13. If a terminal is scheduled such that downlink reception in one subframe immediately precedes a subframe of uplink transmission, a guard time is necessary for the terminal to switch from reception to transmission. This is created in such cases by allowing the terminal to skip receiving[6] the last OFDM symbol(s) in the downlink subframe, as illustrated in Figure 9.10.

9.5.2 Time-Division Duplex (TDD)

In the case of TDD operation (lower part of Figure 9.9), there is a single carrier frequency only and uplink and downlink transmissions are separated in the time domain on a cell basis. As seen in the figure, some subframes are allocated for uplink transmissions and some subframes for downlink transmission, with the switch between downlink and uplink occurring in the *special subframe* (subframe 1 and, in some cases, subframe 6). Different asymmetries in terms of the amount of resources – that is, subframes – allocated for uplink and downlink transmission respectively are provided through the seven different downlink/uplink configurations illustrated in Figure 9.11. As seen in the figure, subframes 0 and 5 are always allocated for downlink transmission while subframe 2 is always allocated for uplink transmissions. The remaining subframes (except the special subframe; see below)

---

[6]The impact on the decoding performance can be accounted for by the rate adaptation mechanism. For very high data rates, the performance impact may be somewhat larger due to the downlink interleaving structure, in which case it may be preferable not to schedule an uplink transmission immediately after a downlink reception.

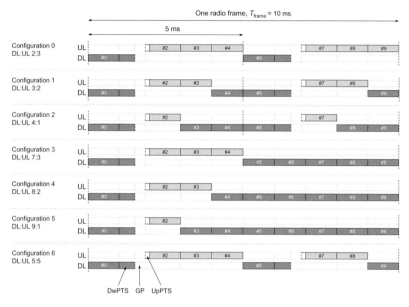

FIGURE 9.11

Different downlink/uplink configurations in the case of TDD.

can then be flexibly allocated for downlink or uplink transmission depending on the configured downlink/uplink configuration. To avoid severe interference between downlink and uplink transmissions in different cells, neighboring cells typically have the same downlink/uplink configuration. This makes it difficult to change the downlink/uplink configuration dynamically, for example on a frame-by-frame basis. Therefore, the current LTE specification assumes that the downlink/uplink configuration is relatively static. It can be changed on a very slow basis though, in order to adapt to changing traffic patterns. It could also, in principle, be different in different areas, for example to match different traffic patterns, although inter-cell interference needs to be carefully addressed in this case.

As the same carrier frequency is used for uplink and downlink transmission, both the base station and the terminal need to switch from transmission to reception and vice versa. The switch between downlink and uplink occurs in the special subframe, which is split into three parts: a downlink part (DwPTS), a guard period (GP), and an uplink part (UpPTS).

The DwPTS is in essence treated as a normal downlink subframe,[7] although the amount of data that can be transmitted is smaller due to the reduced length of the DwPTS compared to a normal subframe. The UpPTS, however, is not used for data transmission due to the very short duration. Instead, it can be used for channel sounding or random access. It can also be left empty, in which case it serves as extra guard period.

An essential aspect of any TDD system is the possibility to provide a sufficiently large *guard period* (or guard time), where neither downlink nor uplink transmissions occur. This guard period is necessary for switching from downlink to uplink transmission and vice versa and, as already mentioned, is

---

[7]For the shortest DwPTS duration of three OFDM symbols, DwPTS cannot be used for PDSCH transmission.

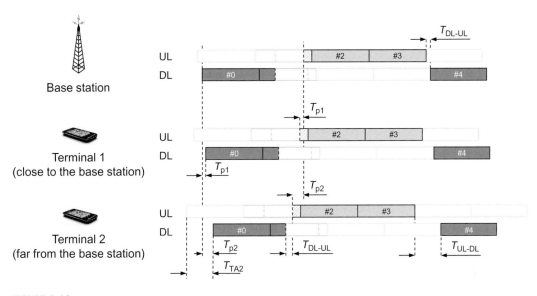

FIGURE 9.12

Timing relation for TDD operation.

obtained from the special subframe. The required length of the guard period depends on several factors. First, it should be sufficiently large to provide the necessary time for the circuitry in base stations and the terminals to switch from downlink to uplink. Switching is typically relatively fast, of the order of 20 microseconds, and in most deployments does not significantly contribute to the required guard time.

Secondly, the guard time should also ensure that uplink and downlink transmissions do not interfere at the base station. This is handled by advancing the uplink timing at the terminals such that, at the base station, the last uplink subframe before the uplink-to-downlink switch ends before the start of the first downlink subframe. The uplink timing of each terminal can be controlled by the base station by using the timing advance mechanism, as will be elaborated upon in Chapter 11. Obviously, the guard period must be large enough to allow the terminal to receive the downlink transmission and switch to transmission before it starts the (timing-advanced) uplink transmission. In essence, some of the guard period of the special subframe is "moved" from the downlink-to-uplink switch to the uplink-to-downlink switch by the timing-advance mechanism. This is illustrated in Figure 9.12. As the timing advance is proportional to the distance to the base station, a larger guard period is required when operating in large cells compared to small cells.

Finally, the selection of the guard period also needs to take interference between base stations into account. In a multi-cell network, inter-cell interference from downlink transmissions in neighboring cells must decay to a sufficiently low level before the base station can start to receive uplink transmissions. Hence, a larger guard period than is motivated by the cell size itself may be required as the last part of the downlink transmissions from distant base stations otherwise may interfere with uplink reception. The amount of guard period depends on the propagation environments, but in some cases the inter-base-station interference is a non-negligible factor when determining the guard period.

Table 9.1 Resulting Guard Period in OFDM Symbols for Different DwPTS and UpPTS Lengths (Normal Cyclic Prefix)

		DwPTS								
		12	11		10		9		3	
GP		1	1	2	2	3	3	4	9	10
UpPTS		1	2	1	2	1	2	1	2	1

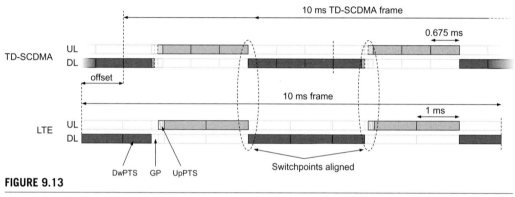

FIGURE 9.13

Coexistence between TD-SCDMA and LTE.

From the above discussion, it is clear that a sufficient amount of configurability of the guard period is needed to meet different deployment scenarios. Therefore, a set of DwPTS/GP/UpPTS configurations is supported as shown in Table 9.1, where each configuration corresponds to a given length of the three fields in the special subframes. The DwPTS/GP/UpPTS configuration used in the cell is signaled as part of the system information.

9.5.3 LTE and TD-SCDMA Coexistence

In addition to supporting a wide range of different guard periods, an important aspect in the design of TDD in LTE was to simplify coexistence with and migration from systems based on the 3GPP TD-SCDMA standard (see Chapter 19 for details on TD-SCDMA). Basically, to handle inter-system interference from two different but co-sited TDD systems operating close in frequency, it is necessary to align the switch-points between the two systems. Since LTE supports configurable lengths of the DwPTS field, the switch-points of LTE and TD-SCDMA (or any other TDD system) can be aligned, despite the different subframe lengths used in the two systems. Aligning the switch-points between TD-SCDMA and LTE is the technical reason for splitting the special subframe into the three fields DwPTS/GP/UpPTS instead of locating the switch-point at the subframe boundary. An example of LTE/TD-SCDMA coexistence is given in Figure 9.13.

The set of possible lengths of DwPTS/GP/UpPTS is selected to support common coexistence deployments, as well as to provide a high degree of guard-period flexibility for the reasons discussed earlier. The UpPTS length is one or two OFDM symbols and the DwPTS length can vary from three[8] to twelve OFDM symbols, resulting in guard periods ranging from one to ten OFDM symbols. The resulting guard period for the different DwPTS and UpPTS configurations supported is summarized in Table 9.1 for the case of normal cyclic prefix. As discussed earlier, the DwPTS can be used for downlink data transmission, while the UpPTS can be used for sounding or random access only, due to its short duration.

---

[8]The smallest DwPTS length is motivated by the location of the primary synchronization signal in the DwPTS (see Chapter 14).

Downlink Physical-Layer Processing

In Chapter 8, the LTE radio-interface architecture was discussed with an overview of the functions and characteristics of the different protocol layers. Chapter 9 then gave an overview of the basic time–frequency structure of LTE transmissions, including the structure of the OFDM time–frequency grid being the fundamental physical resource on both uplink and downlink.

This chapter will provide a more detailed description of the downlink physical-layer functionality, including the transport-channel processing (Section 10.1), downlink reference signals (Section 10.2), details on downlink multi-antenna transmission (Section 10.3), and downlink L1/L2 control signaling (Section 10.4). Chapter 11 will provide a corresponding description for the *uplink* transmission direction. The later chapters will then go further into the details of some specific uplink and downlink functions and procedures.

10.1 TRANSPORT-CHANNEL PROCESSING

As described in Chapter 8, transport channels provide the interface between the MAC layer and the physical layer. As also described, for the LTE downlink there are four different types of transport channels defined, the Downlink Shared Channel (DL-SCH), the Multicast Channel (MCH), the Paging Channel (PCH), and the Broadcast Channel (BCH). This section provides a description of the physical-layer processing applied to DL-SCH transport channels, including the mapping to the physical resource – that is, to the resource elements of the OFDM time–frequency grid. DL-SCH is the main downlink transport-channel type in LTE and is used for transmission of user-specific higher-layer information, both user data and dedicated control information, as well as part of the downlink system information. The DL-SCH physical-layer processing is to a large extent applicable also to MCH and PCH transport channels, although with some additional constraints. On the other hand, the physical-layer processing, and the transmission structure in general, for the BCH is quite different. BCH transmission is described in Chapter 14 as part of the discussion on LTE system information.

10.1.1 Processing Steps

The different steps of the DL-SCH physical layer processing are outlined in Figure 10.1. In the case of carrier aggregation – that is, transmission on multiple component carriers in parallel to the same terminal – the transmissions on the different carriers correspond to separate transport channels with separate and more or less independent physical-layer processing. The transport-channel processing outlined in Figure 10.1 and the discussion below is thus valid also in the case of carrier aggregation.

4G LTE/LTE-Advanced for Mobile Broadband.
© 2011 Erik Dahlman, Stefan Parkvall & Johan Sköld. Published by Elsevier Ltd. All rights reserved.

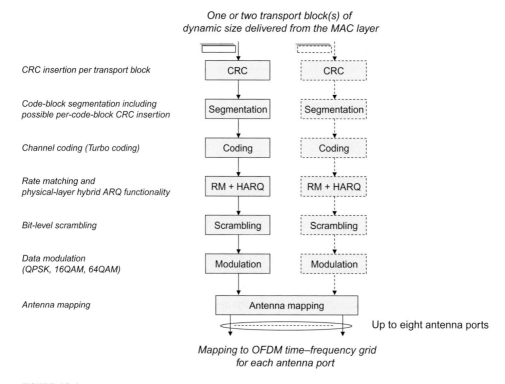

One or two transport block(s) of
dynamic size delivered from the MAC layer

CRC insertion per transport block	CRC / CRC
Code-block segmentation including possible per-code-block CRC insertion	Segmentation / Segmentation
Channel coding (Turbo coding)	Coding / Coding
Rate matching and physical-layer hybrid ARQ functionality	RM + HARQ / RM + HARQ
Bit-level scrambling	Scrambling / Scrambling
Data modulation (QPSK, 16QAM, 64QAM)	Modulation / Modulation
Antenna mapping	Antenna mapping

Up to eight antenna ports

Mapping to OFDM time–frequency grid
for each antenna port

FIGURE 10.1

Physical-layer processing for DL-SCH.

Within each *Transmission Time Interval* (TTI), corresponding to one subframe of length 1 ms, up to two transport blocks of dynamic size are delivered to the physical layer and transmitted over the radio interface for each component carrier. The number of transport blocks transmitted within a TTI depends on the configuration of the multi-antenna transmission scheme, as described in Section 10.3.

- In the case of no *spatial multiplexing* there is at most a single transport block in a TTI.
- In the case of spatial multiplexing, with transmission on multiple *layers* in parallel to the same terminal, there are two transport blocks within a TTI.[1]

10.1.1.1 *CRC Insertion Per Transport Block*

In the first step of the physical-layer processing, a 24-bit CRC is calculated for and appended to each transport block. The CRC allows for receiver-side detection of errors in the decoded transport block. The corresponding error indication can, for example, be used by the downlink hybrid-ARQ protocol as a trigger for requesting retransmissions.

---

[1]This is true for initial transmissions. In the case of hybrid-ARQ retransmissions there may also be cases when a single transport block is transmitted over multiple layers, as discussed, for example, in Section 10.3.2.

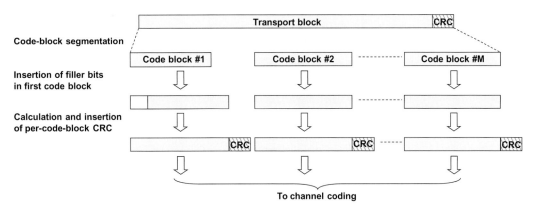

FIGURE 10.2

Code-block segmentation and per-code-block CRC insertion.

10.1.1.2 *Code-Block Segmentation and Per-Code-Block CRC Insertion*

The LTE Turbo-coder internal interleaver is only defined for a limited number of code-block sizes, with a maximum block size of 6144 bits. If the transport block, including the transport-block CRC, exceeds this maximum code-block size, *code-block segmentation*, illustrated in Figure 10.2, is applied before the Turbo coding. Code-block segmentation implies that the transport block is segmented into smaller *code blocks*, the sizes of which should match the set of code-block sizes supported by the Turbo coder.

In order to ensure that a transport block of arbitrary size can be segmented into code blocks that match the set of available code-block sizes, the specification includes the possibility to insert "dummy" *filler bits* at the head of the first code block. However, the set of transport-block sizes currently defined for LTE has been selected so that filler bits are not needed.

As can be seen in Figure 10.2, code-block segmentation also implies that an additional CRC (also of length 24 bits, but different compared to the transport-block CRC described above) is calculated for and appended to each code block. Having a CRC per code block allows for early detection of correctly decoded code blocks and correspondingly early termination of the iterative decoding of that code block. This can be used to reduce the terminal processing effort and corresponding energy consumption. In the case of a single code block no additional code-block CRC is applied.

One could argue that, in case of code-block segmentation, the transport-block CRC is redundant and implies unnecessary overhead as the set of code-block CRCs should indirectly provide information about the correctness of the complete transport block. However, code-block segmentation is only applied to large transport blocks for which the relative extra overhead due to the additional transport-block CRC is small. The transport-block CRC also adds additional error-detection capabilities and thus further reduces the risk for undetected errors in the decoded transport block.

Information about the transport-block size is provided to the terminal as part of the scheduling assignment transmitted on the PDCCH control channel, as described in Section 10.4.4. Based on this information, the terminal can determine the code-block size and number of code blocks. The terminal receiver can thus, based on the information provided in the scheduling assignment, straightforwardly undo the code-block segmentation and recover the decoded transport blocks.

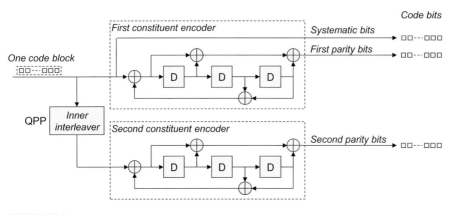

FIGURE 10.3

LTE Turbo encoder.

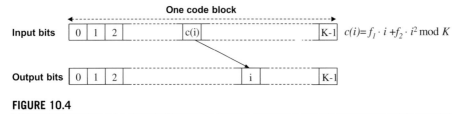

FIGURE 10.4

Principles of QPP-based interleaving.

10.1.1.3 *Channel Coding*

Channel coding for DL-SCH (as well as for PCH and MCH) is based on Turbo coding [53], with encoding according to Figure 10.3. The encoding consists of two rate-1/2, eight-state constituent encoders, implying an overall code rate of 1/3, in combination with QPP-based[2] interleaving [69]. As illustrated in Figure 10.4, the QPP interleaver provides a mapping from the input (non-interleaved) bits to the output (interleaved) bits according to the function:

$$c(i) = f_1 \cdot i + f_2 \cdot i^2 \bmod K,$$

where i is the index of the bit at the output of the interleaver, $c(i)$ is the index of the same bit at the input of the interleaver, and K is the code-block/interleaver size. The values of the parameters f_1 and f_2 depend on the code-block size K. The LTE specification lists all supported code-block sizes, ranging from a minimum of 40 bits to a maximum of 6144 bits, together with the associated values for the parameters f_1 and f_2. Thus, once the code-block size is known, the Turbo-coder inner interleaving, as well as the corresponding de-interleaving at the receiver side, can straightforwardly be carried out.

A QPP-based interleaver is *maximum contention free* [70], implying that the decoding can be parallelized without the risk for contention when the different parallel processes are accessing the

---

[2]QPP = Quadrature Permutation Polynomial.

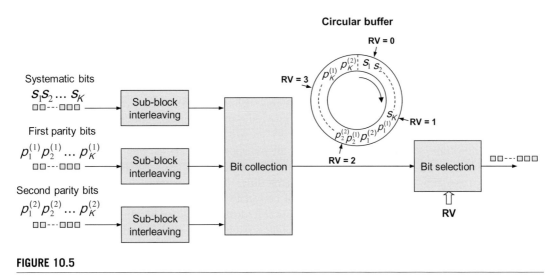

FIGURE 10.5

Rate matching and hybrid-ARQ functionality.

interleaver memory. For the very high data rates supported by LTE, the improved possibilities for parallel processing offered by QPP-based interleaving can substantially simplify the Turbo-encoder/decoder implementation.

10.1.1.4 *Rate Matching and Physical-Layer Hybrid-ARQ Functionality*

The task of the rate-matching and physical-layer hybrid-ARQ functionality is to extract, from the blocks of code bits delivered by the channel encoder, the exact set of code bits to be transmitted within a given TTI/subframe.

As illustrated in Figure 10.5, the outputs of the Turbo encoder (systematic bits, first parity bits, and second parity bits) are first separately interleaved. The interleaved bits are then inserted into what can be described as a circular buffer with the systematic bits inserted first, followed by alternating insertion of the first and second parity bits.

The bit selection then extracts consecutive bits from the circular buffer to an extent that matches the number of available resource elements in the resource blocks assigned for the transmission. The exact set of bits to extract depends on the *redundancy version* (RV) corresponding to different starting points for the extraction of coded bits from the circular buffer. As can be seen, there are four different alternatives for the redundancy version. The transmitter/scheduler selects the redundancy version and provides information about the selection as part of the scheduling assignment (see Section 10.4.4).

Note that the rate-matching and hybrid-ARQ functionality operates on the full set of code bits corresponding to one transport block and not separately on the code bits corresponding to a single code block.

10.1.1.5 *Bit-Level Scrambling*

LTE downlink scrambling implies that the block of code bits delivered by the hybrid-ARQ functionality is multiplied (*exclusive-or* operation) by a bit-level *scrambling sequence*. Without downlink scrambling, the channel decoder at the terminal could, at least in principle, be equally matched to an interfering signal as to the target signal, thus being unable to properly suppress the interference. By applying different scrambling sequences for neighboring cells, the interfering signal(s) after descrambling is (are) randomized, ensuring full utilization of the processing gain provided by the channel code. Thus, the bit scrambling essentially serves the same purpose as the scrambling applied at chip level after the direct-sequence spreading in DS-CDMA-based systems such as WCDMA/HSPA. Fundamentally, channel coding can be seen as "advanced" spreading providing processing gain similar to direct-sequence spreading but also additional coding gain.

In LTE, downlink scrambling is applied to all transport channels as well as to the downlink L1/L2 control signaling. For all downlink transport-channel types except MCH, as well as for the L1/L2 control signaling, the scrambling sequences differ between neighboring cells (*cell-specific scrambling*) to ensure interference randomization between the cells. This is achieved by having the scrambling sequences depend on the physical-layer cell identity (Chapter 14). In contrast, in the case of MBSFN-based transmission using MCH, the same scrambling should be applied to all cells taking part in the MBSFN transmission – that is, all cells within the so-called *MBSFN area* (see Chapter 15).

10.1.1.6 *Data Modulation*

The downlink data modulation transforms the block of scrambled bits to a corresponding block of complex modulation symbols. The set of modulation schemes supported for the LTE downlink includes QPSK, 16QAM, and 64QAM, corresponding to two, four, and six bits per modulation symbol respectively.

10.1.1.7 *Antenna Mapping*

The antenna mapping jointly processes the modulation symbols corresponding to the one or two transport blocks and maps the result to different *antenna ports*. The antenna mapping can be configured in different ways corresponding to different multi-antenna transmission schemes, including transmit diversity, beam-forming, and spatial multiplexing. As indicated in Figure 10.1, LTE supports transmission using up to eight antenna ports depending on the exact multi-antenna transmission scheme. More details about LTE downlink multi-antenna transmission are provided in Section 10.3.

Note that the antenna ports referred to above do not necessarily correspond to specific physical antennas. Rather, an antenna port is a more general concept introduced, for example, to allow for beam-forming using multiple physical antennas without the terminal being aware of the beam-forming carried out at the transmitter side.

At least for the downlink, an antenna port can be seen as corresponding to the transmission of a *reference signal* (Section 10.2). Any data transmission from the antenna port can then rely on that reference signal for channel estimation for coherent demodulation. Thus, if the same reference signal is transmitted from multiple physical antennas, these physical antennas correspond to a *single* antenna port. Similarly, if two different reference signals are transmitted from the same set of physical antennas, this corresponds to *two separate antenna ports*.

It should be noted that the LTE specification actually has a somewhat more general definition of an antenna port, essentially just saying that two received signals can be assumed to have experienced the same overall channel, including any joint processing at the transmitter side, if and only if they have been transmitted on the same antenna port.

10.1.1.8 *Resource-Block Mapping*

The resource-block mapping takes the symbols to be transmitted on each antenna port and maps them to the resource elements of the set of resource blocks assigned by the MAC scheduler for the transmission. As described in Chapter 9, each resource block consists of 84 resource elements (twelve sub-carriers during seven OFDM symbols).[3] However, some of the resource elements within a resource block will not be available for the transport-channel transmission as they are occupied by:

- different types of downlink reference signals, as described in Section 10.2;
- downlink L1/L2 control signaling (one, two, or three OFDM symbols at the head of each subframe), as described in Section 10.4[4].

Furthermore, as will be described in Chapter 14, within some resource blocks, additional resource elements are reserved for the transmission of *synchronization signals* as well as for the transmission of the BCH transport channel.

In the TDD special subframe (Section 9.5.2), mapping is limited to the DwPTS.

10.1.2 Localized and Distributed Resource Mapping

As already discussed in Chapter 7, when deciding what set of resource blocks to use for transmission to a specific terminal, the network may take the downlink channel conditions in both the time and frequency domains into account. Such time/frequency-domain channel-dependent scheduling, taking channel variations – for example, due to frequency-selective fading – into account, may significantly improve system performance in terms of achievable data rates and overall cell throughput.

However, in some cases downlink channel-dependent scheduling is not suitable to use or is not practically possible:

- For low-rate services such as voice, the feedback signaling associated with channel-dependent scheduling may lead to extensive relative overhead.
- At high mobility (high terminal speed), it may be difficult or even practically impossible to track the instantaneous channel conditions to the accuracy required for channel-dependent scheduling to be efficient.

In such situations, an alternative means to handle radio-channel frequency selectivity is to achieve frequency diversity by distributing a downlink transmission in the frequency domain.

One way to distribute a downlink transmission in the frequency domain, and thereby achieve frequency diversity, is to assign multiple non-frequency-contiguous resource blocks for the transmission to a terminal. LTE allows for such *distributed resource-block allocation* by means of *resource allocation types 0 and 1* (see Section 10.4.4). However, although sufficient in many cases, distributed resource-block allocation by means of these resource-allocation types has certain drawbacks:

- For both types of resource allocations, the minimum size of the allocated resource can be as large as four resource-block pairs and may thus not be suitable when resource allocations of smaller sizes are needed.
- In general, both these resource-allocation methods are associated with a relatively large PDCCH payload.

---

[3] 72 resource elements in the case of extended cyclic prefix.
[4] In MBSFN subframes, the control region is limited to a maximum of two OFDM symbols.

In contrast, *resource-allocation type 2* (Section 10.4.4) always allows for the allocation of a single resource-block pair and is also associated with a relatively small PDCCH payload size. However, resource allocation type 2 only allows for the allocation of resource blocks that are contiguous in the frequency domain. In addition, regardless of the type of resource allocation, frequency diversity by means of distributed resource-block allocation will only be achieved in the case of resource allocations larger than one resource-block pair.

In order to provide the possibility for distributed resource-block allocation in the case of resource-allocation type 2, as well as to allow for distributing the transmission of a single resource-block pair in the frequency domain, the notion of a *Virtual Resource Block* (VRB) has been introduced for LTE.

What is being provided in the resource allocation is the resource allocation in terms of VRB pairs. The key to distributed transmission then lies in the mapping from VRB pairs to *Physical Resource Block* (PRB) pairs – that is, to the actual physical resource used for transmission.

The LTE specification defines two types of VRBs: *localized* VRBs and *distributed* VRBs. In the case of localized VRBs, there is a direct mapping from VRB pairs to PRB pairs, as illustrated in Figure 10.6. However, in the case of distributed VRBs, the mapping from VRB pairs to PRB pairs is more elaborate in the sense that:

- consecutive VRBs are not mapped to PRBs that are consecutive in the frequency domain; and
- even a single VRB pair is distributed in the frequency domain.

The basic principle of distributed transmission is outlined in Figure 10.7 and consists of two steps:

- A mapping from VRB pairs to PRB pairs such that consecutive VRB pairs are not mapped to frequency-consecutive PRB pairs (first step of Figure 10.7). This provides frequency diversity between consecutive VRB pairs. The spreading in the frequency domain is done by means of a block-based "interleaver" operating on resource-block pairs.

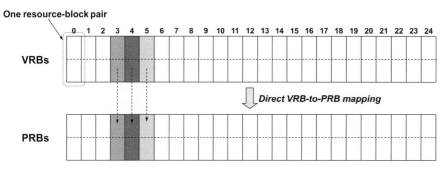

FIGURE 10.6

VRB-to-PRB mapping in the case of localized VRBs. A cell bandwidth corresponding to 25 resource blocks is assumed.

- A split of each resource-block pair such that the two resource blocks of the resource-block pair are transmitted with a certain frequency gap in between (second step of Figure 10.7). This also provides frequency diversity for a single VRB pair. This step can be seen as the introduction of frequency hopping on a slot basis.

Whether the VRBs are localized (and thus mapped according to Figure 10.6) or distributed (mapped according to Figure 10.7) is indicated on the associated PDCCH in the case of type 2 resource allocation. Thus, it is possible to dynamically switch between distributed and localized transmission and also mix distributed and localized transmission for different terminals within the same subframe.

The exact size of the frequency gap in Figure 10.7 depends on the overall downlink cell bandwidth according to Table 10.1. These gaps have been chosen based on two criteria:

1. The gap should be of the order of half the downlink cell bandwidth in order to provide good frequency diversity also in the case of a single VRB pair.
2. The gap should be a multiple of P^2, where P is the size of a *resource-block group* as defined in Section 10.4.4 and used for resource allocation types 0 and 1. The reason for this constraint is to ensure a smooth coexistence in the same subframe between distributed transmission as described above and transmissions based on downlink allocation types 0 and 1.

Due to the constraint that the gap size should be a multiple of P^2, the gap size will in most cases deviate from exactly half the cell bandwidth. In these cases, not all resource blocks within the cell bandwidth can be used for distributed transmission. As an example, for a cell bandwidth corresponding to 25 resource blocks (the example in Figure 10.7) and a corresponding gap size equal to 12 according to Table 10.1, the 25th resource-block pair cannot be used for distributed transmission.

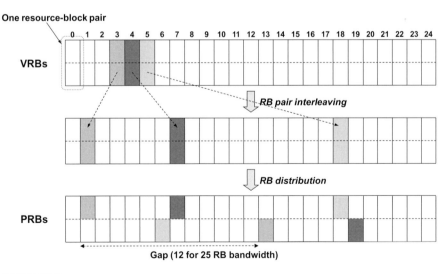

FIGURE 10.7

VRB-to-PRB mapping in the case of distributed VRBs. A cell bandwidth corresponding to 25 resource blocks is assumed.

Table 10.1 Gap Size for Different Cell Bandwidths (Number of Resource Blocks)

	Bandwidth									
	6	7–8	9–10	11	12–19	20–26	27–44	45–63	64–79	80–110
P	1	1	1	2	2	2	3	3	4	4
Gap size	3	4	5	4	8	12	18	27	32	48

Table 10.2 Second Gap Size for Different Cell Bandwidths (Only Applicable for Cell Bandwidths of 50 RBs and Beyond)

	Bandwidth	
	50–63	64–110
Gap size	9	16

As another example, for a cell bandwidth corresponding to 50 resource blocks (gap size equal to 27 according to Table 10.1), only 46 resource blocks would be available for distributed transmission.

In addition to the gap size outlined in Table 10.1, for wider cell bandwidths (50 RBs and beyond), there is a possibility to use a second, smaller frequency gap with a size of the order of one-fourth of the cell bandwidth (see Table 10.2). The use of the smaller gap enables restriction of the distributed transmission to only a part of the overall cell bandwidth. Selection between the larger gap according to Table 10.1 and the smaller gap according to Table 10.2 is indicated by an additional bit in the resource allocation on PDCCH.

10.2 DOWNLINK REFERENCE SIGNALS

Downlink reference signals are predefined signals occupying specific resource elements within the downlink time–frequency grid. The LTE specification includes several types of downlink reference signals that are transmitted in different ways and used for different purposes by the receiving terminal:

- *Cell-specific reference signals* (CRS) are transmitted in every downlink subframe and in every resource block in the frequency domain, thus covering the entire cell bandwidth. The cell-specific reference signals can be used by the terminal for channel estimation for coherent demodulation of any downlink physical channel except for PMCH and for PDSCH in the case of *transmission modes 7, 8, or 9*. As described in Section 10.3, these transmission modes correspond to so-called *non-codebook-based precoding*. The cell-specific reference signals can also be used by the terminal to acquire *channel-state information* (CSI). Finally, terminal measurements on cell-specific reference signals are used as the basis for cell-selection and handover decisions.

- *Demodulation reference signals* (DM-RS), also sometimes referred to as *UE-specific reference signals*, are specifically intended to be used by terminals for channel estimation for PDSCH in the case of transmission modes 7, 8, or 9. The label "*UE-specific*" relates to the fact that each demodulation reference signal is intended for channel estimation by a single terminal. That specific reference signal is then only transmitted within the resource blocks assigned for PDSCH transmission to that terminal.

- *CSI reference signals* (CSI-RS) are specifically intended to be used by terminals to acquire *channel-state information* (CSI) in the case when demodulation reference signals are used for channel estimation.[5] CSI-RS have a significantly lower time/frequency density, thus implying less overhead, compared to the cell-specific reference signals.

- *MBSFN reference signals* are intended to be used for channel estimation for coherent demodulation in the case of MCH transmission using so-called *MBSFN* (see Chapter 15 for more details on MCH transmission).

- *Positioning reference signals* were introduced in LTE release 9 to enhance *LTE positioning functionality*, more specifically to support the use of terminal measurements on multiple LTE cells to estimate the geographical position of the terminal. The positioning reference symbols of a certain cell can be configured to correspond to empty resource elements in neighboring cells, thus enabling high-SIR conditions when receiving neighbor-cell positioning reference signals.

10.2.1 Cell-Specific Reference Signals

Cell-specific reference signals, introduced in the first release of LTE (release 8), are the most basic downlink reference signals in LTE. There can be one, two, or four cell-specific reference signals in a cell, defining one, two, or four corresponding antenna ports.

10.2.1.1 Structure of a Single Reference Signal

Figure 10.8 illustrates the structure of a single cell-specific reference signal. As can be seen, it consists of *reference symbols* of predefined values inserted within the first and third last[6] OFDM symbol of each slot and with a frequency-domain spacing of six subcarriers. Furthermore, there is a frequency-domain staggering of three subcarriers for the reference symbols within the third last OFDM symbol. Within each resource-block pair, consisting of 12 subcarriers during one 1 ms subframe, there are thus eight reference symbols.

In general, the values of the reference symbols vary between different reference-symbol positions and also between different cells. Thus, a cell-specific reference signal can be seen as a *two-dimensional cell-specific sequence*. The period of this sequence equals one 10 ms frame. Furthermore, regardless of the cell bandwidth, the reference-signal sequence is defined assuming the maximum possible LTE carrier bandwidth corresponding to 110 resource blocks in the frequency domain. Thus, the basic reference-signal sequence has a length of 8880 symbols.[7] For cell bandwidths less than the maximum possible value, only the reference symbols within that bandwidth are actually transmitted. The reference symbols in the center part of the band will therefore be the same, regardless of the actual cell bandwidth. This allows for the terminal to estimate the channel corresponding to the center

---

[5] More specifically, CSI-RS are only used in the case of transmission mode 9.
[6] This corresponds to the fifth and fourth OFDM symbols of the slot for normal and extended cyclic prefixes respectively.
[7] Four reference symbols per resource block, 110 resource blocks per slot, and 20 slots per frame.

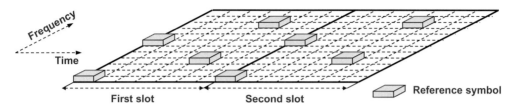

FIGURE 10.8

Structure of cell-specific reference signal within a pair of resource blocks.

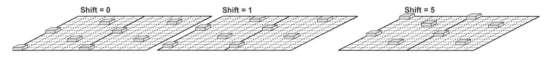

FIGURE 10.9

Different reference-signal frequency shifts.

part of the carrier, where, for example, the basic system information of the cell is transmitted on the BCH transport channel, without knowing the cell bandwidth. Information about the actual cell bandwidth, measured as number of resource blocks, is then provided on the BCH.

There are 504 different reference-signal sequences defined for LTE, where each sequence corresponds to one of 504 different *physical-layer cell identities*. As will be described in more detail in Chapter 14, during the so-called *cell-search procedure* the terminal detects the physical-layer identity of the cell as well as the cell frame timing. Thus, from the cell-search procedure, the terminal knows the reference-signal sequence of the cell (given by the physical-layer cell identity) as well as the start of the reference-signal sequence (given by the frame timing).

The set of reference-symbol positions outlined in Figure 10.8 is only one of six possible *frequency shifts* of the reference symbols, as illustrated in Figure 10.9. The frequency shift to use in a cell depends on the physical-layer identity of the cell such that each shift corresponds to 84 different cell identities. Thus, the six different frequency shifts jointly cover all 504 different cell identities. By properly assigning physical-layer cell identities to different cells, different reference-signal frequency shifts may be used in neighboring cells. This can be beneficial, for example, if the reference symbols are transmitted with higher energy compared to other resource elements, also referred to as *reference-signal power boosting*, in order to improve the reference-signal SIR. If reference signals of neighboring cells were transmitted using the same time/frequency resource, the boosted reference symbols of one cell would be interfered by equally boosted reference symbols of all neighboring cells,[8] implying no gain in the reference-signal SIR. However, if different frequency shifts are used for the reference-signal transmissions of neighboring cells, the reference symbols of one cell will at least partly be interfered by non-reference symbols of neighboring cells, implying an improved reference-signal SIR in the case of reference-signal boosting.

---

[8]This assumes that the cell transmissions are frame-timing aligned.

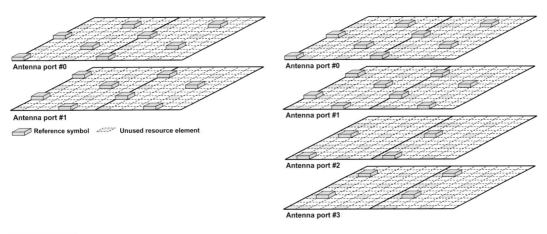

Antenna port #0

Antenna port #1

Antenna port #0

Antenna port #1

Antenna port #2

Antenna port #3

Reference symbol   Unused resource element

FIGURE 10.10

Structure of cell-specific reference signals in the case of multiple reference signals. Two reference signals corresponding to two antenna ports (left). Four reference signals corresponding to four antenna ports (right).

10.2.1.2 *Multiple Reference Signals*

Figure 10.10 illustrates the reference-signal structure in the case of multiple, more specifically two and four, cell-specific reference signals, and corresponding multiple antenna ports, within a cell:[9]

- In the case of two reference signals within a cell (left part of Figure 10.10), the second reference signal is frequency multiplexed with the first reference signal, with a frequency-domain offset of three subcarriers.
- In the case of four reference signals (right part of Figure 10.10), the third and fourth reference signals are frequency multiplexed and transmitted within the *second* OFDM symbol of each slot, thus being time multiplexed with the first and second reference signals.

Obviously, the reference-symbol density for the third and fourth reference signals is lower, compared to the density of the first and second reference signals. The reason for this is to reduce the reference-signal overhead in the case of four reference signals. More specifically, while the first and second reference signals each correspond to a relative overhead of approximately 5% (four reference symbols within a resource block consisting of a total of 84 resource elements), the relative overhead of the third and fourth reference signals is only half of that or approximately 2.5%. This obviously has an impact on the potential of the terminal to track very fast channel variations. However, this can be justified based on an expectation that, for example, high-order spatial multiplexing will mainly be applied to scenarios with low mobility.

It can also be noted that in a resource element carrying reference signals for a certain transmission port, nothing is being transmitted on the antenna ports corresponding to the other reference signals. Thus, a cell-specific reference signal is not interfered by transmissions on other antenna ports. Multi-antenna transmission schemes, such as spatial multiplexing, to a large extent rely on good channel

---

[9] It is not possible to configure a cell with three cell-specific reference signals.

estimates to suppress interference between the different layers at the receiver side. However, in the channel estimation itself there is obviously no such suppression. Reducing the interference to the reference signals of an antenna port is therefore important in order to allow for good channel estimation, and corresponding good interference suppression, at the receiver side.

Note that, in MBSFN subframes, only the reference signals in the two first OFDM symbols of the subframe, corresponding to the control region of the MBSFN subframe, are actually transmitted. Thus, there is no transmission of cell-specific reference signals within the MBSFN part of the MBSFM subframe.

10.2.2 Demodulation Reference Signals

In contrast to cell-specific reference signals, a demodulation reference signal (DM-RS) is intended for a specific terminal and is only transmitted in the resource blocks assigned for transmission to that terminal. Demodulation reference signals are intended to be used for channel estimation for PDSCH transmissions for the case when cell-specific reference signals are not to be used – that is, for non-codebook-based precoding (see Section 10.3.3).

Demodulation reference signals were introduced in the first release of LTE (release 8). However, use of demodulation reference signals was then limited to single-layer transmission – that is, no spatial multiplexing – with at most one reference signal for each terminal. In LTE release 9, transmission based on demodulation reference signals was extended to support also dual-layer transmission, requiring up to two simultaneous reference signals (one for each layer). Transmission based on demodulation reference signals was then further extended in LTE release 10 to support up to eight-layer transmission, corresponding to up to eight reference signals.

Actually, the DM-RS structure introduced in LTE release 9 was not a straightforward extension of the release-8 demodulation reference signals but rather a new structure, supporting both single-layer and dual-layer transmission. Already at the time of LTE release 9, it was relatively clear that the LTE radio-access technology should be further extended to support up to eight-layer spatial multiplexing in release 10. It was also quite clear that this extension would be difficult to achieve based on the release-8 DM-RS structure. Rather than extending the release-8 structure to support two reference signals and then introduce a completely new structure for release 10, it was decided instead to introduce a new, more future-proof structure already in release 9. Here we will focus on the DM-RS structure introduced in LTE release 9, including the release-10 extension to support up to eight simultaneous reference signals.

Figure 10.11 illustrates the DM-RS structure for the case of one or two reference signals.[10] As can be seen, there are 12 reference symbols within a resource-block pair. In contrast to cell-specific reference signals, for which the reference symbols of one antenna port correspond to unused resource elements for the other antenna ports (see Figure 10.10), in the case of two demodulation reference signals all 12 reference symbols in Figure 10.11 are transmitted *for both reference signals* – that is, *on both antenna ports*. Interference between the reference signals is instead handled by applying mutually orthogonal patterns, also referred to as *orthogonal cover codes* (OCC), to pairs of consecutive reference symbols, as illustrated in the lower right corner of the figure.

---

[10]In the case of TDD, the DM-RS structure is slightly modified in the DwPTS due to the shorter duration of the DwPTS compared with normal downlink subframes.

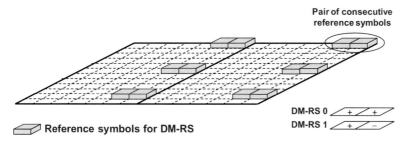

FIGURE 10.11

Structure of demodulation reference signals (DM-RS) for the case of one or two reference signals, including orthogonal cover codes to separate the two reference signals.

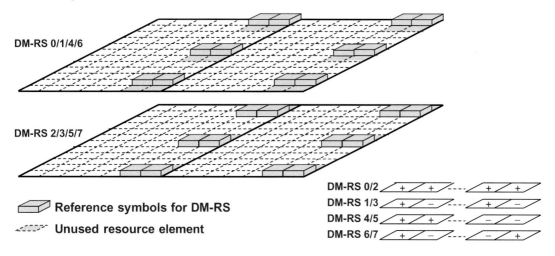

FIGURE 10.12

Demodulation-reference-signal structure for the case of more than two reference signals.

In addition to the mutually orthogonal patterns, one can also apply a pseudo-random sequence to the reference symbols. This sequence is the same for both reference signals and thus does not impact the orthogonality between the transmitted reference signals. Rather, the pseudo-random sequence is intended to separate different demodulation reference signals targeting different terminals in the case of so-called MU-MIMO transmission, as described in Section 10.3.4.1.

Figure 10.12 illustrates the extended DM-RS structure introduced in LTE release 10 to support more than two reference signals. In this case, there are 24 reference symbols within a resource-block pair. The reference signals are frequency multiplexed in groups of four reference signals while, within each group, the reference signals are separated by means of mutually orthogonal patterns covering four reference symbols (two pairs of consecutive reference symbols). It should be noted that orthogonality between the full set of eight reference signals requires that the channel does not vary over the set of reference symbols over which the orthogonal pattern spans. As the four reference symbols are not consecutive in time, this implies somewhat stronger limitations on the amount of channel variations that can be tolerated without losing the reference-signal orthogonality. However, more than four demodulation reference signals is only applicable to the case of spatial multiplexing with more than four layers, a transmission configuration that is typically anyway only applicable to low-mobility scenarios. Also note that the orthogonal patterns are selected such that, for four or less reference signals, orthogonality is achieved already over pairs of reference symbols. Thus, for three and four reference signals, the constraints on channel variations are the same as for two reference signals according to Figure 10.11.

When demodulation reference signals are transmitted within a resource block, PDSCH mapping to the time–frequency grid of the resource block will be modified to avoid the resource elements in which the reference signals are transmitted. Although this modified mapping is not "understood" by earlier-release terminals not supporting demodulation reference signals, this is not a problem as demodulation reference signals will only be transmitted in resource blocks that are scheduled for PDSCH transmission to terminals of later releases supporting demodulation reference signals and thus "understanding" the modified PDSCH mapping. As will be seen below, the situation is somewhat different in the case of CSI-RS.

As the number of transmitted layers may vary dynamically, the number of transmitted demodulation reference signals may also vary. Thus, the transmission may dynamically change between the DM-RS structures in Figures 10.11 and 10.12. The terminal is informed about the number of transmitted layers (the "transmission rank") as part of the scheduling assignment and will thus know the DM-RS structure and associated PDSCH mapping for each subframe.

10.2.3 CSI Reference Signals

Support for CSI reference signals (CSI-RS) was introduced in LTE release 10. CSI-RS is intended to be used by terminals to acquire channel-state information when demodulation reference signals are used for channel estimation. More specifically, CSI-RS are to be used in transmission mode 9.[11]

As described in Section 10.2.1, the cell-specific reference signals, already available in the first release of LTE, can also be used to acquire channel-state information. There are at least two reasons why CSI-RS was introduced in release 10 as a complement to the already supported cell-specific reference signals:

- In LTE release 8 there could be at most four cell-specific reference signals in a cell. As already mentioned, multi-antenna transmission was extended in LTE release 10 to support downlink spatial multiplexing with up to eight layers, corresponding to eight transmit antennas at the base station. Thus, there was a need to extend the LTE CSI capabilities beyond what was possible with the release 8 cell-specific reference signals.
- The time-frequency density of the cell-specific reference signals was selected to allow for channel estimation for coherent demodulation even with the most extreme channel conditions, including very fast channel variations in both the time and frequency domains. As a consequence, each

---

[11]The reason why CSI-RS are not used for transmission modes 7 and 8 is simply that these transmission modes were introduced in LTE releases 8 and 9 respectively, and CSI-RS was not introduced until LTE release 10, together with the introduction of transmission mode 9.

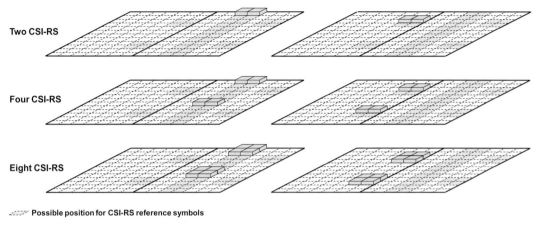

Two CSI-RS

Four CSI-RS

Eight CSI-RS

Possible position for CSI-RS reference symbols

FIGURE 10.13

Examples of reference-signal positions for different number of CSI-RS within a cell. In the case of a single CSI-RS, the same structure as for two CSI-RS is used.

cell-specific reference signal implies a relatively high overhead (approximately 5% and 2.5% for the first/second and third/fourth reference signals respectively). On the other hand, in the case of a reference signal only targeting CSI, such high time-frequency density, and correspondingly relatively large overhead, is not necessary. Thus, rather than extending the cell-specific reference signals it was found to be more efficient to introduce a new type of reference signals (CSI-RS) only targeting CSI, with a flexible and, in general, lower time/frequency density and corresponding lower overhead per reference signal.

10.2.3.1 *CSI-RS Structure*

A cell can be configured with one, two, four, or eight CSI-RS. The exact CSI-RS structure, including the exact set of resource elements used for CSI-RS in a resource block, depends on the number of CSI-RS configured within the cell and may also be different for different cells. More specifically, within a resource-block pair there are 40 *possible* positions for the reference symbols of CSI-RS and, in a given cell, a subset of the corresponding resource elements is used for CSI-RS transmission, as described below and exemplified in Figure 10.13.[12]

- In the case of two CSI-RS within a cell, the CSI-RS consists of two consecutive reference symbols per resource-block pair, as illustrated in the upper part of Figure 10.13. The two CSI-RS are separated by applying mutually *orthogonal cover codes* (OCC) to the two reference symbols, similar to demodulation reference signals described in Section 10.2.2. Thus, there is a possibility for 20 different CSI-RS configurations, two of which are shown in Figure 10.13, based on the 40 *possible* reference-symbol positions.

---

[12]For TDD, there are additional *possible* locations for reference symbols for CSI-RS beyond what is illustrated in Figure 10.13.

- In the case of four/eight CSI-RS, the CSI-RS are pair-wise frequency multiplexed, as illustrated in the middle/lower part of Figure 10.13. Thus, there is a possibility for ten/five different CSI-RS configurations, each supporting four/eight CSI-RS.

In the case of a single CSI-RS, the same structure as for two CSI-RS (upper part of Figure 10.13) is used, although with only one of the orthogonal cover codes. The overhead of a single CSI-RS, in terms of occupied resource elements, is thus the same as for two CSI-RS.

In the time domain, the CSI-RS can be transmitted with different periods, ranging from a period of 5 ms (every fifth subframe) to 80 ms (every eighth frame). With a 5 ms period, the overhead per CSI-RS is thus roughly 0.12% per CSI-RS (one resource element per CSI-RS per resource-block pair, but only in every fifth subframe), with longer periods implying correspondingly less overhead.

The exact subframe in which CSI-RS is transmitted in a cell can also be configured, allowing for separation of CSI-RS transmissions between cells also in the time domain, in addition to using different sets of resource elements within the same subframe, as illustrated in Figure 10.13.

In subframes in which CSI-RS is transmitted, it is transmitted in every resource block in the frequency domain. In other words, a CSI-RS transmission covers the entire cell bandwidth.

As mentioned above, when demodulation reference signals are transmitted within a resource block, the corresponding resource elements on which the reference signals are transmitted are explicitly avoided when mapping PDSCH symbols to the resource block. This "modified" PDSCH mapping, which is obviously not "understood" by earlier-release terminals, is possible as demodulation reference signals can be assumed to be transmitted only in resource blocks in which terminals supporting such reference signals are scheduled – that is, terminals of release 10 or later.[13] Expressed alternatively, an earlier-release terminal can be assumed never to be scheduled in a resource block in which demodulation reference signals are transmitted and thus in which the modified PDSCH mapping is used.

The situation is somewhat different for CSI-RS. As CSI-RS is transmitted within all resource blocks in the frequency domain, it would imply a strong scheduler constraint to assume that release-8/9 terminals could never be scheduled in a resource block in which CSI-RS is transmitted. If the PDSCH mapping were modified to explicitly avoid the resource elements in which CSI-RS is transmitted, the mapping would not be recognized by a release-8/9 terminal. Instead, in the case of resource blocks scheduled to release-8/9 terminals, the PDSCH is mapped exactly according to release 8 – that is, the mapping is not modified to avoid the resource elements on which CSI-RS is to be transmitted. The CSI-RS is then simply transmitted on top of the corresponding PDSCH symbols.[14] This will obviously impact the PDSCH demodulation performance, as some PDSCH symbols will be highly corrupted. However, the remaining PDSCH symbols will not be impacted and the PDSCH will still be decodable, although with somewhat reduced performance.

On the other hand, if a release-10 terminal is scheduled in a resource block in which CSI-RS is transmitted, the PDSCH mapping is modified to explicitly avoid the resource elements on which the CSI-RS is transmitted, similar to demodulation reference signals. Thus, if CSI-RS is transmitted in a resource block, the PDSCH mapping to that resource block will be somewhat different depending on the release of the terminal being scheduled in the resource block.

---

[13] Partly also for terminals of release 9, but then only for a maximum of two DM-RS.

[14] In practice, the base station would probably not transmit PDSCH in these resource elements in order to avoid interference to the CSI-RS transmission. The key thing is that the mapping of the remaining PDSCH symbols is in line with release 8.

It should be noted that release-8 mapping also has to be used for transmission of, for example, system information and paging messages, as such transmissions must be possible to receive also by release-8/9 terminals.

10.2.3.2 *Muted CSI-RS*

Figure 10.13 illustrated the set of resource elements that can *potentially* be used for CSI-RS, a subset of which is used to create the different CSI-RS within a cell. In normal operation, the remaining resource elements highlighted in Figure 10.13 are used for PDSCH transmission within the cell.

However, it is also possible to additionally configure one or several subsets of the resource elements outlined in Figure 10.13 as *muted CSI-RS*. A muted CSI-RS has the same structure as a normal (non-muted) CSI-RS except that nothing is actually transmitted on the corresponding resource elements. A muted CSI-RS can thus be seen as a normal (non-muted) CSI-RS *with zero power*.

The intention with muted CSI-RS is to be able to create "transmission holes" corresponding to actual CSI-RS transmissions in other (neighboring) cells. This serves two purposes:

1. To make it possible for a terminal to receive CSI-RS of neighboring cells, without being severely interfered by transmissions in its own cell. CSI estimation on neighboring cells would be of interest if support for different multi-cell-transmission techniques such as *CoMP* (see Chapter 20) were introduced in future LTE releases.
2. To reduce the interference to CSI-RS transmissions in other cells. This is especially applicable to so-called *heterogeneous network deployments* where overlapping cell layers with cells of substantially different power are to coexist in the same spectrum. Interference handling in heterogeneous deployments is further discussed in Chapter 13.

There can be one or multiple sets of muted CSI-RS in a cell. In the case of interference avoidance to neighboring cells (corresponding to point 2 above), there is typically sufficient with a single set of muted CSI-RS as one should typically avoid interference to a single lower-power cell layer. When muted CSI-RS are used to remove own-cell interference in order to be able to better receive CSI-RS from neighboring cells (point 1 above), multiple sets of muted CSI-RS are typically needed as different neighboring cells can be assumed to use mutually different CSI-RS configurations.

10.3 MULTI-ANTENNA TRANSMISSION

As illustrated in Figure 10.14, multi-antenna transmission in LTE can, in general, be described as a mapping from the output of the data modulation to the different antennas ports. The input to the antenna mapping thus consists of the modulation symbols (QPSK, 16QAM, 64QAM) corresponding to the one or two transport blocks. To be more specific, there is one transport block per TTI except for spatial multiplexing, in which case there may be two transport blocks per TTI (see also Section 10.3.2).

The output of the antenna mapping is a set of symbols for each antenna port. The symbols of each antenna port are subsequently applied to the OFDM modulator – that is, mapped to the basic OFDM time–frequency grid corresponding to that antenna port.

The different multi-antenna transmission schemes correspond to different so-called *transmission modes*. There are currently nine different transmission modes defined for LTE. They differ in terms of the specific structure of the antenna mapping of Figure 10.14 but also in terms of what reference

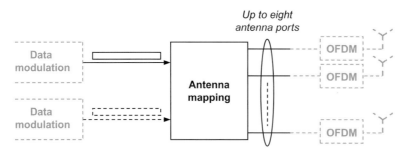

FIGURE 10.14

General structure for LTE downlink multi-antenna transmission. Modulation symbols corresponding to one or two transport blocks mapped to up to eight antenna ports.

signals are assumed to be used for demodulation (cell-specific reference signals or demodulation reference signals respectively) and the type of CSI feedback they rely on. Transmission mode 1 corresponds to single-antenna transmission while the remaining transmission modes correspond to different multi-antenna transmission schemes, including transmit diversity, beam-forming, and spatial multiplexing. Actually, LTE supports both beam-forming and spatial multiplexing as part of more general *antenna precoding*. There are two approaches to downlink antenna precoding, which differ depending on whether cell-specific reference signals or demodulation reference signals are used for channel estimation. These two approaches can also be referred to as *codebook-based precoding* and *non-codebook-based precoding* respectively, the reason for these specific names being further clarified below.

The list below summarizes the currently defined transmission modes and the associated multi-antenna transmission schemes. The different multi-antenna transmission schemes are described in more detail in the sections below.

- **Transmission mode 1:** Single-antenna transmission.
- **Transmission mode 2:** Transmit diversity.
- **Transmission mode 3:** *Open-loop* codebook-based precoding in the case of more than one layer, transmit diversity in the case of rank-one transmission.
- **Transmission mode 4:** *Closed-loop* codebook-based precoding.
- **Transmission mode 5:** Multi-user-MIMO version of transmission mode 4.
- **Transmission mode 6:** Special case of closed-loop codebook-based precoding limited to single-layer transmission.
- **Transmission mode 7:** Release-8 non-codebook-based precoding supporting only single-layer transmission.
- **Transmission mode 8:** Release-9 non-codebook-based precoding supporting up to two layers.
- **Transmission mode 9:** Release-10 non-codebook-based precoding supporting up to eight layers.

Strictly speaking, transmission modes are only applicable for DL-SCH transmission. Thus, a certain transmission mode should not be seen as identical to a certain multi-antenna transmission configuration. Rather, a certain multi-antenna transmission scheme is applied to DL-SCH transmission according to the list above when the corresponding transmission mode is configured for DL-SCH transmission. However, the same multi-antenna transmission scheme may also be applied to other types of transmissions, such as transmission of BCH and L1/L2 control signaling.[15] However, this does not mean that the corresponding transmission mode is applied to such transmissions.

It should also be mentioned that, although a certain multi-antenna transmission scheme can be seen as being associated with a certain transmission mode, for transmission modes 3–9 there is a possibility for fall-back to transmit diversity without implying that the configured transmission mode is changed.

10.3.1 Transmit Diversity

Transmit diversity can be applied to any downlink physical channel. However, it is especially applicable to transmissions that cannot be adapted to varying channel conditions by means of link adaptation and/or channel-dependent scheduling, and thus for which diversity is more important. This includes transmission of the BCH and PCH transport channels, as well as L1/L2 control signaling. Actually, as already mentioned, transmit diversity is the *only* multi-antenna transmission scheme applicable to these channels. Transmit diversity is also used for transmission of DL-SCH when transmission mode 2 is configured. Furthermore, as also already mentioned, transmit diversity is a fall-back "mode" for transmission mode 3 and higher. More specifically, a scheduling assignment using DCI format 1A (see Section 10.4.4) implies the use of transmit diversity regardless of the configured transmission mode.

Transmit diversity assumes the use of cell-specific reference signals for channel estimation. Thus, a transmit-diversity signal is always transmitted on the same antenna ports as the cell-specific reference signals. Actually, if a cell is configured with two cell-specific reference signals, transmit diversity for two antenna ports *must be used* for BCH and PCH, as well as for the L1/L2 control signaling. Similarly, if four cell-specific reference signals are configured for the cell, transmit diversity for four antenna ports has to be used for the transmission of these channels. In this way, a terminal does not have to be explicitly informed about what multi-antenna transmission scheme is used for these channels. Rather, this is given implicitly from the number of cell-specific reference signals configured for a cell.[16]

10.3.1.1 *Transmit Diversity for Two Antenna Ports*

In the case of two antenna ports, LTE transmit diversity is based on *Space-Frequency Block Coding* (SFBC), as described in Chapter 5. As can be seen from Figure 10.15, SFBC implies that two consecutive modulation symbols S_i and S_{i+1} are mapped directly to frequency-adjacent resource elements on the first antenna port. On the second antenna port the frequency-swapped and transformed symbols $-S_{i+1}^*$ and S_i^* are mapped to the corresponding resource elements, where "*" denotes complex conjugate.

Figure 10.15 also indicates how the antenna ports on which a transmit-diversity signal is being transmitted correspond to the cell-specific reference signals, more specifically CRS 0 and CRS 1 in

---

[15] Actually, only single-antenna transmission and transmit diversity is applicable to BCH and L1/L2 control signaling.

[16] Actually, the situation is partly the opposite, i.e. the terminal blindly detects the number of antenna ports used for BCH transmission and, from that, decides on the number of cell-specific reference signals configured within the cell.

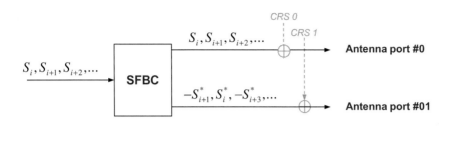

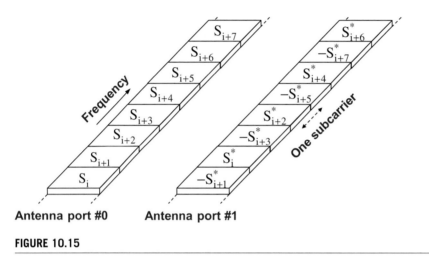

FIGURE 10.15

Transmit diversity for two antenna ports – SFBC.

the case of two antenna ports. Note that one should not interpret this such that the CRS is specifically transmitted for this transmit-diversity signal. The cell-specific reference signals are always transmitted and can be jointly used by multiple transmissions within a cell.

10.3.1.2 *Transmit Diversity for Four Antenna Ports*
In the case of four antenna ports, LTE transmit diversity is based on a combination of SFBC and *Frequency-Switched Transmit Diversity* (FSTD). As can be seen in Figure 10.16, combined SFBC/FSTD implies that pairs of modulation symbols are transmitted by means of SFBC with transmission alternating between pairs of antenna ports (antenna ports 0 and 2 and antenna ports 1 and 3 respectively). For the resource elements where transmission is on one pair of antenna ports, there is no transmission on the other pair of antenna ports. Thus, combined SFBC/FSTD in some sense operates on groups of four modulation symbols and corresponding groups of four frequency-consecutive resource elements on each antenna port. As mentioned in Section 10.4.1, this is the reason for the use of *resource-element groups*, each consisting of four resource elements, when defining the mapping of the L1/L2 control signaling to the physical resource.

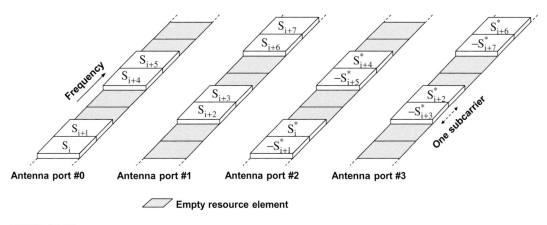

FIGURE 10.16

Transmit diversity for four antenna ports – combined SFBC/FSTD.

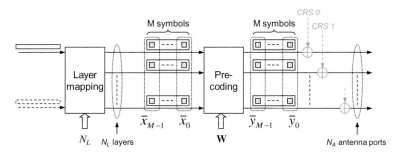

FIGURE 10.17

The basic structure of LTE codebook-based antenna precoding. The figure also indicates how cell-specific reference signals (CRS) are applied after precoding.

10.3.2 Codebook-Based Precoding

The basic processing for codebook-based precoding is illustrated in Figure 10.17. The modulation symbols corresponding to one or two transport blocks are first mapped to N_L *layers*. The number of layers may range from a minimum of one layer up to a maximum number of layers equal to the number of antenna ports.[17] The layers are then mapped to the antenna ports by means of the *precoder functionality*. As codebook-based precoding relies on the cell-specific reference signals for channel estimation, and there are at most four cell-specific reference signals in a cell, codebook-based precoding allows for a maximum of four antenna ports and, as a consequence, a maximum of four layers.

The mapping to layers is outlined in Figure 10.18 for the case of an initial transmission. There is one transport block in the case of a single layer ($N_L = 1$) and two transport blocks for two or more

---

[17]In practice, the number of layers is also limited by, and should not exceed, the number of receive antennas available at the terminal.

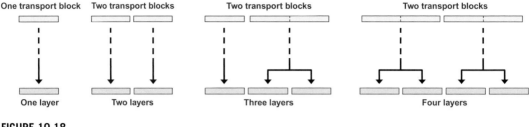

FIGURE 10.18

Transport-block-to-layer mapping for codebook-based antenna precoding (initial transmission).

layers ($N_L > 1$). In the case of a hybrid-ARQ retransmission, if only one of two transport blocks needs to be retransmitted and that transport block was mapped to two layers for the initial transmission, the retransmission may also be carried out on two layers. Thus, in the case of a retransmission, a single transport block may also be transmitted using two layers.

The mapping to layers is such that the number of modulation symbols on each layer is the same and equal to the number of symbols to be transmitted on each antenna port. Thus, in the case of three layers, there should be twice as many modulation symbols corresponding to the second transport block (mapped to the second and third layers) compared to the first transport block (mapped to the first layer). This is ensured by the selection of an appropriate transport-block size in combination with the rate-matching functionality. In the case of four layers, the first transport block is mapped to the first and second layers while the second transport block is mapped to the third and fourth layers. In this case, the number of modulation symbols should thus be the same for the two transport blocks. For one transport block being mapped to two layers, the modulated symbols corresponding to the transport block are mapped to the layers in an alternating fashion – that is, every second modulation symbol is mapped to the first and second layer respectively.

In the case of multi-antenna precoding the number of layers is also often referred to as the *transmission rank*.[18] The transmission rank can vary dynamically, for example based on the number of layers that can be supported by the channel. The latter is sometimes also referred to as the *channel rank*.

After layer mapping, a set of N_L symbols (one symbol from each layer) is linearly combined and mapped to the antenna ports. This combination/mapping can be described by a *precoder matrix* $\mathbf{W}$ of size $N_A \times N_L$, where N_A is the number of antenna ports. More specifically, the vector $\bar{y}_i$ of size N_A, consisting of one symbol for each antenna port, is given by $\bar{y}_i = \mathbf{W} \cdot \bar{x}_i$, where the vector $\bar{x}_i$ of size N_L consists of one symbol from each layer. As the number of layers can vary dynamically, also the number of columns of the precoder matrix will vary dynamically. Specifically, in the case of a single layer, the precoder matrix $\mathbf{W}$ is a vector of size $N_A \times 1$ that provides beam-forming for a single modulation symbol.

Figure 10.17 also indicates how the cell-specific reference signals are applied after antenna precoding. Channel estimation based on the cell-specific reference signals will thus reflect the channel for each antenna port *not including the precoding*. As a consequence, the terminal receiver must have

---

[18] In the LTE specification, transmit diversity is actually also described as transmission using *multiple layers*. However, transmit diversity is still a *single-rank* transmission scheme.

explicit knowledge about what precoding has been applied at the transmitter side in order to properly process the received signal and recover the different layers. Once again, the figure should not be interpreted such that cell-specific reference signals are inserted specifically for a given PDSCH transmission.

There are two operational modes for codebook-based precoding, *closed-loop operation* and *open-loop operation*. These two modes differ in terms of the exact structure of the precoder matrix and how the matrix is selected by the network and made known to the terminal.

10.3.2.1 *Closed-Loop Operation*

In case of closed-loop precoding it is assumed that the network selects the precoder matrix based on feedback from the terminal.

Based on measurements on the cell-specific reference signals, the terminal selects a suitable transmission rank and corresponding precoder matrix. Information about the selected rank and precoder matrix is then reported to the network in the form of a *Rank Indication* (RI) and a *Precoder-Matrix Indication* (PMI), as described in Section 11.4. It is important to understand though that the RI and PMI are only recommendations and the network does not need to follow the RI/PMI provided by the terminal when selecting the actual transmission rank and precoder matrix to be used for transmission to the terminal. When not following the terminal recommendation, the network must explicitly inform the terminal what precoder matrix is used for the downlink transmission. On the other hand, if the network uses the precoder matrix recommended by the terminal, only a confirmation that the network is using the recommended matrix is signaled.

To limit the signaling on both uplink and downlink only a limited set of precoder matrices, also referred to as the *codebook*, is defined for each transmission rank for a given number of antenna ports. Both the terminal (when reporting PMI) and the network (when selecting the actual precoder matrix to use for the subsequent downlink transmission to the terminal) should select a precoder matrix from the corresponding codebook. Thus, for terminal PMI reporting, as well as when the network informs the terminal about the actual precoder matrix used for the downlink transmission, only the index of the selected matrix needs to be signaled.

As LTE supports multi-antenna transmission using two and four antenna ports, codebooks are defined for:

- Two antenna ports and one and two layers, corresponding to precoder matrices of size 2×1 and 2×2 respectively.
- Four antenna ports and one, two, three, and four layers, corresponding to precoder matrices of size 4×1, 4×2, 4×3, and 4×4 respectively.

As an example, the precoder matrices specified for the case of two antenna ports are illustrated in Table 10.3. As can be seen, there are four 2×1 precoder matrices for single-layer transmission and three 2×2 precoder matrices for two-layer transmission. In the same way, sets of 4×1, 4×2, 4×3, and 4×4 matrices are defined for the case of four antenna ports and one, two, three, and four layers respectively. It should be pointed out that the first rank-2 (2×2) matrix in Table 10.3 is not used in closed-loop operation but only for *open-loop precoding*, as described in the next section.

Even if the network is following the precoder-matrix recommendation provided by the terminal, the network may, for different reasons, decide to use a lower rank for the transmission, so-called *rank*

Table 10.3 Precoder Matrices for Two Antenna Ports and One and Two Layers (the First 2 × 2 Matrix Is Only Used for Open-Loop Precoding)

One layer	$\frac{1}{\sqrt{2}}\begin{bmatrix} +1 \\ +1 \end{bmatrix}$	$\frac{1}{\sqrt{2}}\begin{bmatrix} +1 \\ -1 \end{bmatrix}$	$\frac{1}{\sqrt{2}}\begin{bmatrix} +1 \\ +j \end{bmatrix}$	$\frac{1}{\sqrt{2}}\begin{bmatrix} +1 \\ -j \end{bmatrix}$
Two layers	$\frac{1}{\sqrt{2}}\begin{bmatrix} +1 & 0 \\ 0 & +1 \end{bmatrix}$	$\frac{1}{2}\begin{bmatrix} +1 & +1 \\ +1 & -1 \end{bmatrix}$	$\frac{1}{2}\begin{bmatrix} +1 & +1 \\ +j & -j \end{bmatrix}$	

override. In that case the network will use a subset of the columns of the recommended precoder matrix. The network precoder confirmation will then include explicit information about the set of columns being used or, equivalently, about the set of layers being transmitted.

There is also a possibility to apply closed-loop precoding strictly limited to single-layer (rank-1) transmission. This kind of multi-antenna transmission is associated with *transmission mode 6*. The reason for defining an additional transmission mode limited to single-layer transmission rather than relying on the general closed-loop precoding associated with transmission mode 4 is that, by strictly limiting to single-layer transmission, the signaling overhead on both downlink and uplink can be reduced. Transmission mode 6 can, for example, be configured for terminals with low SINR for which multi-layer transmission would anyway not apply.

10.3.2.2 *Open-Loop Operation*

Open-loop precoding does not rely on any detailed precoder recommendation being reported by the terminal and does not require any explicit network signaling of the actual precoder used for the downlink transmission. Instead, the precoder matrix is selected in a predefined and deterministic way known to the terminal in advance. One use of open-loop precoding is in high-mobility scenarios where accurate feedback is difficult to achieve due to the latency in the PMI reporting.

The basic transmission structure for open-loop precoding is aligned with the general codebook-based precoding outlined in Figure 10.17 and only differs from closed-loop precoding in the structure of the precoding matrix **W**.

In the case of open-loop precoding, the precoder matrix can be described as the product of two matrices **W′** and **P**, where **W′** and **P** are of size $N_A \times N_L$ and $N_L \times N_L$ respectively:

$$\mathbf{W} = \mathbf{W'} \cdot \mathbf{P}. \tag{10.1}$$

In the case of two antenna ports, the matrix **W′** is the normalized 2 × 2 identity matrix:[19]

$$\mathbf{W'} = \frac{1}{\sqrt{2}}\begin{bmatrix} +1 & 0 \\ 0 & +1 \end{bmatrix}. \tag{10.2}$$

---

[19] As non-codebook-based precoding is not used for rank-1 transmission (see below), there is no need for any matrix **W′** of size 2 × 1.

In the case of four antenna ports, $\mathbf{W}'$ is given by cycling through four of the defined $4 \times N_L$ precoder matrices and is different for consecutive resource elements.

The matrix $\mathbf{P}$ can be expressed as $\mathbf{P} = \mathbf{D}_i \cdot \mathbf{U}$, where $\mathbf{U}$ is a constant matrix of size $N_L \times N_L$ and $\mathbf{D}_i$ is a matrix of size $N_L \times N_L$ that varies between subcarriers (indicated by the index i). As an example, the matrices $\mathbf{U}$ and $\mathbf{D}_i$ for the case of two layers ($N_L = 2$) are given by:

$$\mathbf{U} = \frac{1}{\sqrt{2}}\begin{bmatrix} 1 & 1 \\ 1 & e^{-j2\pi/2} \end{bmatrix}, \quad \mathbf{D}_i = \begin{bmatrix} 1 & 0 \\ 0 & e^{-j2\pi i/2} \end{bmatrix}. \tag{10.3}$$

The basic idea with the matrix $\mathbf{P}$ is to average out any differences in the channel conditions as seen by the different layers.

Similar to closed-loop precoding, the transmission rank for open-loop precoding can also vary dynamically down to a minimum of two layers. Transmission mode 3, associated with open-loop precoding, also allows for rank-1 transmission. In that case, transmit diversity as described in Section 10.3.1 is used – that is, SFBC for two antenna ports and combined SFBC/FSTD for four antenna ports.

10.3.3 Non-Codebook-Based Precoding

Similar to codebook-based precoding, non-codebook-based precoding is only applicable to DL-SCH transmission. Non-codebook-based precoding was introduced in LTE release 9 but was then limited to a maximum of two layers. The extension to eight layers was then introduced as part of release 10. The release-9 scheme, associated with transmission mode 8, is a subset of the extended release-10 scheme (transmission mode 9).

There is also a release-8 non-codebook-based precoding defined, associated with transmission mode 7. Transmission mode 7 relies on the release-8 demodulation reference signals mentioned but not described in detail in Section 10.2.2 and only supports single-layer transmission. In this description we will focus on non-codebook-based precoding corresponding to transmission modes 8 and 9.

The basic principles for non-codebook-based precoding can be explained based on Figure 10.19 (where the precoder is intentionally shaded; see below). As can be seen, this figure is very similar to the corresponding figure illustrating codebook-based precoding (Figure 10.17), with layer mapping of modulation symbols corresponding to one or two transport blocks followed by precoding. The layer mapping also follows the same principles as that of codebook-based precoding (see Figure 10.18) but is extended to support up to eight layers. In particular, at least for an initial transmission, there are two transport blocks per TTI except for the case of a single layer, in which case there is only one transport block within the TTI. Similar to codebook-based precoding, for hybrid-ARQ retransmissions there may in some cases be a single transport block also in the case of multi-layer transmission.

The main difference in Figure 10.19 compared to Figure 10.17 (codebook-based precoding) is the presence of demodulation reference signals (DM-RS) before the precoding. The transmission of precoded reference signals allows for demodulation and recovery of the transmitted layers at the receiver side *without explicit receiver knowledge of the precoding applied at the transmitter side*. Put simply, channel estimation based on precoded demodulation reference signals will reflect the channel experienced by the layers, *including the precoding*, and can thus be used directly for coherent demodulation of the different layers. There is thus no need to signal any precoder-matrix information

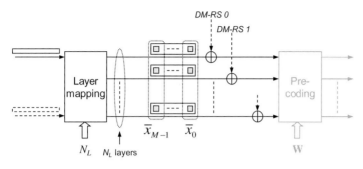

FIGURE 10.19

Basic principles for non-codebook-based antenna precoding.

to the terminal, which only needs to know the number of layers – that is, the transmission rank. As a consequence, the network can select an arbitrary precoder and there is no need for any explicit codebook to select from. This is the reason for the term *non-codebook-based* precoding. It should be noted though that non-codebook-based precoding may still rely on codebooks for the terminal feedback, as described below.

The possibility to select an arbitrary precoder matrix for the transmission is also the reason why the precoder is shaded in Figure 10.19. The precoder part of Figure 10.19 is not visible in the LTE specification and, strictly speaking, in the case of non-codebook-based precoding the antenna mapping defined according to Figure 10.14 consists of only the layer mapping. This also means that the antenna ports defined in Figure 10.14 correspond to the different layers in Figure 10.19 or, expressed differently, precoding occurs *after* the antenna ports.

Still, there must be a way for the network to select a suitable precoder matrix for the transmission. There are essentially two ways by which this can be done in the case of non-codebook-based precoding.

The network may estimate the uplink channel state, for example based on transmission of uplink sounding reference signals as described in the next chapter, and rely on uplink/downlink channel reciprocity when selecting the precoder matrix to use for the downlink transmission. This is especially of interest for TDD operation for which the use of the same frequency for uplink and downlink transmission typically leads to a higher degree of downlink/uplink channel reciprocity. It should be noted though that if the terminal uses multiple receive antennas, it also has to transmit on multiple antennas in order for uplink measurements to fully reflect the downlink channel state.

Alternatively, the network may rely on terminal feedback for precoder-matrix selection. For transmission modes 8 and 9, this feedback is actually very similar to the corresponding feedback for closed-loop codebook-based precoding. In particular, the feedback is based on the same precoder matrices as for codebook-based precoding (see Table 10.3 for an example) but extended to handle up to eight antenna ports. Furthermore, for transmission mode 9, the terminal measurements should be based on CSI-RS, as described in Section 10.2.3, rather than the cell-specific reference signals.

Thus, despite the name, non-codebook-based precoding may also use defined codebooks. However, in contrast to codebook-based precoding, the codebooks are then only used for the terminal PMI reporting and not for the actual downlink transmission.

10.3.4 **Downlink Multi-User MIMO**

Spatial multiplexing implies the transmission of multiple layers – that is, multiple parallel transmissions on the same time–frequency resource, to the same terminal. The presence of multiple antennas at both the transmitter and receiver sides in combination with transmitter and/or receiver signal processing is then used to suppress interference between the different layers.

Spatial multiplexing is also often referred to as *Multi-Input-Multi-Output* (MIMO) transmission, reflecting the fact that the channel in the case of spatial multiplexing can be seen as having *multiple inputs*, corresponding to the multiple transmit antennas, and *multiple outputs*, corresponding to the multiple receive antennas. The more specific term *Single-User MIMO* (SU-MIMO) is also often used for reasons that will become obvious below.

The term *Multi-User* MIMO (MU-MIMO) is, in 3GPP, used to denote transmission to *different* terminals using the same time–frequency resource, in practice relying on multiple antennas at the transmitter (network) side to separate the two transmissions.

In principle, one could realize MU-MIMO as a direct extension to spatial multiplexing, with the different layers simply being intended for different terminals. The set of terminals would demodulate and perhaps also decode the full set of layers. The data on a layer not intended for a specific terminal would then just be discarded by that terminal after demodulation/decoding.

However, such an approach would imply that all terminals involved in the MU-MIMO transmission would need to know about the full set of layers being transmitted. It would also imply that one would need to have exactly the same resource assignment for all terminals involved in the MU-MIMO transmission. All terminals would also need to include the full set of receive antennas necessary to receive the overall multi-layer transmission.

Instead, 3GPP has chosen an MU-MIMO approach that does not require terminal knowledge about the presence of the other transmissions, allows for only partly overlapping resource assignments, and, at least in principle, does not require the presence of multiple receive antennas at the mobile terminal.[20] There are two approaches to MU-MIMO specified in 3GPP, one being an integrated part of transmission modes 8 and 9 corresponding to non-codebook-based precoding, and one being based on codebook-based precoding but associated with a special transmission mode, *transmission mode 5*.

10.3.4.1 *MU-MIMO within Transmission Modes 8/9*

The only thing that is needed to support MU-MIMO as part of transmission modes 8 and 9 – that is, within the concept of non-codebook precoding – is to be able to assign different demodulation reference signals for the transmissions to the different terminals. As discussed in Section 10.2.2, for transmission mode 8 it is possible to define two orthogonal reference signals by means of two mutually orthogonal patterns, or *orthogonal cover codes* (OCC). In the case of spatial multiplexing, these OCC and corresponding reference signals are used for the transmission of the two layers. However, they can also be assigned to different users to enable MU-MIMO transmission within transmission mode 8.

In addition, as also mentioned in Section 10.2.2, a pseudo-random sequence is applied to the demodulation reference signals. This sequence is not terminal specific in the sense that each terminal

---

[20]Note, though, that the LTE performance requirements in general assume the presence of at least two receive antennas at the mobile terminal.

has its own sequence. Rather, there are only two sequences available and information on what pseudo-random sequence is used for a certain demodulation reference signal to a certain terminal is provided in the scheduling assignment together with information about the OCC.

In total, this allows for four different reference signals that can be assigned for four different transmissions being transmitted in parallel to different terminals using MU-MIMO. This assumes no spatial multiplexing on a per-link basis – that is, the transmission to each terminal consists of a single layer. One can also combine MU-MIMO with spatial multiplexing by transmitting two layers to a single terminal, in which case the number of multiplexed terminals is reduced correspondingly as the OCC are then used to separate the two layers transmitted to a single terminal. Obviously, the transmission of multiple layers to the same terminal requires the use of multiple receive antennas at the terminal.

Finally, although MU-MIMO does not rely on any terminal knowledge of the other, parallel, transmissions, nothing prevents a terminal from assuming that there are transmissions on the other "layers", corresponding to the other demodulation reference signals, and applying receiver-side signal processing to suppress the potential interference from those *possibly* present transmissions. Such suppression would then, once again, be based on the presence of multiple receive antennas at the mobile terminal.

10.3.4.2 *MU-MIMO Based on CRS*

The above-described MU-MIMO transmission is part of transmission modes 8 and 9 and thus became available in LTE release 9, with further extension in release 10. However, already in LTE release 8, MU-MIMO was possible by a minor modification of transmission mode 4 – that is, closed-loop codebook-based beam-forming, leading to *transmission mode 5*. The only difference between transmission modes 4 and 5 is the signaling of an additional power offset between PDSCH and the cell-specific reference signals (CRS).

In general, for transmission modes relying on CRS (as well as when relying on DM-RS) for channel estimation the terminal will use the reference signal as a phase reference but also as a power/amplitude reference for the demodulation of signals transmitted by means of higher-order modulation (16QAM and 64QAM). Thus, for proper demodulation of higher-order modulation, the terminal needs to know the power offset between the CRS and the PDSCH.

The terminal is informed about this power offset by means of higher-layer signaling. However, what is then provided is the offset between CRS power and the overall PDSCH power, including all layers. In the case of spatial multiplexing, the overall PDSCH power has to be divided between the different layers, and it is the relation between the CRS power and the per-layer PDSCH power that is relevant for demodulation.

In the case of pure spatial multiplexing (no MU-MIMO) – that is, transmission modes 3 and 4 – the terminal knows about the number of layers and thus indirectly about the offset between the CRS power and the per-layer PDSCH power.

In the case of MU-MIMO, the total available power will typically also be divided between the transmissions to the different terminals, with less PDSCH power being available for each transmission. However, terminals are not aware of the presence of parallel transmissions to other terminals and are thus not aware of any per-PDSCH power reduction. For this reason, transmission mode 5 includes the explicit signaling of an *additional power offset* of $-3\,dB$ to be used by the terminal in addition to the CRS/PDSCH power offset signaled by higher layers.

Transmission mode 5 is limited to single-rank transmission and, in practice, limited to two users being scheduled in parallel, as there is only a single $-3\,dB$ offset defined.

Note that the above power-offset signaling is not needed for MU-MIMO based on demodulation reference signals (transmission modes 8 and 9) as, in this case, each transmission has its own set of reference signals. The power per reference signal will thus scale with the number of layers and transmissions, similar to the PDSCH power and the reference-signal to per-layer PDSCH power ratio will remain constant.

10.4 DOWNLINK L1/L2 CONTROL SIGNALING

To support the transmission of downlink and uplink transport channels, there is a need for certain *associated downlink control signaling*. This control signaling is often referred to as *downlink L1/L2 control signaling*, indicating that the corresponding information partly originates from the physical layer (Layer 1) and partly from Layer 2 MAC. Downlink L1/L2 control signaling consists of downlink scheduling assignments, including information required for the terminal to be able to properly receive, demodulate, and decode the DL-SCH[21] on a component carrier, uplink scheduling grants informing the terminal about the resources and transport format to use for uplink (UL-SCH) transmission, and hybrid-ARQ acknowledgements in response to UL-SCH transmissions. In addition, the downlink control signaling can also be used for the transmission of power-control commands for power control of uplink physical channels, as well as for certain special purposes such as MBSFN notifications.

As illustrated in Figure 10.20, the downlink L1/L2 control signaling is transmitted within the first part of each subframe. Thus, each subframe can be said to be divided into a *control region* followed by a *data region*, where the control region corresponds to the part of the subframe in which the L1/L2 control signaling is transmitted. To simplify the overall design, the control region always occupies an integer number of OFDM symbols, more specifically one, two, or three OFDM symbols (for narrow cell bandwidths, 10 resource blocks or less, the control region consists of two, three, or four OFDM symbols to allow for a sufficient amount of control signaling). In the case of carrier aggregation, there is one control region per component carrier.

The size of the control region expressed in number of OFDM symbols, or, equivalently, the start of the data region, can be dynamically varied on a per-subframe basis independently for each component carrier.[22] Thus, the amount of radio resources used for control signaling can be dynamically adjusted to match the instantaneous traffic situation. For a small number of users being scheduled in a subframe, the required amount of control signaling is small and a larger part of the subframe can be used for data transmission (larger data region).

The reason for transmitting the control signaling at the beginning of the subframe is to allow for terminals to decode downlink scheduling assignments as early as possible. Processing of the data region – that is, demodulation and decoding of the DL-SCH transmission – can then begin before the end of the subframe. This reduces the delay in the DL-SCH decoding and thus the overall downlink transmission delay. Furthermore, by transmitting the L1/L2 control channel at the beginning of the subframe – that is, by allowing for early decoding of the L1/L2 control information – terminals that

---

[21] L1/L2 control signaling is also needed for the reception, demodulation, and decoding of the PCH transport channel.

[22] The start of the data region is semi-statically configured when using cross-carrier scheduling (see Section 10.4.1).

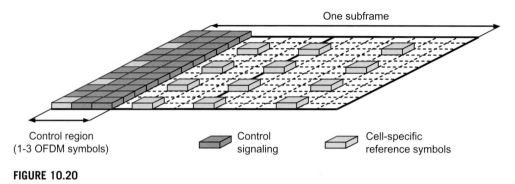

One subframe

Control region (1-3 OFDM symbols) · Control signaling · Cell-specific reference symbols

FIGURE 10.20

LTE time–frequency grid illustrating the split of the subframe into (variable-sized) control and data regions.

are not scheduled in the subframe may power down the receiver circuitry for a part of the subframe, allowing for a reduced terminal power consumption.

The downlink L1/L2 control signaling consists of four different physical-channel types:

- The *Physical Control Format Indicator Channel* (PCFICH), informing the terminal about the size of the control region (one, two, or three OFDM symbols). There is one and only one PCFICH on each component carrier or, equivalently, in each cell.
- The *Physical Downlink Control Channel* (PDCCH), used to signal downlink scheduling assignments and uplink scheduling grants. Each PDCCH typically carries signaling for a single terminal, but can also be used to address a group of terminals. Multiple PDCCHs can exist in each cell.
- The *Physical Hybrid-ARQ Indicator Channel* (PHICH), used to signal hybrid-ARQ acknowledgements in response to uplink UL-SCH transmissions. Multiple PHICHs can exist in each cell.
- The *Relay Physical Downlink Control Channel* (R-PDCCH), used for relaying. A detailed discussion can be found in Chapter 16 in conjunction with the overall description of relays; at this stage it suffices to note that the R-PDCCH is not transmitted in the control region.

The maximum size of the control region is normally three OFDM symbols (four in the case of narrow cell bandwidths), as mentioned above. However, there are a few exceptions to this rule. When operating in TDD mode, the control region in subframes 1 and 6 is restricted to at most two OFDM symbols since, for TDD, the primary synchronization signal (see Chapter 14) occupies the third OFDM symbol in those subframes. Similarly, for MBSFN subframes (see Chapter 9), the control region is restricted to a maximum of two OFDM symbols.

10.4.1 Physical Control Format Indicator Channel

The PCFICH indicates the size of the control region in terms of the number of OFDM symbols – that is, indirectly where in the subframe the data region starts. Correct decoding of the PCFICH information is thus essential. If the PCFICH is incorrectly decoded, the terminal will neither know how to process the control channels nor where the data region starts for the corresponding subframe.[23]

---

[23] Theoretically, the terminal could blindly try to decode all possible control channel formats and, from which format that was correctly decoded, deduce the starting position of the data region, but this can be a very complex procedure.

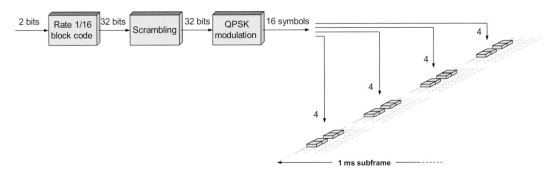

FIGURE 10.21

Overview of the PCFICH processing.

The PCFICH consists of two bits of information, corresponding to the three[24] control-region sizes of one, two, or three OFDM symbols (two, three or four for narrow bandwidths), which are coded into a 32-bit codeword. The coded bits are scrambled with a cell- and subframe-specific scrambling code to randomize inter-cell interference, QPSK modulated, and mapped to 16 resource elements. As the size of the control region is unknown until the PCFICH is decoded, the PCFICH is always mapped to the first OFDM symbol of each subframe.

The mapping of the PCFICH to resource elements in the first OFDM symbol in the subframe is done in groups of four resource elements, with the four groups being well separated in frequency to obtain good diversity. Furthermore, to avoid collisions between PCFICH transmissions in neighboring cells, the location of the four groups in the frequency domain depends on the physical-layer cell identity.

The transmission power of the PCFICH is under control of the eNodeB. If necessary for coverage in a certain cell, the power of the PCFICH can be set higher than for other channels by "borrowing" power from, for example, simultaneously transmitted PDCCHs. Obviously, increasing the power of the PCFICH to improve the performance in an interference-limited system depends on the neighboring cells not increasing their transmit power on the interfering resource elements. Otherwise, the interference would increase as much as the signal power, implying no gain in received SIR. However, as the PCFICH-to-resource-element mapping depends on the cell identity, the probability of (partial) collisions with PCFICH in neighboring cells in synchronized networks is reduced, thereby improving the performance of PCFICH power boosting as a tool to control the error rate.

The overall PCFICH processing is illustrated in Figure 10.21.

To describe the mapping of the PCFICH, and L1/L2 control signaling in general, to resource elements, some terminology is required. As mentioned above, the mapping is specified in terms of groups of four resource elements, so-called *resource-element groups*. To each resource-element group, a *symbol quadruplet* consisting of four (QPSK) symbols is mapped. The main motivation behind this, instead of simply mapping the symbols one by one, is the support of transmit diversity. As discussed in Section 10.3, transmit diversity with up to four antenna ports is specified for L1/L2

---

[24]The fourth combination is reserved for future use.

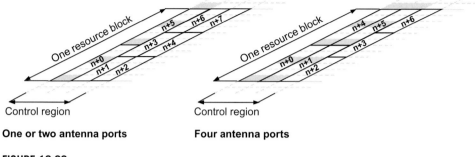

One or two antenna ports   **Four antenna ports**

FIGURE 10.22

Numbering of resource-element groups in the control region (assuming a size of three OFDM symbols).

control signaling. Transmit diversity for four antenna ports is specified in terms of groups of four symbols (resource elements) and, consequently, the L1/L2 control-channel processing is also defined in terms of symbol quadruplets.

The definition of the resource-element groups assumes that reference symbols corresponding to two antenna ports are present in the first OFDM symbol, regardless of the actual number of antenna ports configured in the cell. This simplifies the definition and reduces the number of different structures to handle. Thus, as illustrated in Figure 10.22, in the first OFDM symbol there are two resource-element groups per resource block, as every third resource element is reserved for reference signals (or non-used resource elements corresponding to reference symbols on the other antenna port). As also illustrated in Figure 10.22, in the second OFDM symbol (if part of the control region) there are two or three resource-element groups depending on the number of antenna ports configured. Finally, in the third OFDM symbol (if part of the control region) there are always three resource-element groups per resource block. Figure 10.22 also illustrates how resource-element groups are numbered in a time-first manner within the size of the control region.

Returning to the PCFICH, four resource-element groups are used for the transmission of the 16 QPSK symbols. To obtain good frequency diversity the resource-element groups should be well spread in frequency and cover the full downlink cell bandwidth. Therefore, the four resource-element groups are separated by one-fourth of the downlink cell bandwidth in the frequency domain, with the starting position given by physical-layer cell identity. This is illustrated in Figure 10.23, where the PCFICH mapping to the first OFDM symbol in a subframe is shown for three different physical-layer cell identities in the case of a downlink cell bandwidth of eight resource blocks. As seen in the figure, the PCFICH mapping depends on the physical-layer cell identity to reduce the risk of inter-cell PCFICH collisions. The cell-specific shifts of the reference symbols, described in Section 10.2.1, are also seen in the figure.

For terminals capable of carrier aggregation, there is one PCFICH per component carrier. Independent signaling of the control-region size on the different component carriers is used, implying that the control region may be of different size on different component carriers. Hence, in principle the terminal needs to receive the PCFICH on each of the component carriers it is scheduled upon. Furthermore, as different component carriers may have different physical-layer cell identities, the location and scrambling may differ across component carriers.

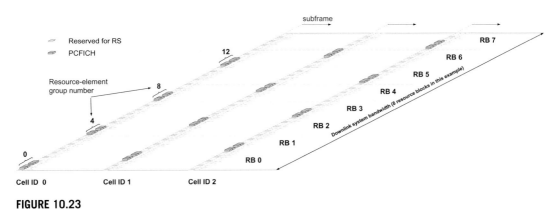

FIGURE 10.23

Example of PCFICH mapping in the first OFDM symbol for three different physical-layer cell identities.

If cross-carrier scheduling is used (see Section 10.4.6) – that is, control signaling related to a certain PDSCH transmission is transmitted on a component carrier other than the PDSCH itself – the terminal needs to know the starting position for the data region on the carrier upon which the PDSCH is transmitted. Using the PCFICH on the component carrier carrying the PDSCH would be possible in principle, although it would increase the probability of incorrectly decoding the PDSCH since there are two PCFICH instances, one for the PDCCH decoding on one component carrier and one for PDSCH reception on the other component carrier. This would be problematic, especially since the main use of cross-carrier scheduling is enhanced support of heterogeneous deployments (see Chapter 13), where some of the component carriers may be subject to strong interference. Therefore, for cross-carrier scheduled transmissions, the start of the data region is not obtained from the PCFICH on that component carrier, but is configured on a semi-static basis. The semi-statically configured value may differ from the value signaled on the PCFICH on the component carrier carrying the PDSCH transmission.

10.4.2 Physical Hybrid-ARQ Indicator Channel

The PHICH is used for transmission of hybrid-ARQ acknowledgements in response to UL-SCH transmission. There is one PHICH transmitted per received transport block and TTI – that is, when uplink spatial multiplexing is used on a component carrier, two PHICHs are used to acknowledge the transmission, one per transport block.

For proper operation of the hybrid-ARQ protocol, as discussed in Chapter 12, the error rate of the PHICH should be sufficiently low. The operating point of the PHICH is not specified and is up to the network operator to decide upon, but typically ACK-to-NAK and NAK-to-ACK error rates of the order of 10^{-2} and 10^{-3}–10^{-4} respectively are targeted. The reason for the asymmetric error rates is that an NAK-to-ACK error would imply loss of a transport block at the MAC level, a loss that has to be recovered by RLC retransmissions with the associated delays, while an ACK-to-NAK error only implies an unnecessary retransmission of an already correctly decoded transport block. To meet these error-rate targets without excessive power, it is beneficial to control the PHICH transmission power as a function of the radio-channel quality of the terminal to which the PHICH is directed. This has influenced the design of the PHICH structure.

In principle, a PHICH could be mapped to a set of resource elements exclusively used by this PHICH. However, taking the dynamic PHICH power setting into account, this could result in significant variations in transmission power between resource elements, which can be challenging from an RF implementation perspective. Therefore, it is preferable to spread each PHICH on multiple resource elements to reduce the power differences while at the same time providing the energy necessary for accurate reception. To fulfill this, a structure where several PHICHs are code multiplexed on to a set of resource elements is used in LTE. The hybrid-ARQ acknowledgement (one single bit of information per transport block) is repeated three times, followed by BPSK modulation on either the I or the Q branch and spreading with a length-four orthogonal sequence. A set of PHICHs transmitted on the same set of resource elements is called a PHICH group, where a PHICH group consists of eight PHICHs in the case of normal cyclic prefix. An individual PHICH can thus be uniquely represented by a single number from which the number of the PHICH group, the number of the orthogonal sequence within the group, and the branch, I or Q, can be derived.

For extended cyclic prefix, which is typically used in time-dispersive environments, the radio channel may not be flat over the frequency spanned by a length-four sequence. A non-flat channel would negatively impact the orthogonality between the sequences. Hence, for extended cyclic prefix, orthogonal sequences of length two are used for spreading, implying only four PHICHs per PHICH group. However, the general structure remains the same as for the normal cyclic prefix.

After forming the composite signal representing the PHICHs in a group, cell-specific scrambling is applied and the 12 scrambled symbols are mapped to three resource-element groups. Similarly to the other L1/L2 control channels, the mapping is described using resource-element groups to be compatible with the transmit diversity schemes defined for LTE.

The overall PHICH processing is illustrated in Figure 10.24.

The requirements on the mapping of PHICH groups to resource elements are similar to those for the PCFICH, namely to obtain good frequency diversity and to avoid collisions between neighboring cells in synchronized networks. Hence, each PHICH group is mapped to three resource-element groups, separated by approximately one-third of the downlink cell bandwidth. In the first OFDM

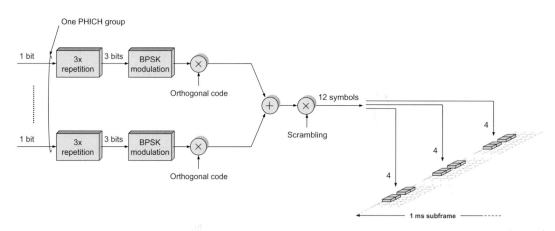

FIGURE 10.24

PHICH structure.

symbol in the control region, resources are first allocated to the PCFICH, the PHICHs are mapped to resource elements not used by the PCFICH, and finally, as will be discussed further below, the PDCCHs are mapped to the remaining resource elements.

Typically, the PHICH is transmitted in the first OFDM symbol only, which allows the terminal to attempt to decode the PHICH even if it failed decoding of the PCFICH. This is advantageous as the error requirements on the PHICH typically are stricter than for PCFICH. However, in some propagation environments, having a PHICH duration of a single OFDM symbol would unnecessarily restrict the coverage. To alleviate this, it is possible to semi-statically configure a PHICH duration of three OFDM symbols. In this case, the control region is three OFDM symbols long in all subframes to fulfill the general principle of separating the control region from data in the time domain only. Obviously, the value transmitted on the PCFICH will be fixed (and can be ignored) in this case. For narrow bandwidths, where the control region can be up to four OFDM symbols long, there is still a need to use the PCFICH to differentiate between a size-three and size-four control region.

As a general principle, LTE transmits the PHICH on the same component carrier that was used for the scheduling grant for the corresponding uplink data transmission. Not only is this principle general in the sense that it can handle symmetric as well as asymmetric carrier aggregation scenarios, it is also beneficial from a terminal power consumption perspective as the terminal only need monitor the component carriers it monitors for uplink scheduling grants (especially as the PDCCH may override the PHICH to support adaptive retransmissions, as discussed in Chapter 12).

The association between PUSCH transmission in an uplink subframe and the corresponding downlink PHICH resource is discussed in Chapter 12. At this stage, it is sufficient to note that, for TDD configuration zeros, uplink transmissions in multiple subframes may need to be acknowledged in one downlink subframe. Consequently, the amount of resources for PHICH may differ between subframes in the case of TDD.

The PHICH configuration is part of the system information transmitted on the PBCH; one bit indicates whether the duration is one or three OFDM symbols and two bits indicate the amount of resources in the control region reserved for PHICHs, expressed as a fraction of the downlink cell bandwidth in terms of resource blocks. Having the possibility to configure the amount of PHICH resources is useful as the PHICH capacity depends on, for example, whether the network uses multi-user MIMO or not. The PHICH configuration must reside on the PBCH, as it needs to be known in order to properly process the PDCCHs for reception of the part of the system information on the DL-SCH. For TDD, the PHICH information provided on the PBCH is not sufficient for the terminal to know the exact set of resources used by PHICH, as there is also a dependency on the uplink/downlink allocation, provided as part of the system information transmitted on the PDSCH. In order to receive the system information on the DL-SCH, which contains the uplink/downlink allocation, the terminal therefore has to blindly process the PDCCHs under different PHICH configuration hypotheses.

10.4.3 Physical Downlink Control Channel

The PDCCH is used to carry downlink control information (DCI) such as scheduling decisions and power-control commands. More specifically, the DCI includes:

- Downlink scheduling assignments, including PDSCH resource indication, transport format, hybrid-ARQ information, and control information related to spatial multiplexing (if applicable). A downlink scheduling assignment also includes a command for power control of the PUCCH used for transmission of hybrid-ARQ acknowledgements in response to downlink scheduling assignments.

- Uplink scheduling grants, including PUSCH resource indication, transport format, and hybrid-ARQ-related information. An uplink scheduling grant also includes a command for power control of the PUSCH.
- Power-control commands for a set of terminals as a complement to the commands included in the scheduling assignments/grants.

The different types of control information, both between the groups above as well as within the groups, correspond to different DCI message sizes. For example, supporting spatial multiplexing with non-contiguous allocation of resource blocks in the frequency domain requires a larger scheduling message in comparison with an uplink grant allowing for frequency-contiguous allocations only. The DCI is therefore categorized into different *DCI formats*, where a format corresponds to a certain message size and usage. The DCI formats are summarized in Table 10.4, sorted in approximate order of the relative size. The actual message size depends on, among other factors, the cell bandwith as, for larger bandwidths, a larger number of bits is required to indicate the resource-block allocation. The number of cell-specific reference signals in the cell and whether cross-carrier scheduling is configured or not will also affect the absolute size of most DCI formats. Hence, a given DCI format may have different sizes depending on the overall configuration of the cell. This will be discussed further below; at this stage it suffices to note that formats 0, 1A, 3, and 3A have the same message size.[25]

Table 10.4 DCI Formats

Size	Uplink Grant		Downlink Assignment		Power Control
			Usage		
Small	–	1C	Special purpose compact assignment		–
	0	Single layer	1A	Contiguous allocations only	3, 3A
...	–		1B	Codebook-based beam-forming using CRS	–
	–		1D	Multi-user MIMO using CRS	–
	4	Spatial multiplexing	–		–
	–		1	Flexible allocations	–
	–		2A	Open-loop spatial multiplexing using CRS	–
	–		2B	Dual-layer transmission using DM-RS	–
	–		2C	Multi-layer transmission using DM-RS	–
Large	–		2	Closed-loop spatial multiplexing using CRS	–

[25]The smaller of DCI formats 0 and 1A is padded to ensure the same payload size. Which of 0 and 1A is padded depends on the uplink and downlink cell bandwidths; in the case of identical uplink and downlink bandwidths in a cell there is a single bit of padding in format 0.

One PDCCH carries one DCI message with one of the formats above. As multiple terminals can be scheduled simultaneously, on both downlink and uplink, there must be a possibility to transmit multiple scheduling messages within each subframe. Each scheduling message is transmitted on a separate PDCCH, and consequently there are typically multiple simultaneous PDCCH transmissions within each cell. Furthermore, to support different radio-channel conditions, link adaptation can be used, where the code rate of the PDCCH is selected to match the radio-channel conditions.

For carrier aggregation, scheduling assignments/grants are transmitted individually per component carrier. PDCCH transmission for carrier aggregation is further elaborated upon in Section 10.4.6.

10.4.4 Downlink Scheduling Assignment

Downlink scheduling assignments are valid for the same subframe in which they are transmitted. The scheduling assignments use one of the DCI formats 1, 1A, 1B, 1C, 1D, 2, 2A, 2B, or 2C and the DCI formats used depend on the transmission mode configured (the relation between transmission modes and DCI formats is further described in Section 10.4.9). The reason for supporting multiple formats with different message sizes for the same purpose is to allow for a trade-off in control-signaling overhead and scheduling flexibility. Parts of the contents are the same for the different DCI formats, as seen in Table 10.5, but obviously there are also differences due to the different capabilities.

DCI format 1 is the basic downlink assignment format in the absence of spatial multiplexing (transmission modes 1, 2, and 7). It supports non-contiguous allocations of resource blocks and the full range of modulation-and-coding schemes.

DCI format 1A, also known as the "compact" downlink assignment, supports allocation of frequency-contiguous resource blocks only and can be used in all transmission modes. Contiguous allocations reduce the payload size of the control information with a somewhat reduced flexibility in resource allocations. The full range of modulation-and-coding schemes is supported.

DCI format 1B is used to support codebook-based beam-forming described in Section 10.3.2, with a low control-signaling overhead (transmission mode 6). The content is similar to DCI format 1A with the addition of bits for signaling of the precoding matrix. As codebook-based beam-forming can be used to improve the data rates for cell-edge terminals, it is important to keep the related DCI message size small so as not to unnecessarily limit the coverage.

DCI format 1C is used for various special purposes such as random-access response, paging, and transmission of system information. Common for these applications is simultaneous reception of a relatively small amount of information by *multiple* users. Hence, DCI format 1C supports QPSK only, has no support for hybrid-ARQ retransmissions, and does not support closed-loop spatial multiplexing. Consequently, the message size for DCI format 1C is very small, which is beneficial for coverage and efficient transmission of the type of system messages for which it is intended. Furthermore, as only a small number of resource blocks can be indicated, the size of the corresponding indication field in DCI format 1C is independent of the cell bandwidth.

DCI format 1D is used to support multi-user MIMO (transmission mode 5) scheduling of one codeword with precoder information. To support dynamic sharing of the transmission power between the terminals sharing the same resource block in multi-user MIMO, one bit of power offset information is included in DCI format 1D, as described in Section 10.3.4.2.

DCI format 2 is an extension for DCI format 1 to support closed-loop spatial multiplexing (transmission mode 4). Thus, information about the number of transmission layers and the index of the

Table 10.5 DCI Formats Used for Downlink Scheduling

Field		DCI Format								
		1	1A	1B	1C	1D	2	2A	2B	2C
Resource information	Carrier indicator	•	•	•		•	•	•	•	•
	Resource block assignment type	0/1	2	2	2'	2	0/1	0/1	0/1	0/1
HARQ process number		•	•	•		•	•	•	•	•
First transport block	MCS	•	•	•	•	•	•	•	•	•
	RV	•	•	•		•	•	•	•	•
	NDI	•	•	•		•	•	•	•	•
Second transport block	MCS						•	•	•	•
	RV						•	•	•	•
	NDI						•	•	•	•
Multi-antenna information	PMI confirmation			•						
	Precoding information			•		•	•	•		
	Transport block swap flag						•	•		
	Power offset					•				
	DM-RS scrambling								•	
	# Layers/DM-RS scrambling									•
Downlink assignment index		•	•	•		•	•	•	•	•
PUCCH power control		•	•	•		•	•	•	•	•
Flag for 0/1A differentiation			•							
Padding (only if needed)		(•)	(•)	(•)		(•)	(•)	(•)	(•)	(•)
Identity		•	•	•	•	•	•	•	•	•

precoder matrix used are jointly encoded in the precoding information field. Some of the fields in DCI format 1 have been duplicated to handle the two transport blocks transmitted in parallel in the case of spatial multiplexing.

DCI format 2A is similar to DCI format 2 except that it supports open-loop spatial multiplexing (transmission mode 3) instead of closed-loop spatial multiplexing. The precoder information field is used to signal the number of transmission layers only, hence the field has a smaller size than in DCI format 2. Furthermore, since DCI format 2A is used for scheduling of multi-layer transmissions only, the precoder information field is only necessary in the case of four transmit antenna ports (a two-antenna-port setup cannot be used for more than two transmission layers).

DCI format 2B was introduced in release 9 in order to support dual-layer spatial multiplexing in combination with beam-forming using DM-RS (transmission mode 8). Since scheduling with DCI format 2B relies on DM-RS, precoding/beam-forming is transparent to the terminal and there is no need to signal a precoder index. The number of layers can be controlled by disabling one of the transport blocks. Two different scrambling sequences for the DM-RS can be used, as described in Section 10.2.2.

DCI format 2C was introduced in release 10 and is used to support spatial multiplexing using DM-RS (transmission mode 9). To some extent, it can be seen as a generalization of format 2B to support spatial multiplexing of up to eight layers. DM-RS scrambling and the number of layers are jointly signaled by a single three-bit field.

Many information fields in the different DCI formats are, as already mentioned, common among several of the formats, while some types of information exist only in certain formats. The information in the DCI formats used for downlink scheduling can be organized into different groups, as shown in Table 10.5, with the fields present varying between the DCI formats. The contents of the different DCI formats are explained in more detail below:

- Resource information, consisting of:
    - Carrier indicator (0 or 3 bit). This field is only present if cross-carrier scheduling is enabled via RRC signaling and is used to indicate the component carrier the downlink control information relates to (see Section 10.4.6).
    - Resource-block allocation. This field indicates the resource blocks on one component carrier upon which the terminal should receive the PDSCH. The size of the field depends on the cell bandwidth and on the DCI format, more specifically on the resource indication type, as discussed in Section 10.4.4.1. Resource allocation types 0 and 1, which are the same size, support non-contiguous resource-block allocations, while resource allocation type 2 has a smaller size but supports contiguous allocations only. DCI format 1C uses a restricted version of type 2 in order to further reduce control signaling overhead.
- Hybrid-ARQ process number (3 bit for FDD, 4 bit for TDD), informing the terminal about the hybrid-ARQ process to use for soft combining. Not present in DCI format 1C.
- For the first (or only) transport block:[26]
    - Modulation-and-coding scheme (5 bit), used to provide the terminal with information about the modulation scheme, the code rate, and the transport-block size, as described further below. DCI format 1C has a restricted size of this field as only QPSK is supported.

---

[26] A transport block can be disabled by setting the modulation-and-coding scheme to zero and the redundancy version to one in the DCI.

- New-data indicator (1 bit), used to clear the soft buffer for initial transmissions. Not present in DCI format 1C as this format does not support hybrid ARQ.
- Redundancy version (2 bit).
- For the second transport block (only present in DCI format supporting spatial multiplexing):
  - Modulation-and-coding scheme (5 bit).
  - New-data indicator (1 bit).
  - Redundancy version (2 bit).
- Multi-antenna information. The different DCI formats are intended for different multi-antenna schemes and which of the fields below that are included depends on the DCI format shown in Table 10.5.
  - PMI confirmation (1 bit), present in format 1B only. Indicates whether the eNodeB uses the (frequency-selective) precoding matrix recommendation from the terminal or if the recommendation is overridden by the information in the PMI field.
  - Precoding information, provides information about the index of the precoding matrix used for the downlink transmission and, indirectly, about the number of transmission layers.
  - Transport block swap flag (1 bit), indicating whether the two codewords should be swapped prior to being fed to the hybrid-ARQ processes. Used for averaging the channel quality between the codewords.
  - Power offset between the PDSCH and cell-specific reference signals used to support dynamic power sharing between multiple terminals for multi-user MIMO.
  - Reference-signal scrambling sequence, used to control the generation of quasi-orthogonal reference-signal sequences, as discussed in Section 10.2.2.
  - Number of layers and reference-signal scrambling sequence (jointly encoded information in release 10).
- Downlink assignment index (2 bit), informing the terminal about the number of downlink transmissions for which a single hybrid-ARQ acknowledgement should be generated according to Section 12.1.3. Present for TDD only.
- Transmit-power control for PUCCH (2 bit). For scheduling of a secondary component carrier in the case of carrier aggregation, these bits are reused as *acknowledgement resource indicator* (ARI) – see Chapter 11.
- DCI format 0/1A indication (1 bit), used to differentiate between DCI formats 1A and 0 as the two formats have the same message size. This field is present in DCI formats 0 and 1A only. DCI formats 3 and 3A, which have the same size, are separated from DCI formats 0 and 1A through the use of a different RNTI.
- Padding; the smaller of DCI formats 0 and 1A is padded to ensure the same payload size irrespective of the uplink and downlink cell bandwidths. Padding is also used to ensure that the DCI size is different for different DCI formats (this is rarely required in practice as the payload sizes are different due to the different amounts of information). Finally, padding is used to avoid certain DCI sizes that may cause ambiguous decoding.[27]

---

[27] For a small set of specific payload sizes, the control signaling may be correctly decoded at an aggregation level other than the one used by the transmitter. Since the PHICH resource is derived from the first CCE used for the PDCCH, this may result in incorrect PHICH being monitored by the terminal. To overcome this, padding is used if necessary to avoid the problematic payload sizes.

- Identity (RNTI) of the terminal for which the PDSCH transmission is intended (16 bit). As described in Section 10.4.8, the identity is not explicitly transmitted but implicitly included in the CRC calculation. There are different RNTIs defined depending on the type of transmission (unicast data transmission, paging, power-control commands, etc.).

10.4.4.1 *Signaling of Downlink Resource-Block Allocations*

Focusing on the signaling of resource-block allocations, there are three different possibilities, types 0, 1, and 2, as indicated in Table 10.5. Resource-block allocation types 0 and 1 both support non-contiguous allocations of resource blocks in the frequency domain, whereas type 2 supports contiguous allocations only. A natural question is why multiple ways of signaling the resource-block allocations are supported, a question whose answer lies in the number of bits required for the signaling. The most flexible way of indicating the set of resource blocks the terminal is supposed to receive the downlink transmission upon is to include a bitmap with size equal to the number of resource blocks in the cell bandwidth. This would allow for an arbitrary combination of resource blocks to be scheduled for transmission to the terminal but would, unfortunately, also result in a very large bitmap for the larger cell bandwidths. For example, in the case of a downlink cell bandwidth corresponding to 100 resource blocks, the downlink PDCCH would require 100 bits for the bitmap alone, to which the other pieces of information need to be added. Not only would this result in a large control-signaling overhead, but it could also result in downlink coverage problems as more than 100 bits in one OFDM symbol correspond to a data rate exceeding 1.4 Mbit/s. Consequently, there is a need for a resource allocation scheme requiring a smaller number of bits while keeping sufficient allocation flexibility.

In resource allocation type 0, the size of the bitmap has been reduced by pointing not to individual resource blocks in the frequency domain, but to groups of contiguous resource blocks, as shown at the top of Figure 10.25. The size of such a group is determined by the downlink cell bandwidth; for the smallest bandwidths there is only a single resource block in a group, implying that an arbitrary set of resource blocks can be scheduled, whereas for the largest cell bandwidths groups of four resource blocks are used (in the example in Figure 10.25, the cell bandwidth is 25 resource blocks, implying a group size of two resource blocks). Thus, the bitmap for the system with a downlink cell bandwidth of 100 resource blocks is reduced from 100 to 25 bits. A drawback is that the scheduling granularity is reduced; single resource blocks cannot be scheduled for the largest cell bandwidths using allocation type 0.

However, also in large cell bandwidths, frequency resolution of a single resource block is sometimes useful, for example to support small payloads. Resource allocation type 1 addresses this by dividing the total number of resource blocks in the frequency domain into dispersed subsets, as shown in the middle of Figure 10.25. The number of subsets is given from the cell bandwidth with the number of subsets in type 1 being equal to the group size in type 0. Thus, in Figure 10.25, there are two subsets, whereas for a cell bandwidth of 100 resource blocks there would have been four different subsets. Within a subset, a bitmap indicates the resource blocks in the frequency domain upon which the downlink transmission occurs.

To inform the terminal whether resource allocation type 0 or 1 is used, the resource allocation field includes a flag for this purpose, denoted "type" to the left in Figure 10.25. For type 0, the only additional information is the bitmap discussed above. For type 1, on the other hand, in addition to the bitmap

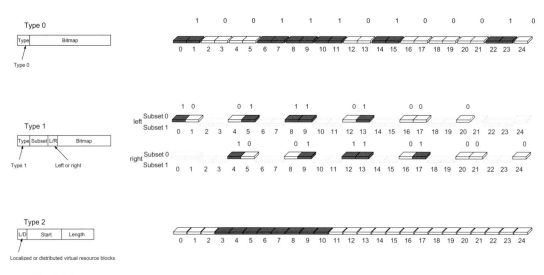

FIGURE 10.25

Illustration of resource-block allocation types (cell bandwidth corresponding to 25 resource blocks used in this example).

itself, information about the subset for which the bitmap relates is also required. As one of the requirements in the design of resource allocation type 1 was to maintain the same number of bits in the allocation as for type 0 without adding unnecessary overhead,[28] the bitmap in resource allocation type 1 is smaller than in type 0 to allow for the signaling of the subset number. However, a consequence of a smaller bitmap is that not all resource blocks in the subset can be addressed simultaneously. To be able to address all resources with the bitmap, there is a flag indicating whether the bitmap relates to the "left" or "right" part of the resource blocks, as depicted in the middle part of Figure 10.25.

Unlike the other two types of resource-block allocation signaling, type 2 does not rely on a bitmap. Instead, it encodes the resource allocation as a start position and length of the resource-block allocation. Thus, it does not support arbitrary allocations of resource blocks but only frequency-contiguous allocations, thereby reducing the number of bits required for signaling the resource-block allocation. The number of bits required for resource-signaling type 2 compared to type 0 or 1 is shown in Figure 10.26 and, as shown, the difference is fairly large for the larger cell bandwidths.

All three resource-allocation types refer to *virtual* resource blocks (see Section 10.1.1.8 for a discussion of resource-block types). For resource-allocation types 0 and 1, the virtual resource blocks are of localized type and the virtual resource blocks are directly mapped to physical resource blocks. For resource-allocation type 2, on the other hand, both localized and distributed virtual resource blocks are supported. One bit in the resource allocation field indicates whether the allocation signaling refers to localized or distributed resource blocks.

---

[28] Allowing different sizes would result in an increase in the number of blind decoding attempts required in the terminal.

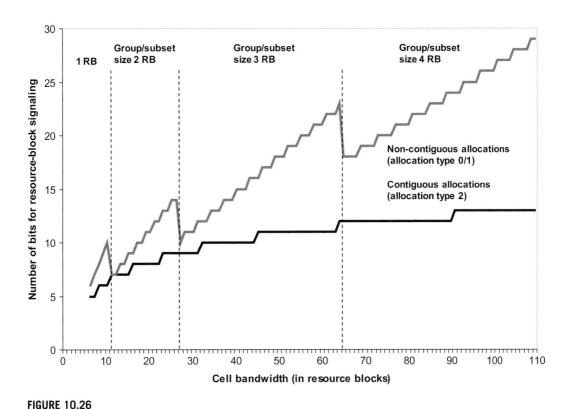

FIGURE 10.26

Number of bits used for downlink resource-allocation signaling for downlink allocation types 0/1 and 2.

10.4.4.2 *Signaling of Transport-Block Sizes*

Proper reception of a downlink transmission requires, in addition to the set of resource blocks, knowledge about the modulation scheme and the transport-block size, information (indirectly) provided by a five-bit field in the different DCI formats. Of the 32 combinations, 29 are used to signal the modulation-and-coding scheme whereas three are reserved, the purpose of which is described later. Together, the modulation-and-coding scheme and the number of resource blocks assigned provide the transport-block size on the DL-SCH. Thus, the possible transport-block sizes can be described as a table with 29 rows and 110 columns, one column for each number of resource blocks possible to transmit upon (the number of columns follows from the maximum downlink component carrier bandwidth of 110 resource blocks).

Each modulation-and-coding scheme represents a particular combination of modulation scheme and channel-coding rate or, equivalently, a certain spectral efficiency measured in the number of information bits per modulation symbol. Although the 29-by-110 table of transport-block sizes in principle could be filled directly from the modulation-and-coding scheme and the number of resource blocks, this would result in arbitrary transport-block sizes, which is not desirable. First, as all the higher-layer protocol layers are byte aligned, the resulting transport-block sizes should be an integer number of bytes. Secondly, common payloads, for example RRC signaling messages and VoIP, should be possible to transmit without padding. Aligning with the QPP interleaver sizes is also beneficial, as this would avoid the use of filler bits (see Section 10.1.1.3). Finally, the same transport-block size should ideally appear for several different resource-block allocations, as this allows the number of resource blocks to be changed between retransmission attempts, providing increased scheduling flexibility. Therefore, a "mother table" of transport-block sizes is first defined, fulfilling the requirements above. Each entry in the 29-by-110 table is picked from the mother table such that the resulting spectral efficiency is as close as possible to the spectral efficiency of the signaled modulation-and-coding scheme. The mother table spans the full range of transport-block sizes possible, with an approximately constant worst-case padding.

From a simplicity perspective, it is desirable if the transport-block sizes do not vary with the configuration of the system. The set of transport-block sizes is therefore independent of the actual number of antenna ports and the size of the control region.[29] The design of the table assumes a control region of three OFDM symbols and two antenna ports, the "reference configuration". If the actual configuration is different, the resulting code rate for the DL-SCH will be slightly different as a result of the rate-matching procedure. However, the difference is small and of no practical concern. Also, if the actual size of the control region is smaller than the three-symbol assumption in the reference configuration, the spectral efficiencies will be somewhat smaller than the range indicated by the modulation-and-coding scheme signaled as part of the DCI. Thus, information about the modulation scheme used is obtained directly from the modulation-and-coding scheme, whereas the exact code rate and rate matching is obtained from the implicitly signaled transport-block size together with the number of resource elements used for DL-SCH transmission.

For bandwidths smaller than the maximum of 110 resource blocks, a subset of the table is used. More specifically, in case of a cell bandwidth of N resource blocks, the first N columns of the table are used. Also, in the case of spatial multiplexing, a single transport block can be mapped to up to four layers. To support the higher data rates this facilitates, the set of supported transport-block sizes needs to be extended beyond what is possible in the absence of spatial multiplexing. The additional entries are in principle obtained by multiplying the sizes with the number of layers to which a transport block is mapped and adjusting the result to match the QPP interleaver size.

The 29 combinations of modulation-and-coding schemes each represent a reference spectral efficiency in the approximate range of 0.1–5.9 bits/s/symbol.[30] There is some overlap in the combinations in the sense that some of the 29 combinations represent the same spectral efficiency. The reason is that the best combination for realizing a specific spectral efficiency depends on the channel properties; sometimes higher-order modulation with a low code rate is preferable over lower-order modulation with a higher code rate, sometimes the opposite is true. With the overlap, the eNodeB can select the best combination, given the propagation scenario. As a consequence of the overlap, two of the rows in the 29-by-110 table are duplicates and result in the same spectral efficiency but with different modulation schemes, and there are only 27 unique rows of transport-block sizes.

Returning to the three reserved combinations in the modulation-and-coding field mentioned at the beginning, those entries can be used for retransmissions only. In the case of a retransmission, the

---

[29] For DwPTS, the transport-block size is scaled by a factor of 0.75 compared to the values found in the table, motivated by the DwPTS having a shorter duration than a normal subframe.

[30] The exact values vary slightly with the number of resource blocks allocated due to rounding.

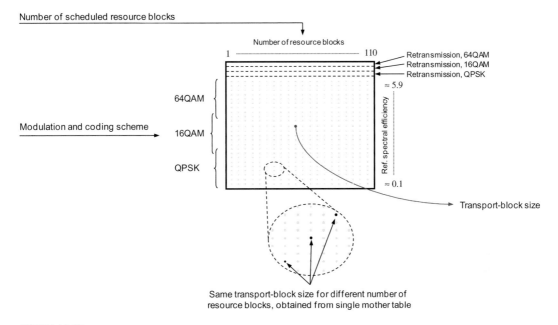

FIGURE 10.27

Computing the transport-block size.

transport-block size is, by definition, unchanged and fundamentally there is no need to signal this piece of information. Instead, the three reserved values represent the modulation scheme, QPSK, 16QAM or 64QAM, which allows the scheduler to use an (almost) arbitrary combination of resource blocks for the retransmission. Obviously, using any of the three reserved combinations assumes that the terminal properly received the control signaling for the initial transmission; if this is not the case, the retransmission should explicitly indicate the transport-block size.

The derivation of the transport-block size from the modulation-and-coding scheme and the number of scheduled resource blocks is illustrated in Figure 10.27.

10.4.5 Uplink Scheduling Grants

Uplink scheduling grants use one of DCI formats 0 or 4, where DCI format 4 was added in release 10 to support uplink spatial multiplexing. The basic resource-allocation scheme for the uplink is single-cluster allocations where the resource blocks are contiguous in the frequency domain, although release 10 added support for multi-cluster transmissions of up to two clusters on a single component carrier.

DCI format 0 is used for scheduling uplink transmissions not using spatial multiplexing on one component carrier. It has the same size control-signaling message as the "compact" downlink assignment (DCI format 1A). A flag in the message is used to inform the terminal whether the message is an uplink scheduling grant (DCI format 0) or a downlink scheduling assignment (DCI format 1A).

Table 10.6 DCI Formats for Uplink Scheduling Grants

Field		DCI Format	
		0	**4**
Resource information	Carrier indicator	•	•
	Multi-cluster flag	•	•
	Resource block assignment	0	1
First transport block	MCS/RV	•	•
	NDI	•	•
Second transport block	MCS/RV		•
	NDI		•
DMRS phase rotation		•	•
Precoding information			•
CSI request		•	•
SRS request			•
Uplink index/DAI		•	•
PUSCH power control		•	•
Flag for 0/1A differentiation		•	
Padding (only if needed)		(•)	(•)
Identity		•	•

DCI format 4 is used for uplink transmissions using spatial multiplexing on one component carrier. Consequently, the size of DCI format 4 is larger than that of DCI format 0, as additional information fields are required.

Many information fields are common to the two DCI formats, but there are also differences, as shown in Table 10.6. The contents of the different DCI formats are explained in more detail below:

- Resource information, consisting of:
  - Carrier indicator (0 or 3 bit). This field is only present if cross-carrier scheduling is enabled via RRC signaling and is used to indicate the uplink component carrier the grant relates to (see Section 10.4.6).
  - Multi-cluster flag (1 bit), indicating whether one or two clusters of resource blocks are used for the uplink transmission. This flag is not present in releases prior to release 10. In previous releases, the downlink bandwidth is, in practice, always at least as large as the uplink bandwidth, implying that one padding bit is used for DCI format 0 in those releases to align with the size of format 1A. The padding bit could therefore be replaced by the multi-cluster flag in release 10 without sacrificing backwards compatibility. In DCI format 4, supported in release 10 only, the multi-cluster flag is always present.
  - Resource-block allocation, including hopping indication. This field indicates the resource blocks upon which the terminal should transmit the PUSCH using uplink resource-allocation type 0 (DCI format 0) or type 1 (DCI format 4), as described in Section 10.4.5.1. The size of

the field depends on the cell bandwidth. For single-cluster allocations, uplink frequency hopping, as described in Chapter 11, can be applied to the uplink PUSCH transmission.

- For the first (or only) transport block:
  - Modulation-and-coding scheme including redundancy version (5 bit), used to provide the terminal with information about the modulation scheme, the code rate, and the transport-block size. The signaling of the transport-block size uses the same transport-block table as for the downlink – that is, the modulation-and-coding scheme together with the number of scheduled resource blocks provides the transport-block size. However, as the support of 64QAM in the uplink is not mandatory for all terminals, terminals not capable of 64QAM use 16QAM when 64QAM is indicated in the modulation-and-coding field. The use of the three reserved combinations is slightly different than for the downlink; the three reserved values are used for implicit signaling of the redundancy version, as described below. A transport block can be disabled by signaling a specific combination of modulation-and-coding scheme and number of resource blocks. Disabling one transport block is used when retransmitting a single transport block only.
  - New-data indicator (1 bit), used to indicate to the terminal whether transmission of a new transport block or retransmission of the previous transport block is granted.
- For the second transport block (DCI format 4 only):
  - Modulation-and-coding scheme including redundancy version (5 bit).
  - New-data indicator (1 bit).
- Phase rotation of the uplink demodulation reference signal[31] (3 bit), used to support multi-user MIMO, as described in Chapter 11. By assigning different reference-signal phase rotations to terminals scheduled on the same time–frequency resources, the eNodeB can estimate the uplink channel response from each terminal and suppress the inter-terminal interference by the appropriate processing.
- Precoding information, used to signal the precoder to use for the uplink transmission.
- Channel-state request flag (1 or 2 bit). The network can explicitly request an aperiodic channel-state report to be transmitted on the UL-SCH by setting this bit(s) in the uplink grant. In the case of carrier aggregation, two bits are used to indicate which downlink component carrier the CSI should be reported for (see Chapter 12).
- SRS request (2 bit), used to trigger aperiodic sounding using one of up to three preconfigured settings, as discussed in Chapter 11.
- Uplink index/DAI (2 bit). This field is present only when operating in TDD. For uplink–downlink configuration 0 (uplink-heavy configuration), it is used as an uplink index to signal for which uplink subframe(s) the grant is valid, as described in Chapter 13. For other uplink–downlink configurations, it is used as downlink assignment index to indicate the number of downlink transmissions the eNodeB expects hybrid-ARQ acknowledgement for.
- Transmit-power control for PUSCH (2 bit).
- DCI format 0/1A indication (1 bit), used to differentiate between DCI formats 1A and 0 as the two formats have the same message size. This field is present in DCI formats 0 and 1A only.

---

[31] In release 10, this field can also be used to select the orthogonal cover sequence used for uplink reference signals (see Chapter 11 for details).

- Padding; the smaller of DCI formats 0 and 1A is padded to ensure the same payload size irrespective of the uplink and downlink cell bandwidths. Padding is also used to ensure that the DCI size is different for different DCI formats (this is rarely required in practice as the payload sizes are different due to the different amounts of information). Finally, padding is used to avoid certain DCI sizes that may cause ambiguous decoding.
- Identity (RNTI) of the terminal for which the grant is intended (16 bit). As described in Section 10.4.8, the identity is not explicitly transmitted but implicitly included in the CRC calculation.

There is no explicit signaling of the redundancy version in the uplink scheduling grants. This is motivated by the use of a synchronous hybrid-ARQ protocol in the uplink; retransmissions are normally triggered by a negative acknowledgement on the PHICH and not explicitly scheduled as for downlink data transmissions. However, as described in Chapter 12, there is a possibility to explicitly schedule retransmissions. This is useful in a situation where the network will explicitly move the retransmission in the frequency domain by using the PDCCH instead of the PHICH. Three values of the modulation-and-coding field are reserved to mean redundancy versions one, two, and three. If one of those values is signaled, the terminal should assume that the same modulation and coding as the original transmission is used. The remaining entries are used to signal the modulation-and-coding scheme to use and also imply that redundancy version zero should be used. The difference in usage of the reserved values compared to the downlink scheduling assignments implies that the modulation scheme, unlike the downlink case, cannot change between uplink (re)transmission attempts.

10.4.5.1 *Signaling of Uplink Resource-Block Allocations*

The basic uplink resource-allocation scheme is single-cluster allocations – that is, allocations contiguous in the frequency domain – but release 10 also provides the possibility for multi-cluster uplink transmissions.

Single-cluster allocations use uplink resource-allocation type 0, which is identical to downlink resource allocation type 2 described in Section 10.4.4.1 except that the single-bit flag indicating localized/distributed transmission is replaced by a single-bit hopping flag. The resource allocation field in the DCI provides the set of virtual resource blocks to use for uplink transmission. The set of physical resource blocks to use in the two slots of a subframe is controlled by the hopping flag, as described in Chapter 11.

Multi-cluster allocations with up to two clusters were introduced in release 10, using uplink resource-allocation type 1. In resource-allocation type 1, the starting and ending positions of two clusters of frequency-contiguous resource blocks are encoded into an index. Uplink resource-allocation type 1 does not support frequency hopping (diversity is achieved through the use of two clusters instead). Indicating two clusters of resources naturally requires additional bits compared to the single-cluster case. At the same time, the total number of bits used for resource-allocation type 1 should be identical to that of type 0 (this is similar to the situation for the downlink allocation types 0 and 1; without aligning the sizes a new DCI format with a corresponding negative impact on the number of blind decodings is necessary). Since frequency hopping is not supported for allocation type 1, the bit otherwise used for the hopping flag can be reused for extending the resource-allocation field. However, despite the extension of the resource-allocation field by one bit, the number of bits is not sufficient to provide a single-resource-block resolution in the two clusters for all bandwidths. Instead, similar to downlink resource allocation type 0, groups of resource blocks are used and the starting and ending positions of the two clusters are

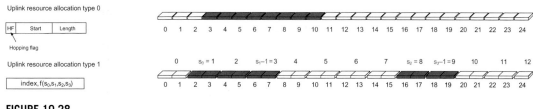

FIGURE 10.28

Illustration of uplink resource-block allocation types (uplink bandwidth corresponding to 25 resource blocks used in this example).

given in terms of group numbers. The size of such a group is determined by the uplink carrier bandwidth in a similar way as for the downlink; for the smallest bandwidths there is only a single resource block in a group, implying that an arbitrary (as long as the limit of at most two clusters is observed) set of resource blocks can be scheduled, whereas for the largest cell bandwidths groups of four resource blocks are used (in the example in Figure 10.28, the cell bandwidth is 25 resource blocks, implying a group size of two resource blocks).

10.4.6 Carrier Aggregation and Cross-Carrier Scheduling

Carrier aggregation, where a terminal receives or transmits on multiple component carriers, is an integral part of LTE from release 10 onwards. Clearly, for a terminal supporting multiple component carriers, the terminal needs to know to which component carrier a certain DCI relates. This information can either be implicit or explicit, depending on whether cross-carrier scheduling is used or not, as illustrated in Figure 10.29. Enabling cross-carrier scheduling is done individually via RRC signaling on a per-terminal and per-component-carrier basis.

In the absence of cross-carrier scheduling, downlink scheduling assignments are valid for the component carrier upon which they are transmitted. Similarly, for uplink grants, there is an association between downlink and uplink component carriers such that each uplink component carrier has an associated downlink component carrier. The association is provided as part of the system information. Thus, from the uplink–downlink association, the terminal will know to which uplink component carrier the downlink control information relates to.

In the presence of cross-carrier scheduling, where downlink PDSCH or uplink PUSCH is transmitted on an (associated) component carrier other than the PDCCH, the carrier indicator in the PDCCH provides information about the component carrier used for the PDSCH or PUSCH.

Whether cross-carrier scheduling is used or not is, as already mentioned, configured using higher-layer signaling; if cross-carrier scheduling is not configured then no carrier indication field is included in the DCI. Thus, most of the DCI formats come in two "flavors", with and without the carrier indication field, and which "flavor" the terminal is supposed to monitor is determined by enabling/disabling support for cross-carrier scheduling.

To signal which component carrier a grant relates to, the component carriers are numbered. The primary component carrier is always given the number zero, while the different secondary component carriers are assigned a unique number each through UE-specific RRC signaling. Hence, even if the

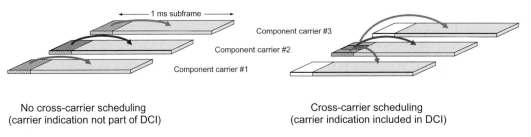

No cross-carrier scheduling
(carrier indication not part of DCI)

Cross-carrier scheduling
(carrier indication included in DCI)

FIGURE 10.29

Cross-carrier scheduling.

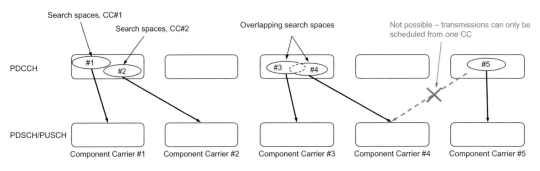

FIGURE 10.30

Example of scheduling of multiple component carriers (search spaces are discussed in Section 10.4.9).

terminal and the eNodeB may have different understandings of the component carrier numbering during a brief period of reconfiguration, at least transmissions on the primary component carrier can be scheduled.

Irrespective of whether cross-carrier scheduling is used or not, PDSCH/PUSCH on a component carrier can only be scheduled from *one* component. Thus, for each PDSCH/PUSCH component carrier there is an associated component carrier, configured via UE-specific RRC signaling, where the corresponding DCI can be transmitted. Figure 10.30 illustrates one example hereof, where PDSCH/PUSCH transmissions on component carrier 1 are scheduled using PDCCHs transmitted on component carrier 1. In this case, as cross-carrier scheduling is not used, there is no carrier indicator in the corresponding DCI formats. PDSCH/PUSCH transmissions on component carrier 2 are cross-carrier scheduled from PDCCHs transmitted on component carrier 1. Hence, the DCI formats in the UE-specific search space for component carrier 2 include the carrier indicator.

Note also, as transmissions on a component carrier can be scheduled by PDCCHs on *one* component carrier only, component carrier 4 cannot be scheduled by PDCCHs on component carrier 5 as the semi-static association between the component carriers used for PDCCH transmission and the actual data transmission has associated data on component carrier 4, with PDCCHs on component carrier 3 in this example.

10.4.7 **Power-Control Commands**

As a complement to the power-control commands provided as part of the downlink scheduling assignments and the uplink scheduling grants, there is the potential to transmit a power-control command using DCI formats 3 (single-bit command per terminal) or 3A (two-bit command per terminal). The main motivation for DCI format 3/3A is to support power control for semi-persistent scheduling. The power-control message is directed to a group of terminals using an RNTI specific for that group. Each terminal can be allocated two power-control RNTIs, one for PUCCH power control and the other for PUSCH power control. Although the power-control RNTIs are common to a group of terminals, each terminal is informed through RRC signaling which bit(s) in the DCI message it should follow. No carrier indicator is used for formats 3 and 3A.

10.4.8 **PDCCH Processing**

Having discussed the contents of L1/L2 control signaling in terms of the different DCI formats, the transmission of the DCI message on a PDCCH can be described. The processing of downlink control signaling is illustrated in Figure 10.31. A CRC is attached to each DCI message payload. The identity of the terminal (or terminals) addressed – that is, the RNTI – is included in the CRC calculation and not explicitly transmitted. Depending on the purpose of the DCI message (unicast data transmission, power-control command, random-access response, etc.), different RNTIs are used; for normal unicast data transmission, the terminal-specific C-RNTI is used.

Upon reception of DCI, the terminal will check the CRC using its set of assigned RNTIs. If the CRC checks, the message is declared to be correctly received and intended for the terminal. Thus, the identity of the terminal that is supposed to receive the DCI message is implicitly encoded in the CRC and not explicitly transmitted. This reduces the amount of bits necessary to transmit on the PDCCH as, from a terminal point of view, there is no difference between a corrupt message whose CRC will not check and a message intended for another terminal.

After CRC attachment, the bits are coded with a rate-1/3 tail-biting convolutional code and rate-matched to fit the amount of resources used for PDCCH transmission. Tail-biting convolutional coding is similar to conventional convolutional coding with the exception that no tail bits are used. Instead, the convolutional encoder is initialized with the last bits of the message prior to the encoding process. Thus, the starting and ending states in the trellis in an MLSE (Viterbi) decoder are identical.

After the PDCCHs to be transmitted in a given subframe have been allocated to the desired resource elements (the details of which are given below), the sequence of bits corresponding to all the PDCCH resource elements to be transmitted in the subframe, including the unused resource elements, is scrambled by a cell- and subframe-specific scrambling sequence to randomize inter-cell interference, followed by QPSK modulation and mapping to resource elements.

To allow for simple yet efficient processing of the control channels in the terminal, the mapping of PDCCHs to resource elements is subject to a certain structure. This structure is based on so-called *Control-Channel Elements* (CCEs), which in essence is a convenient name for a set of 36 useful resource elements (nine resource-element groups as defined in Section 10.4.1). The number of CCEs, one, two, four, or eight, required for a certain PDCCH depends on the payload size of the control information (DCI payload) and the channel-coding rate. This is used to realize link adaptation for the PDCCH; if the channel conditions for the terminal to which the PDCCH is intended are

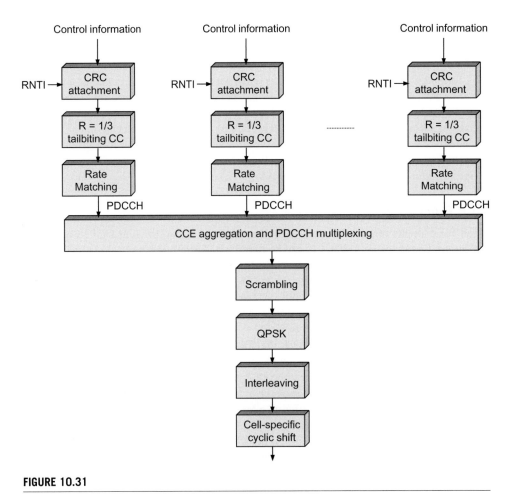

FIGURE 10.31

Processing of L1/L2 control signaling.

disadvantageous, a larger number of CCEs needs to be used compared to the case of advantageous channel conditions. The number of CCEs used for a PDCCH is also referred to as the aggregation level.

The number of CCEs available for PDCCHs depends on the size of the control region, the cell bandwidth, the number of downlink antenna ports, and the amount of resources occupied by PHICH. The size of the control region can vary dynamically from subframe to subframe as indicated by the PCFICH, whereas the other quantities are semi-statically configured. The CCEs available for PDCCH transmission can be numbered from zero and upward, as illustrated in Figure 10.32. A specific PDCCH can thus be identified by the numbers of the corresponding CCEs in the control region.

As the number of CCEs for each of the PDCCHs may vary and is not signaled, the terminal has to blindly determine the number of CCEs used for the PDCCH it is addressed upon. To reduce the complexity of this process somewhat, certain restrictions on the aggregation of contiguous CCEs have been specified. For example, an aggregation of eight CCEs can only start on CCE numbers evenly

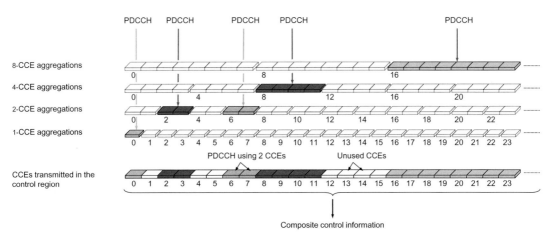

FIGURE 10.32

CCE aggregation and PDCCH multiplexing.

divisible by 8, as illustrated in Figure 10.32. The same principle is applied to the other aggregation levels. Furthermore, some combinations of DCI formats and CCE aggregations that result in excessively high channel-coding rates are removed.

The sequence of CCEs should match the amount of resources available for PDCCH transmission in a given subframe – that is, the number of CCEs varies according to the value transmitted on the PCFICH. In many cases, not all the PDCCHs that can be transmitted in the control region are used. Nevertheless, unused PDCCHs are part of the interleaving and mapping process in the same way as any other PDCCH. At the terminal, the CRC will not check for those "dummy" PDCCHs. Preferably, the transmission power is set to zero for those unused PDCCHs; the power can be used by other control channels.

The mapping of the modulated composite control information is, for the same reason as for the other control channels, described in terms of symbol quadruplets being mapped to resource-element groups. Thus, the first step of the mapping stage is to group the QPSK symbols into symbol quadruplets, each consisting of four consecutive QPSK symbols. In principle, the sequence of quadruplets could be mapped directly to the resource elements in sequential order. However, this would not exploit all the frequency diversity available in the channel and diversity is important for good performance. Furthermore, if the same CCE-to-resource-element mapping is used in all neighboring cells, a given PDCCH will persistently collide with one and the same PDCCH in the neighboring cells assuming a fixed PDCCH format and inter-cell synchronization. In practice, the number of CCEs per PDCCH varies in the cell as a function of the scheduling decisions, which gives some randomization to the interference, but further randomization is desirable to obtain a robust control-channel design. Therefore, the sequence of quadruplets is first interleaved using a block interleaver to allow exploitation of the frequency diversity, followed by a cell-specific cyclic shift to randomize the interference between neighboring cells. The output from the cell-specific shift is mapped to resource-element groups in a time-first manner, as illustrated in Figure 10.22, skipping resource-element groups used for PCFICH and PHICH. Time-first mapping

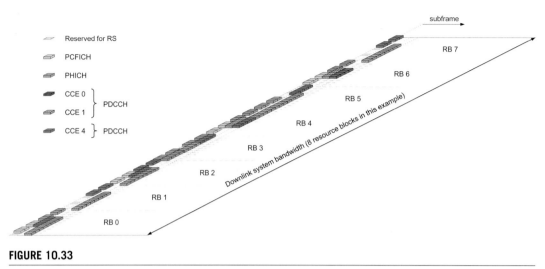

FIGURE 10.33

Example of mapping of PCFICH, PHICH, and PDCCH.

preserves the interleaving properties; with frequency-first over multiple OFDM symbols, resource-element groups that are spread far apart after the interleaving process may end up close in frequency, although on different OFDM symbols.

The interleaving operation described above, in addition to enabling exploitation of the frequency diversity and randomizing the inter-cell interference, also serves the purpose of ensuring that each CCE spans virtually all the OFDM symbols in the control region. This is beneficial for coverage as it allows flexible power balancing between the PDCCHs to ensure good performance for each of the terminals addressed. In principle, the energy available in the OFDM symbols in the control region can be balanced arbitrarily between the PDCCHs. The alternative of restricting each PDCCH to a single OFDM symbol would imply that power cannot be shared between PDCCHs in different OFDM symbols.

Similarly to the PCFICH, the transmission power of each PDCCH is under the control of the eNodeB. Power adjustments can therefore be used as a complementary link adaptation mechanism in addition to adjusting the code rate. Relying on power adjustments alone might seem a tempting solution but, although possible in principle, it can result in relatively large power differences between resource elements. This may have implications on the RF implementation and may violate the out-of-band emission masks specified. Hence, to keep the power differences between the resource elements reasonable, link adaptation through adjusting the channel code rate, or equivalently the number of CCEs aggregated for a PDCCH, is necessary. The two mechanisms for link adaptation, power adjustments and different code rates, complement each other.

To summarize and to illustrate the mapping of PDCCHs to resource elements in the control region, consider the example shown in Figure 10.33. In this example, the size of the control region in the subframe considered equals three OFDM symbols. Two downlink antenna ports are configured (but, as explained above, the mapping would be identical in the case of a single antenna port). One PHICH group is configured and three resource-element groups are therefore used by the PHICHs. The cell identity is assumed to be identical to zero in this case.

The mapping can then be understood as follows: First, the PCFICH is mapped to four resource-element groups, followed by allocating the resource-element groups required for the PHICH. The resource-element groups left after the PCFICH and PHICH are used for the different PDCCHs in the system. In this particular example, one PDCCH is using CCE numbers 0 and 1, while another PDCCH is using CCE number 4. Consequently, there is a relatively large amount of unused resource-element groups in this example; either they can be used for additional PDCCHs or the power otherwise used for the unused CCEs could be allocated to the PDCCHs in use (as long as the power difference between resource elements is kept within the limits set by the RF requirements). Furthermore, depending on the inter-cell interference situation, fractional loading of the control region may be desirable, implying that some CCEs are left unused to reduce the average inter-cell interference.

10.4.9 Blind Decoding of PDCCHs

As described above, each PDCCH supports multiple formats and the format used is a priori unknown to the terminal. Therefore, the terminal needs to blindly detect the format of the PDCCHs. The CCE structure described in the previous section helps in reducing the number of blind decoding attempts, but is not sufficient. Hence, it is required to have mechanisms to limit the CCE aggregations that the terminal is supposed to monitor. Clearly, from a scheduling point of view, restrictions in the allowed CCE aggregations are undesirable as they may influence the scheduling flexibility and require additional processing at the transmitter side. At the same time, requiring the terminal to monitor all possible CCE aggregations, also for the larger cell bandwidths, is not attractive from a terminal-complexity point of view. To impose as few restrictions as possible on the scheduler while at the same time limit the maximum number of blind decoding attempts in the terminal,[32] LTE defines so-called *search spaces*, which describe the set of CCEs the terminal is supposed to monitor for scheduling assignments/grants relating to a certain component carrier. Search spaces are discussed below, starting with the case of a single component carrier and later extended to the case of multiple component carriers.

A search space is a set of candidate control channels formed by CCEs on a given aggregation level, which the terminal is supposed to attempt to decode. As there are multiple aggregation levels, corresponding to one, two, four, and eight CCEs, a terminal has multiple search spaces. In each subframe, the terminals will attempt to decode all the PDCCHs that can be formed from the CCEs in each of its search spaces. If the CRC checks, the content of the control channel is declared as valid for this terminal and the terminal processes the information (scheduling assignment, scheduling grants, etc.). Clearly, the network can only address a terminal if the control information is transmitted on a PDCCH formed by the CCEs in one of the terminal's search spaces. For example, terminal A in Figure 10.34 cannot be addressed on a PDCCH starting at CCE number 20, whereas terminal B can. Furthermore, if terminal A is using CCEs 16–23, terminal B cannot be addressed on aggregation level 4 as all CCEs in its level-4 search space are blocked by the use for the other terminals. From this it can be intuitively understood that for efficient utilization of the CCEs in the system, the search spaces should differ between terminals. Each terminal in the system therefore has a *terminal-specific* search space at each aggregation level.

---

[32] In release 8/9, the number of blind decoding attempts is 44 per subframe, while for release 10 with uplink spatial multiplexing the number is 60 (assuming a single component carrier).

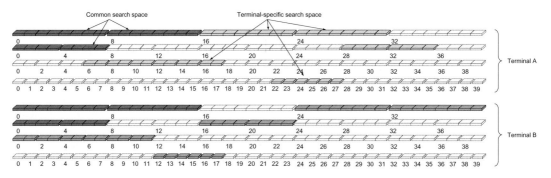

FIGURE 10.34

Principal illustration of search spaces in two terminals.

As the terminal-specific search space is typically smaller than the number of PDCCHs the network could transmit at the corresponding aggregation level, there must be a mechanism determining the set of CCEs in the terminal-specific search space for each aggregation level. One possibility would be to let the network configure the terminal-specific search space in each terminal. However, this would require explicit signaling to each of the terminals and possibly reconfiguration at handover. Instead, the terminal-specific search spaces in LTE are defined without explicit signaling through a function of the terminal identity and implicitly the subframe number. Dependence on the subframe number results in the terminal-specific search spaces being time varying, which helps resolve blocking between terminals. If a given terminal cannot be scheduled in a subframe as all the CCEs that the terminal is monitoring have already been used for scheduling other terminals in the same subframe, the time-varying definition of the terminal-specific search spaces is likely to resolve the blocking in the next subframe.

In several situations, there is a need to address a group of, or all, terminals in the system. One example hereof is dynamic scheduling of system information, another is transmission of paging messages, both described in Chapter 14. Transmission of explicit power-control commands to a group of terminals is a third example. To allow all terminals to be addressed at the same time, LTE has defined *common search spaces* in addition to the terminal-specific search spaces. A common search space is, as the name implies, common, and all terminals in the cell monitor the CCEs in the common search spaces for control information. Although the motivation for the common search space is primarily transmission of various system messages, it can be used to schedule individual terminals as well. Thus, it can be used to resolve situations where scheduling of one terminal is blocked due to lack of available resources in the terminal-specific search space. Unlike unicast transmissions, where the transmission parameters of the control signaling can be tuned to match the channel conditions of a specific terminal, system messages typically need to reach the cell border. Furthermore, the data rate of the associated DL-SCH transmission is typically modest. Consequently, the common search spaces are only defined for aggregation levels of four and eight CCEs and only for the smallest DCI formats, 0/1A/3/3A and 1C. There is no support for DCI formats with spatial multiplexing in the common search space. This helps to reduce the number of blind decoding attempts in the terminal used for monitoring the common search space.

Table 10.7 Downlink DCI Formats Monitored in Different Search Spaces for C-RNTI

Mode	Monitored DCI Formats in Search Space			Description
	Common	UE-Specific	UE-Specific	
1	1A	1A	1	Single antenna transmission
2	1A	1A	1	Transmit diversity
3	1A	1A	2A	Open-loop spatial multiplexing
4	1A	1A	2	Closed-loop spatial multiplexing
5	1A	1A	1D	Multi-user MIMO
6	1A	1A	1B	Single layer codebook-based precoding
7	1A	1A	1	Single-layer transmission using DM-RS
8	1A	1A	2B	Dual-layer transmission using DM-RS
9	1A	1A	2C	Multi-layer transmission using DM-RS

Figure 10.34 illustrates the terminal-specific and common search spaces in two terminals in a certain subframe. The terminal-specific search spaces are different in the two terminals and will, as described above, vary from subframe to subframe. Furthermore, the terminal-specific search spaces partially overlap between the two terminals in this subframe (CCEs 24–31 on aggregation level 8) but, as the terminal-specific search space varies between subframes, the overlap in the next subframe is most likely different.

The downlink DCI formats to decode in the terminal-specific search space depend on the *transmission mode* configured for the terminal. Transmission modes are described in Section 10.3 and, in principle, correspond to different multi-antenna configurations. As an example, there is no need to attempt to decode DCI format 2 when the terminal has not been configured for spatial multiplexing, which helps to reduce the number of blind decoding attempts. Configuration of the transmission mode is done via RRC signaling. As the exact subframe number when this configuration takes effect in the terminal is not specified and may vary depending on, for example, RLC retransmissions, there is a (short) period when the network and the terminal may have different understandings of which transmission mode is configured. Therefore, in order not to lose the possibility of communicating with the terminal, it is necessary to have at least one DCI format that is decoded irrespective of the transmission mode. For downlink transmissions, DCI format 1A serves this purpose and the network can therefore always transmit data to the terminal using this DCI format. Another function of format 1A is to reduce overhead for transmissions when full flexibility in resource-block assignment is not needed.

In Table 10.7, the DCI monitoring is described for the case of a single component carrier. For multiple component carriers, the above procedures are applied in principle to each of the activated[33] downlink component carriers. Hence, in principle there is one UE-specific search space per aggregation level and per (activated) component carrier upon which PDSCH can be received (or PUSCH transmitted), although there are some carrier-aggregation-specific modifications. Clearly, for terminals configured to use carrier aggregation, this results in an increase in the number of blind decoding

---

[33] Individual component carriers can be activated/deactivated as discussed in Chapter 12.

attempts compared to a terminal not using carrier aggregation, as scheduling assignments/grants for each of the component carriers need to be monitored.

The common search space is only defined for transmissions on the primary component carrier. As the main function of the common search space is to handle scheduling of system information intended for multiple terminals, and such information must be receivable by all terminals in the cell, scheduling in this case uses the common search space. For this reason, the carrier indication field is never present in DCI formats monitored in the common search space.

As mentioned before, there is one UE-specific search space per aggregation level and component carrier[34] used for the PDSCH/PUSCH. This is illustrated in Figure 10.30, where PDSCH/PUSCH transmissions on component carrier 1 are scheduled using PDCCHs transmitted on component carrier 1. No carrier indicator is assumed in the UE-specific search space for component carrier 1 as cross-carrier scheduling is not used. For component carrier 2, on the other hand, a carrier indicator is assumed in the UE-specific search space as component carrier 2 is cross-carrier scheduled from PDCCHs transmitted on component carrier 1.

Search spaces for different component carriers may overlap in some subframes. In Figure 10.30, this happens for the UE-specific search spaces for component carriers 3 and 4. The terminal will handle the two search spaces independently, assuming (in this example) a carrier indicator for component carrier 4 but not for component carrier 3. If the UE-specific and common search spaces relating to different component carriers happen to overlap for some aggregation level when cross-carrier scheduling is configured, the terminal only needs to monitor the common search space. The reason for this is to avoid ambiguities; if the component carriers have different bandwidths a DCI format in the common search space may have the same payload size as another DCI format in the UE-specific search space relating to another component carrier.

---

[34] If the DCI formats for two different component carriers have the same size, then search spaces can be shared such that, in the example in Figure 10.30, component carrier 2 can be scheduled from search spaces 2 *and* 1. This comes "for free" as the DCI formats in search space 1 need to be monitored anyway and the carrier-indicator field points to the component carrier for which the grant/assignment is intended.

Uplink Physical-Layer Processing

This chapter provides a description of the basic physical-layer functionality of the LTE uplink. It essentially follows the same outline as the corresponding downlink description provided in the previous chapter with detailed descriptions regarding transport-channel processing (Section 11.1), the reference-signal structure (Section 11.2), multi-antenna transmission (Section 11.3), and the uplink L1/L2 control-channel structure (Section 11.4). Physical aspects directly related to some specific uplink functions and procedures such as random access will be provided in later chapters. The chapter ends with a discussion on the uplink timing-alignment procedure in Section 11.5.

11.1 TRANSPORT-CHANNEL PROCESSING

This section describes the physical-layer processing applied to the uplink shared channel (UL-SCH) and the subsequent mapping to the uplink physical resource in the form of the basic OFDM time–frequency grid. The UL-SCH is the only uplink transport-channel type in LTE[1] and is used for the transmission of all uplink higher-layer information (both data and control).

11.1.1 Processing Steps

Figure 11.1 outlines the different steps of the UL-SCH physical-layer processing. Similar to the downlink, in the case of uplink carrier aggregation the different component carriers correspond to separate transport channels with separate physical-layer processing. Also, most of the uplink processing steps outlined in Figure 11.1 are identical to the corresponding steps for the downlink transport-channel processing outlined in Section 10.1.

The different steps of the uplink transport-channel processing are summarized below. For a more detailed overview of the different steps, the reader is referred to the corresponding downlink description in Section 10.1.

- *CRC insertion per transport block.* A 24-bit CRC is calculated for and appended to each uplink transport block.
- *Code-block segmentation and per-code-block CRC insertion.* In the same way as for the downlink, code-block segmentation, including per-code-block CRC insertion, is applied for transport blocks larger than 6144 bits.

---

[1] Strictly speaking, the LTE Random-Access Channel is also defined as a transport-channel type, see Chapter 8. However, RACH only includes a layer-1 preamble and carries no actual transport-channel data.

4G LTE/LTE-Advanced for Mobile Broadband.

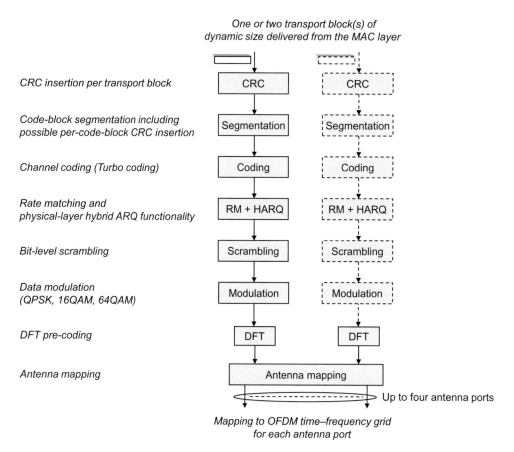

CRC insertion per transport block

Code-block segmentation including
possible per-code-block CRC insertion

Channel coding (Turbo coding)

Rate matching and
physical-layer hybrid ARQ functionality

Bit-level scrambling

Data modulation
(QPSK, 16QAM, 64QAM)

DFT pre-coding

Antenna mapping

FIGURE 11.1

Physical-layer processing for UL-SCH.

- *Channel coding.* Rate-1/3 Turbo coding with QPP-based inner interleaving is also used for the uplink.
- *Rate matching and physical-layer hybrid-ARQ functionality.* The physical-layer part of the uplink rate-matching and hybrid-ARQ functionality is essentially the same as the corresponding downlink functionality, with sub-block interleaving and insertion into a circular buffer followed by bit selection with four redundancy versions. Note that there are some important differences between the downlink and uplink hybrid-ARQ protocols, such as asynchronous versus synchronous operation as described in Chapter 12. However, these differences are not visible in terms of the physical-layer processing.
- *Bit-level scrambling.* The aim of uplink scrambling is the same as for the downlink – that is, to randomize the interference and thus ensure that the processing gain provided by the channel code can be fully utilized.

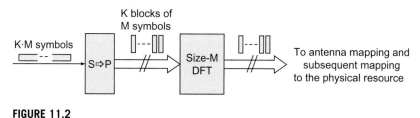

FIGURE 11.2

DFT precoding *K* blocks of *M* modulation symbols.

- *Data modulation.* Similar to the downlink, QPSK, 16QAM, and 64QAM modulation can also be used for uplink transport-channel transmission.
- *DFT precoding.* As illustrated in Figure 11.2, the modulation symbols, in blocks of M symbols, are fed through a size-M DFT, where M corresponds to the number of subcarriers assigned for the transmission. The reason for the precoding is to reduce the cubic metric for the transmitted signal. From an implementation-complexity point of view, the DFT size should preferably be constrained to a power of 2. However, such a constraint would limit the scheduler flexibility in terms of the amount of resources that can be assigned for an uplink transmission and, from a flexibility point of view, all possible DFT sizes should preferably be allowed. For LTE, a middle way has been adopted where the DFT size, and thus also the size of the resource allocation, is limited to products of the integers 2, 3, and 5. Thus, for example, DFT sizes of 60, 72, and 96 are allowed but a DFT size of 84 is not allowed.[2] In this way, the DFT can be implemented as a combination of relatively low-complex radix-2, radix-3, and radix-5 FFT processing.
- *Antenna mapping. The antenna mapping maps the output of the DFT precoder to antenna ports* for subsequent mapping to the physical resource (the OFDM time–frequency grid). In the first releases of LTE (release 8 and 9), only single-antenna transmission was supported for the uplink.[3] However, as part of LTE release 10, multi-antenna transmission, more specifically antenna precoding with up to four antennas ports, is supported in the uplink. More details about LTE uplink multi-antenna transmission are provided in Section 11.3.

11.1.2 Mapping to the Physical Resource

The scheduler assigns a set of resource-block pairs to be used for the uplink transmission, more specifically for transmission of the PUSCH physical channel that carries the UL-SCH transport channel. Each such resource-block pair spans 14 OFDM symbols (one subframe) in time.[4] However, as will be described in Section 11.2.1, two of these symbols are used for *uplink demodulation reference signals* and are thus not available for PUSCH transmission. Furthermore, one additional symbol may be

---

[2] As uplink resource assignments are always done in terms of resource blocks of size 12 subcarriers, the DFT size is always a multiple of 12.

[3] Actually, uplink *antenna selection* has been part of the LTE specification since release 8. However, it is an optional terminal feature that has, as of today, not been implemented in any commercial terminals.

[4] Twelve symbols in the case of extended cyclic prefix.

reserved for the transmission of *sounding reference signals* (Section 11.2.2). Thus, 11 or 12 OFDM symbols are available for PUSCH transmission within each resource-block pair.

Figure 11.3 illustrates how $K \cdot M$ DFT-precoded symbols at the output of the antenna mapping are mapped to the basic OFDM time–frequency grid, where K is the number of available OFDM symbols within a subframe (11 or 12 according to above) and M is the assigned bandwidth in number of sub-carriers. As there are 12 subcarriers within a resource-block pair, $M = N \cdot 12$, where N is the number of assigned resource-block pairs.

Figure 11.3 assumes that the set of assigned resource-block pairs are contiguous in the frequency domain. This is the typical assumption for DFTS-OFDM and was strictly the case for LTE release 8 and 9. Mapping of the DFT-precoded signal to frequency-contiguous resource elements is preferred in order to retain good cubic-metric properties of the uplink transmission. At the same time, such a restriction implies an additional constraint for the uplink scheduler, something which may not always be desirable. Therefore, LTE release 10 introduced the possibility to also assign partly frequency-separated resources for PUSCH transmission. More specifically, in release 10 the assigned uplink resource may consist of a maximum of two frequency-separated *clusters*, as illustrated in Figure 11.4, where each cluster consists of a number of resource-block pairs ($N1$ and $N2$ resource-block pairs respectively). In the case of such *multi-cluster transmission*, a single DFT precoding spans the overall assigned resource in the frequency domain – that is, both clusters. This means that the total assigned bandwidth in number of subcarriers ($M = M1 + M2$) should be aligned with the restrictions on available DFT sizes described above.

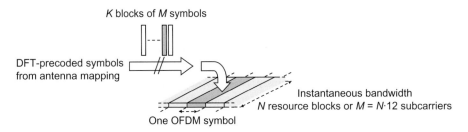

FIGURE 11.3

Mapping to the uplink physical resource.

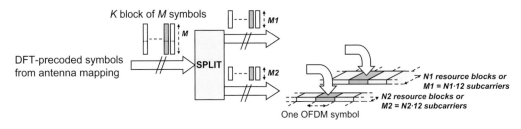

FIGURE 11.4

Uplink multi-cluster transmission.

11.1.3 PUSCH Frequency Hopping

In Chapter 10 it was described how the notion of *virtual resource blocks* (VRBs) in combination with the mapping from VRBs to *physical resource blocks* (PRBs) allowed for downlink distributed transmission – that is, the spreading of a downlink transmission in the frequency domain. As described, downlink distributed transmission consists of two separate steps: (1) a mapping from VRB pairs provided as part of the uplink scheduling grant to PRB pairs such that frequency-consecutive VRB pairs are not mapped to frequency-consecutive PRB pairs and (2) a split of each resource-block pair such that the two resource blocks of the resource-block pair are transmitted with a certain frequency gap in between. This second step can be seen as frequency hopping on a slot basis. In the absence of uplink frequency hopping, each VRB is directly mapped to a PRB.

The notion of VRBs can also be used for the LTE uplink, allowing for frequency-domain-distributed transmission for the uplink. However, in the uplink, where transmission from a terminal should always be over a set of consecutive subcarriers in the absence of multi-cluster transmission, distributing resource-block pairs in the frequency domain, as in the first step of downlink distributed transmission, is not possible. Rather, uplink distributed transmission is similar to the second step of downlink distributed transmission – that is, a frequency separation of the transmissions in the first and second slots of a subframe. Uplink distributed transmission for PUSCH can thus more directly be referred to as *uplink frequency hopping*.

There are two types of uplink frequency hopping defined for PUSCH:

- Sub-band-based hopping according to cell-specific hopping/mirroring patterns.
- Hopping based on explicit hopping information in the scheduling grant.

Uplink frequency hopping is not supported for multi-cluster transmission as, in that case, sufficient diversity can be obtained by the proper location of the two clusters.

11.1.3.1 *Hopping Based on Cell-Specific Hopping/Mirroring Patterns*

To support sub-band-based hopping according to cell-specific hopping/mirroring patterns, a set of consecutive sub-bands of a certain size is defined from the overall uplink frequency band, as illustrated in Figure 11.5. It should be noted that the sub-bands do not cover the total uplink frequency

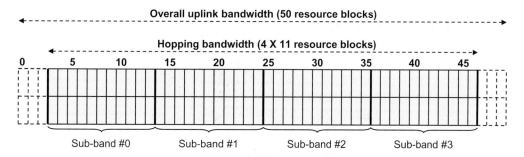

FIGURE 11.5

Definition of sub-bands for PUSCH hopping. The figure assumes a total of four sub-bands, each consisting of 11 resource blocks.

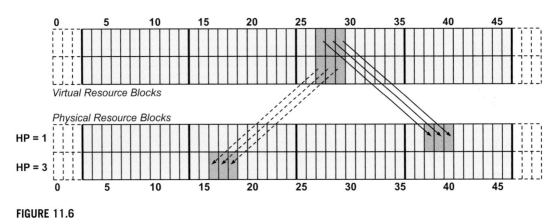

FIGURE 11.6

Hopping according to predefined hopping pattern.

band, mainly due to the fact that a number of resource blocks at the edges of the uplink frequency band are used for transmission of L1/L2 control signaling on the PUCCH. For example, in Figure 11.5 the overall uplink bandwidth corresponds to 50 resource blocks and there are a total of four sub-bands, each consisting of 11 resource blocks. Six resource blocks are not included in the hopping bandwidth and could, for example, be used for PUCCH transmission.

In the case of sub-band-based hopping, the set of virtual resource blocks provided in the scheduling grant are mapped to a corresponding set of physical resource blocks according to a cell-specific hopping pattern. The resource to use for transmission, the *PRBs*, is obtained by shifting the VRBs provided in the scheduling grant by a number of sub-bands according to the hopping pattern, where the hopping pattern can provide different shifts for each slot. As illustrated in Figure 11.6, a terminal is assigned VRBs 27, 28, and 29. In the first slot, the predefined hopping pattern takes the value 1, implying transmission using PRBs one sub-band to the right – that is, PRBs 38, 39, and 40. In the second slot, the predefined hopping pattern takes the value 3, implying a shift of three sub-bands to the right in the figure and, consequently, transmission using resource blocks 16, 17, and 18. Note that the shifting "wraps-around" – that is, in the case of four sub-bands, a shift of three sub-bands is the same as a negative shift of one sub-band. As the hopping pattern is cell specific – that is, the same for all terminals within a cell – different terminals will transmit on non-overlapping physical resources as long as they are assigned non-overlapping virtual resources.

In addition to the hopping pattern, there is also a cell-specific *mirroring pattern* defined in a cell. The mirroring pattern controls, on a slot basis, whether or not mirroring within each sub-band should be applied to the assigned resource. In essence, mirroring implies that the resource blocks within each sub-band are numbered right to left instead of left to right. Figure 11.7 illustrates mirroring in combination with hopping. Here, the mirroring pattern is such that mirroring is not applied to the first slot while mirroring is applied to the second slot.

Both the hopping pattern and the mirroring pattern depend on the physical-layer cell identity and are thus typically different in neighboring cells. Furthermore, the period of the hopping/mirroring patterns corresponds to one frame.

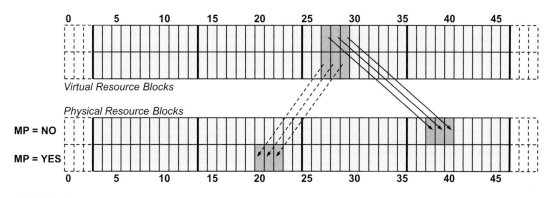

FIGURE 11.7

Hopping/mirroring according to predefined hopping/mirroring patterns. Same hopping pattern as in Figure 11.6.

11.1.3.2 *Hopping Based on Explicit Hopping Information*

As an alternative to hopping/mirroring according to cell-specific hopping/mirroring patterns as described above, uplink slot-based frequency hopping for PUSCH can also be controlled by *explicit hopping information* provided in the scheduling grant. In such a case the scheduling grant includes:

- information about the resource to use for uplink transmission in the first slot, exactly as in the non-hopping case; and
- additional information about the offset of the resource to use for uplink transmission in the second slot, relative to the resource of the first slot.

Selection between hopping according to cell-specific hopping/mirroring patterns as described above or hopping according to explicit information in the scheduling grant can be done dynamically. More specifically, for cell bandwidths less than 50 resource blocks, there is a single bit in the scheduling grant indicating if hopping should be according to the cell-specific hopping/mirroring patterns or should be according to information in the scheduling grant. In the latter case, the hop is always half of the hopping bandwidth. In the case of larger bandwidths (50 resource blocks and beyond), there are two bits in the scheduling grant. One of the combinations indicates that hopping should be according to the cell-specific hopping/mirroring patterns while the three remaining alternatives indicate hopping of 1/2, +1/4, and −1/4 of the hopping bandwidth. Hopping according to information in the scheduling grant for the case of a cell bandwidth corresponding to 50 resource blocks is illustrated in Figure 11.8. In the first subframe, the scheduling grant indicates a hop of one-half the hopping bandwidth. In the second subframe, the grant indicates a hop of one-fourth the hopping bandwidth (equivalent to a negative hop of three-fourths of the hopping bandwidth). Finally, in the third subframe, the grant indicates a negative hop of one-fourth the hopping bandwidth.

Subframe n, hop = +1/2

Subframe n + 1, hop = +1/4

Subframe n + 2, hop = −1/4

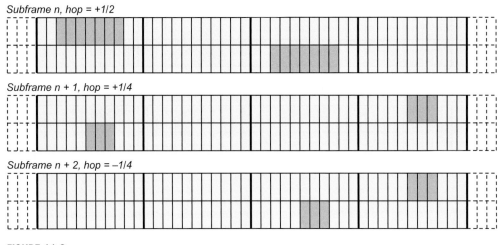

FIGURE 11.8

Frequency hopping according to explicit hopping information.

11.2 UPLINK REFERENCE SIGNALS

Similar to the downlink, reference signals are also transmitted on the LTE uplink. There are two types of reference signals defined for the LTE uplink:

- Uplink *demodulation reference signals* (DM-RS) are intended to be used by the base station for channel estimation for coherent demodulation of the uplink physical channels (PUSCH and PUCCH). Demodulation reference signals are thus only transmitted together with PUSCH or PUCCH and are then transmitted with the same bandwidth as the corresponding physical channel.
- Uplink *sounding reference signals* (SRS) are intended to be used by the base station for channel-state estimation to support uplink channel-dependent scheduling and link adaptation. The SRS can also be used in cases when uplink transmission is needed, although there is no data to transmit. One example is when uplink transmission is needed for the network to be able to control the uplink transmit timing by means of the *uplink-timing-alignment procedure* (Section 11.5). Finally, the uplink SRS can also be used to estimate the downlink channel-state assuming sufficient uplink/downlink reciprocity. This is especially of interest for TDD, where the operation of downlink and uplink on the same carrier frequency typically implies a higher degree of short-term downlink/uplink reciprocity compared to FDD.

11.2.1 Uplink Demodulation Reference Signals

Uplink demodulation reference signals are used for channel estimation for coherent demodulation of the Physical Uplink Shared Channel (PUSCH) to which the UL-SCH transport channel is mapped, as well as for the Physical Uplink Control Channel (PUCCH), which carries different types of L1/L2 control signaling (see Section 11.4). The basic structure for demodulation reference signals is

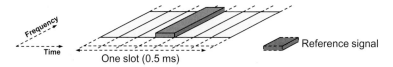

FIGURE 11.9

Transmission of uplink reference signals within a slot in the case of PUSCH transmission.

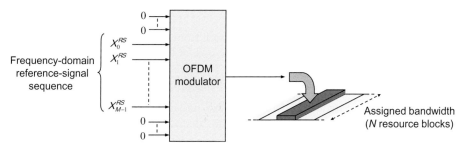

FIGURE 11.10

Generation of uplink reference signal from a frequency-domain reference-signal sequence.

the same for PUSCH and PUCCH transmission, although there are some differences – for example, in terms of the exact set of OFDM symbols in which the reference signals are transmitted.

11.2.1.1 *Basic Principles of Uplink DM-RS*

Due to the importance of low cubic metric and corresponding high power-amplifier efficiency for uplink transmissions, the principles for uplink reference-signal transmission are different compared to the downlink. In essence, transmitting reference signals in parallel with other uplink transmissions from the same terminal is not suitable for the uplink. Instead, certain OFDM symbols are used exclusively for DM-RS transmission – that is, the reference signals are *time multiplexed* with other uplink transmissions from the same terminal. The structure of the reference signal itself then ensures a low cubic metric in these symbols.

More specifically, in the case of PUSCH transmission a demodulation reference signal is transmitted within the fourth symbol of each uplink slot[5] (Figure 11.9). Within each subframe, there are thus two reference-signal transmissions, one in each slot.

In the case of PUCCH transmission, the number of OFDM symbols used for reference-signal transmission in a slot, as well as the exact position of these symbols, differs between different *PUCCH formats,* as described in Section 11.4.

Regardless of the kind of uplink transmission (PUSCH or PUCCH), the basic structure of each reference-signal transmission is the same. As illustrated in Figure 11.10, an uplink reference signal

---

[5]The third symbol in the case of extended cyclic prefix.

can be defined as a *frequency-domain reference-signal sequence* applied to consecutive inputs of the OFDM modulator – that is, to consecutive subcarriers.

In general, there is no reason to estimate the channel outside the frequency band of the corresponding PUSCH/PUCCH transmission that is to be coherently demodulated. The bandwidth of the reference signal, corresponding to the length of the reference-signal sequence in Figure 11.10, should thus be equal to the bandwidth of the corresponding PUSCH/PUCCH transmission measured in number of subcarriers. This means that, for PUSCH transmission, it should be possible to generate reference-signal sequences of different length, corresponding to the possible bandwidths of a PUSCH transmission. Note, however, that the length of the reference-signal sequence will always be a multiple of 12, as uplink resource allocations for PUSCH transmission are always done in terms of resource blocks consisting of 12 subcarriers.

As will be described in Section 11.4, PUCCH transmission from a terminal is always carried out over a single resource block in the frequency domain. In the case of PUCCH transmission, the length of the reference-signal sequence is thus always equal to 12.

Uplink reference signals should preferably have the following properties:

- Limited power variations in the frequency domain to allow for similar channel-estimation quality for all frequencies. Note that this is equivalent to good time-domain cross-correlation properties of the transmitted reference signal.
- Limited power variations in the time domain, leading to low cubic metric of the transmitted signal.

Furthermore, sufficient reference-signal sequences of a given length, corresponding to a certain reference-signal bandwidth, should be available in order to avoid an unreasonable planning effort when assigning reference-signal sequences to cells.

So-called *Zadoff–Chu* sequences [71] have the property of constant power in both the frequency and time domains. The elements of a Zadoff–Chu sequence of length M_{ZC} can be expressed as:

$$X_k^{ZC} = e^{-j\pi u \frac{k(k+1)}{M_{ZC}}} \quad 0 \le k < M_{ZC}, \tag{11.1}$$

where u is the *index* of the Zadoff–Chu sequence within the set of Zadoff–Chu sequences of length M_{ZC}. From the point of view of small power variations in both the frequency and time domains, Zadoff–Chu sequences would thus be excellent as uplink reference-signal sequences. However, there are two reasons why Zadoff–Chu sequences are not suitable for direct use as uplink reference-signal sequences in LTE:

- The number of available Zadoff–Chu sequences of a certain length, corresponding to the number of possible values of the index u in the expression above, equals the number of integers that are relative prime to the sequence length. This implies that to maximize the number of Zadoff–Chu sequences and thus, in the end, to maximize the number of available uplink reference signals, prime-length Zadoff–Chu sequences would be preferred. At the same time, the length of the uplink reference-signal sequences should be a multiple of 12, which is obviously not a prime number.
- For short sequence lengths, corresponding to narrow uplink transmission bandwidths, relatively few reference-signal sequences would be available even if they were based on prime-length Zadoff–Chu sequences.

Instead, for sequence lengths larger than or equal to 36, corresponding to transmission bandwidths larger than or equal to three resource blocks, the reference-signal sequences are defined as *cyclic extensions* of Zadoff–Chu sequences of length M_{ZC}, where M_{ZC} is the largest prime number smaller than or equal to the desired reference-signal sequence length. For example, the largest prime number less than or equal to 36 is 31, implying that reference-signal sequences of length 36 are defined as cyclic extensions of Zadoff–Chu sequences of length 31. The number of available sequences is then equal to 30 – that is, one less than the length of the Zadoff–Chu sequence. For larger sequence lengths, more sequences are available. For example, for a reference-signal sequence length equal to 72, there are 70 sequences available.[6]

For sequence lengths equal to 12 and 24, corresponding to transmission bandwidths of one and two resource blocks respectively, special QPSK-based sequences have instead been found from computer searches and are explicitly listed in the LTE specifications. For each of the two sequence lengths, 30 sequences are available.

11.2.1.2 *Phase-Rotated Reference-Signal Sequences*

In the previous section it was described how different reference-signal sequences (a minimum of 30 sequences for each sequence length) can be derived, primarily by cyclically extending *different prime-length Zadoff–Chu sequences*. Additional reference-signal sequences can be derived by applying different linear phase rotations to the same basic reference-signal sequence, as illustrated in Figure 11.11, where the basic reference-signal sequence is defined according to the above – that is, a cyclically extended Zadoff–Chu sequence for sequence lengths larger than or equal to 36 and a "special" sequence for shorter sequence lengths.

Applying a linear phase rotation in the frequency domain is equivalent to applying a cyclic shift in the time domain. Thus, although being *defined* as different *frequency-domain phase rotations*, in

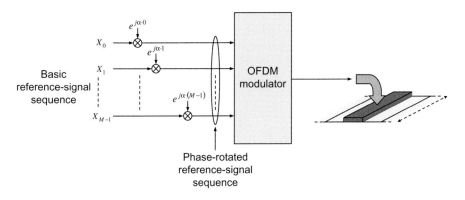

FIGURE 11.11

Generation of uplink reference-signal sequence from linear phase rotation of a basic reference-signal sequence.

---

[6]The largest prime number smaller than or equal to 72 is 71. The number of sequences is then one less than the length of the Zadoff–Chu sequence, i.e. 70.

the LTE specification this is often referred to as applying different *cyclic shifts*[7] to the same basic reference-signal sequence. Here the term "phase rotation" will be used. However, it should be borne in mind that what is referred to here as "phase rotation" is often referred to as "cyclic shift" in the 3GPP-related literature.

Demodulation reference signals defined from different reference-signal sequences typically have relatively low but still non-zero mutual correlation. In contrast, reference signals defined from different phase rotations of the same basic reference-signal sequence can be made completely orthogonal, thus causing no interference to each other, assuming the parameter α in Figure 11.11 takes a value $m\pi/6$, where m ranges from 0 to 11.[8] Up to 12 orthogonal reference signals can thus be defined from each basic reference-signal sequence.

However, to preserve the orthogonality between these reference signals also at the receiver side, the frequency response of the channel should essentially be constant (non-selective) over the span of 12 subcarriers (one resource block). Alternatively expressed, the main part of the channel time disper-sion should not extend beyond the length of the cyclic shift mentioned above. If that is not the case, a subset of the available values for α may be used – for example, only the values $\{0, 2\pi/6, 4\pi/6, \ldots, 10\pi/6\}$ or perhaps even fewer values. Limiting the set of values for α implies orthogonality over a smaller number of subcarriers and thus less sensitivity to channel frequency selectivity.

Another prerequisite for orthogonality between reference signals defined from different phase rotations of the same basic reference-signal sequence is that the reference signals are received well aligned in time. Any timing misalignment between the reference signals will, in the frequency domain, appear as a phase rotation that may counteract the phase rotation applied to separate the reference-signal sequences and cause interference between them.

In general for LTE, uplink transmissions from different terminals within the same cell are typically well time aligned anyway, as this is a prerequisite for retaining the orthogonality between different frequency-multiplexed transmissions. Thus, one possible use of reference signals defined from differ-ent phase rotations of the same basic reference-signal sequence is for the case when multiple terminals within the same cell simultaneously transmit on the uplink *using the same frequency resource*. As will be seen in Section 11.4, this is, for example, the case for uplink PUCCH transmissions.

Multiple reference signals are also needed for PUSCH transmission in the case of spatial multi-plexing (Section 11.3.1) and multi-user MIMO (Section 11.3.2). As discussed in Section 11.2.1.5, using different phase rotations is one approach by which multiple reference signals can be provided in these cases.

11.2.1.3 *Reference-Signal Assignment to Cells*

As described above, the design of the LTE reference-signal sequences allows for at least 30 sequences of each sequence length. To assign reference-signal sequences to cells, these sequences are first grouped into 30 *sequence groups* where each group consists of:

- One reference-signal sequence for each sequence length less than or equal to 60, corresponding to a transmission bandwidth of five resource blocks or less.

---

[7] Not to be confused with the cyclic extension of Zadoff–Chu sequences to generate the basic reference-signal sequences. The cyclic shift discussed here would be a time-domain cyclic shift, equivalent to a frequency-domain phase rotation.

[8] The orthogonality is due to the fact that, for $\alpha = m\pi/6$, there will be an integer number of full-circle rotations over 12 subarriers – that is, over one resource block.

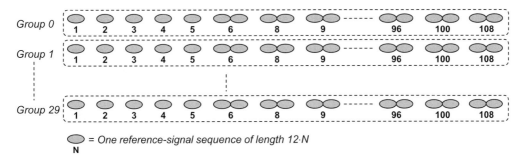

= One reference-signal sequence of length 12·N

FIGURE 11.12

Grouping of reference-signal sequences into sequence groups. The number indicates the corresponding bandwidth in number of resource blocks.

- Two reference-signal sequences for each sequence length larger than or equal to 72,[9] corresponding to a transmission bandwidth of six resource blocks or more.

The grouping of reference-signal sequences into sequence groups is illustrated in Figure 11.12. It can be noted that the sequence groups do not include any reference-signal sequences corresponding to seven resource blocks. As described in Section 11.1.1, the uplink transmission bandwidths, measured in number of resource blocks, should always be a product of the integers 2, 3, and 5.[10] A resource assignment of size seven resource blocks is thus not a valid resource assignment and there is no need to define reference-signal sequences of the corresponding length. For the same reason, the maximum length of the reference-signal sequences corresponds to 108 resource blocks, rather than 110 resource blocks. Similarly, the second largest sequence length corresponds to 100 resource blocks.

In a given time slot, the uplink reference-signal sequences to use within a cell are taken from one specific sequence group. What group to use within a cell may then be the same for all slots (fixed assignment). Alternatively, what group to use within a cell may vary between slots, also referred to as *group hopping*. The system information indicates if there is a fixed group assignment or if group hopping is to be used.

Regardless of whether a fixed group assignment or group hopping is used, there are certain differences between PUCCH and PUSCH transmission.

In the case of a fixed group assignment, the sequence group to use for PUCCH transmission is given by the physical-layer cell identity (see Chapter 14) modulo 30, where the cell identity ranges from 0 to 503. Thus, cell identities 0, 30, 60, ..., 480 correspond to sequence group 0, cell identities 1, 31, 61, ..., 481 correspond to sequence group 1, and so on.

In contrast, what sequence group to use for PUSCH transmission is explicitly signaled as part of the cell system information. The reason for explicitly indicating what sequence group to use for

---

[9]For sequence lengths larger than or equal to 72, there are more than 60 sequences of each length allowing for two sequences per group.
[10]Strictly speaking, the number of subcarriers should be a product of the integers 2, 3, and 5. However, as there are $12 = 3 \cdot 2^2$ subcarriers within one resource block, this is equivalent to a constraint that the number of resource blocks should be a product of the integers 2, 3, and 5.

PUSCH transmission is that it should be possible to use the same sequence group for PUSCH transmission in neighboring cells, despite the fact that such cells typically have different cell identities, resulting in reference-signal orthogonality also *between* cells and a corresponding decrease in interference. In this case, the reference signals for PUSCH transmission would instead be distinguished by different phase rotations, as discussed in Section 11.2.1.2.

In the case of group hopping, an additional cell-specific group-hopping pattern is added to the sequence group. Hence, in the case of reference signals for PUCCH transmission, the sequence group to use in a given slot is given by the value of the group-hopping pattern for that slot plus the cell identity modulo 30. The group-hopping pattern is also given by the cell identity and identical group-hopping patterns are used for PUSCH and PUCCH within a cell.

Thus, to have the same sequence group for PUSCH in two cells:

- The cell identities of the two cells should be different (to have different sequence groups for PUCCH, which should always be the case).
- If group hopping is enabled, the cell identities of the two cells should be such that the same group-hopping pattern is used in the two cells.
- The sequence group for PUSCH (indicated as part of the cell system information) should be the same for the two cells.

11.2.1.4 *Sequence Hopping*

As can be seen from Figure 11.12, for sequence lengths corresponding to six resource blocks and above, there are two reference-signal sequences in each sequence group. What sequence to use may then either be fixed (always the first sequence within the group) or vary between slots, also referred to as *sequence hopping* (not to be mixed with group hopping as described in Section 11.2.1.3). The purpose of sequence hopping is to randomize the inter-cell interference between reference signals. Sequence hopping can only be enabled if group hopping is not enabled – that is, if there is a fixed assignment of groups to cells.

11.2.1.5 *Multiple Demodulation Reference Signals*

Support for uplink multi-antenna transmission, more specifically multi-antenna precoding including spatial multiplexing, was introduced in LTE release 10. As will be seen in Section 11.3.1, in the case of spatial multiplexing there is a need for one demodulation reference signal *per layer*. As LTE release 10 supports the transmission of up to four spatially multiplexed layers in parallel, there is thus a need to be able to transmit up to four demodulation reference signals from the same terminal. Such multiple reference signals can be generated by different means:

- Multiple mutually orthogonal reference signals can be generated by using different phase rotations ("cyclic shifts"), as already described in Section 11.2.1.2.
- Two different reference signals can be created from the same reference signal sequence by applying mutually orthogonal patterns, similar to the downlink referred to as *orthogonal cover codes* (OCC), to the two reference-signal transmissions within a subframe, as illustrated in Figure 11.13.

Multiple orthogonal reference signals, as described above, are also used in the case of uplink multi-user MIMO (Section 11.3.2).

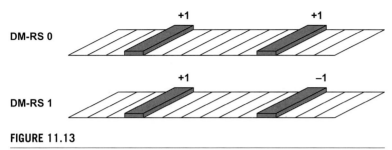

FIGURE 11.13

Generation of multiple demodulation reference signals from orthogonal cover codes.

11.2.2 Uplink Sounding Reference Signals

The demodulation reference signals discussed in Section 11.2.1 are used for channel estimation to allow for coherent demodulation of uplink physical channels (PUSCH or PUCCH). A demodulation reference signal is always transmitted together with and covering the same frequency band as the corresponding physical channel.

In contrast, *sounding reference signals* (SRS) are transmitted on the uplink to allow for the base station to estimate the uplink *channel state* at different frequencies. The channel-state estimates can then, for example, be used by the network scheduler to assign resource blocks of instantaneously good quality for uplink PUSCH transmission (uplink channel-dependent scheduling), as well as to select different transmission parameters such as the instantaneous data rate and different parameters related to uplink multi-antenna transmission. As mentioned earlier, SRS transmission can also be used for uplink timing estimation as well as to estimate downlink channel conditions assuming downlink/uplink channel reciprocity. Thus, an SRS is not necessarily transmitted together with any physical channel and if transmitted together with, for example, PUSCH, the SRS may cover a different, typically larger, frequency span.

There are two types of SRS transmission defined for the LTE uplink: *periodic* SRS transmission, which has been available from the first release of LTE (release 8); and *aperiodic* SRS transmission, introduced in LTE release 10.

11.2.2.1 Periodic SRS Transmission

Periodic SRS transmission from a terminal occurs at regular time intervals, from as often as once every 2 ms (every second subframe) to as infrequently as once every 160 ms (every 16th frame). When SRS is transmitted in a subframe, it occupies the last symbol of the subframe, as illustrated in Figure 11.14. As an alternative, in the case of TDD operation, SRS can also be transmitted within the UpPTS.

In the frequency domain, SRS transmissions should cover the frequency band that is of interest for the scheduler. This can be achieved in two ways:

- By means of a sufficiently wideband SRS transmission that allows for sounding of the entire frequency band of interest with a single SRS transmission, as illustrated in the upper part of Figure 11.15.

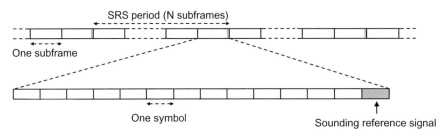

FIGURE 11.14

Transmission of SRS.

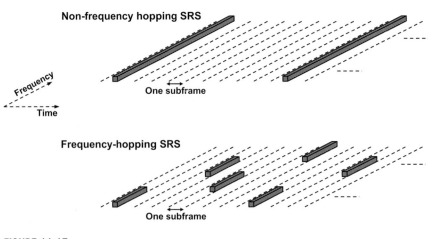

FIGURE 11.15

Non-frequency-hopping (wideband) versus frequency-hopping SRS.

- By means of more narrowband SRS transmission – that is, hopping in the frequency domain – in such a way that a sequence of SRS transmissions jointly covers the frequency band of interest, as illustrated in the lower part of Figure 11.15.

The main benefit of wideband (non-hopping) SRS transmission according to the upper part of Figure 11.15 is that the entire frequency band of interest can be sounded with a single SRS transmission – that is, within a single OFDM symbol. As described below, the entire OFDM symbol in which SRS is transmitted will be unavailable for data transmission in the cell. A single wideband SRS transmission is thus more efficient from a resource-utilization point of view as less OFDM symbols need to be used to sound a given overall bandwidth. However, in the case of a high uplink path loss, wideband SRS transmission may lead to relatively low received power density, which may degrade the channel-state estimation. In such a case it may be preferable to use a more narrowband SRS

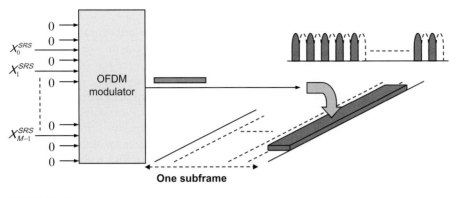

FIGURE 11.16

Generation of SRS from a frequency-domain reference-signal sequence.

transmission, thereby focusing the available transmit power in a more narrow frequency range, and then hop over the total band to be sounded.

In general, different bandwidths of the SRS transmission can be available within a cell. A narrow SRS bandwidth, corresponding to four resource blocks, is always available in all cells, regardless of the uplink cell bandwidth. Up to three additional, more wideband SRS bandwidths may also be configured within the cell. The SRS bandwidths are then always a multiple of four resource blocks. A terminal is then explicitly configured to use one of the SRS bandwidths available in the cell.

If a terminal is transmitting SRS in a certain subframe, the SRS transmission may very well overlap, in the frequency domain, with PUSCH transmissions from other terminals within the cell. To avoid collision between SRS and PUSCH transmissions from different terminals, terminals should in general avoid PUSCH transmission in the OFDM symbols in which SRS transmission may occur. To achieve this, all terminals within a cell are aware of the set of subframes within which SRS *may* be transmitted by *any* terminal within the cell. All terminals should then avoid PUSCH transmission in the last OFDM symbol of those subframes.[11] Information about the set of subframes in which SRS *may be* transmitted within a cell is provided as part of the cell system information.

On a more detailed level, the structure for sounding reference signals (SRS) is similar to that of uplink demodulation reference signals described in Section 11.2.1. More specifically, a sounding reference signal is also defined as a frequency-domain reference-signal sequence derived as a cyclic extension of prime-length Zadoff–Chu sequences. However, in the case of SRS, the reference-signal sequence is mapped to *every second subcarrier*, creating a "comb"-like spectrum, as illustrated in Figure 11.16. Taking into account that the bandwidth of the SRS transmission is always a multiple of four resource blocks, the lengths of the reference-signal sequences for SRS are thus always a multiple of 24. The reference-signal sequence to use for SRS transmission within the cell is taken from the same sequence group as the demodulation reference signals used for channel estimation for PUCCH.

---

[11] SRS transmission in UpPTS can obviously not collide with PUSCH transmission as there is no PUSCH transmission in UpPTS.

Similar to demodulation reference signals, different phase rotations (also, for SRS, typically referred to as "cyclic shifts") can be used to generate different SRS that are orthogonal to each other. By assigning different phase rotations to different terminals, multiple SRS can thus be transmitted in parallel in the same subframe, as illustrated by terminals #1 and #2 in the upper part of Figure 11.17. However, it is then required that the reference signals span the same frequency band.

Another way to allow for SRS to be simultaneously transmitted from different terminals is to rely on the fact that each SRS only occupies every second subcarrier. Thus, SRS transmissions from two terminals can be *frequency multiplexed* by assigning them to different frequency shifts or "combs", as illustrated by terminal #3 in the lower part of Figure 11.17. In contrast to the multiplexing of SRS transmission by means of different "cyclic shifts", frequency multiplexing of SRS transmissions does not require the transmissions to cover identical frequency bands.

To summarize, the following set of parameters defines the characteristics of an SRS transmission:

- SRS transmission bandwidth – that is, the bandwidth covered by a single SRS transmission.
- Hopping bandwidth – that is, the frequency band over which the SRS transmission is frequency hopping.
- Frequency-domain position – that is, the starting point of the SRS transmission in the frequency domain.
- Transmission comb, as illustrated in Figure 11.17.
- Phase rotation (or equivalently cyclic shift) of the reference-signal sequence.
- SRS transmission time-domain period (from 2 to 160 ms) and subframe offset.

A terminal that is to transmit SRS is configured with these parameters by means of higher layer (RRC) signaling. In addition, all terminals within a cell should be informed in what subframes SRS may be transmitted within the cell as, within these subframes, the "SRS symbol" should not be used for PUSCH transmission.

11.2.2.2 *Aperiodic SRS Transmission*

In contrast to periodic SRS, aperiodic SRS are *one-shot* transmissions, triggered by signaling on PDCCH as part of the scheduling grant. The frequency-domain structure of an aperiodic SRS

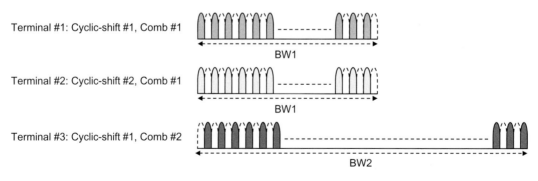

FIGURE 11.17

Multiplexing of SRS transmissions from different terminals.

transmission is identical to that of periodic SRS. Also, in the same way as for periodic SRS transmission, aperiodic SRS are transmitted within the last symbol of a subframe. Furthermore, the time instants when aperiodic SRS may be transmitted are configured per terminal using higher-layer signaling.

The frequency-domain parameters for aperiodic SRS (bandwidth, odd or even "comb", etc.) are configured by higher-layer (RRC) signaling. However, no SRS transmission will actually be carried out until the terminal is explicitly triggered to do so by an explicit *SRS trigger* on PDCCH. When such a trigger is received, a single SRS is transmitted in the next available aperiodic SRS instant configured for the terminal using the configured frequency-domain parameters. Additional SRS transmissions can then be carried out if additional triggers are received.

Three different parameter sets can be configured for aperiodic SRS, for example differing in the frequency position of the SRS transmission and/or the transmission comb. Information on what parameters to use when the SRS is actually transmitted is included in the PDCH information, which consists of two bits, three combinations of which indicate the specific SRS parameter set. The fourth combination simply indicates that no SRS should be transmitted.

11.3 UPLINK MULTI-ANTENNA TRANSMISSION

Downlink multi-antenna transmission was supported by the LTE specification from its first release (release 8). With LTE release 10, support for uplink multi-antenna transmission – that is, uplink transmission relying on multiple transmit antennas at the terminal side – was also introduced for LTE. Uplink multi-antenna transmission can be used to improve the uplink link and/or system performance in different ways:

- To improve the achievable data rates and spectral efficiency for uplink data transmission by allowing for antenna precoding supporting spatial multiplying with up to four layers for the uplink physical data channel PUSCH.
- To improve the uplink control-channel performance by allowing for transmit diversity for the uplink physical control channel PUCCH.

11.3.1 Precoder-Based Multi-Antenna Transmission for PUSCH

In the case of PUSCH transmission, multi-antenna transmission is based on antenna precoding, as illustrated in Figure 11.18. The structure of the uplink antenna precoding is very similar to that of downlink antenna precoding (Section 10.3), including the presence of precoded demodulation reference signals (one per layer) similar to downlink non-codebook-based precoding (Figure 10.19 in Section 10.3.3). Uplink antenna precoding supports transmission using up to four antenna ports, allowing for spatial multiplexing with up to four layers.

The principles for mapping of the modulation symbols to layers are also the same as for the downlink. For an initial transmission, there is one transport block in the case of a single layer and two transport blocks for more than one layer, as illustrated in Figure 11.19. Similar to the downlink, in the case of a hybrid-ARQ retransmission, a single transport block may also be transmitted on multiple layers in some cases.

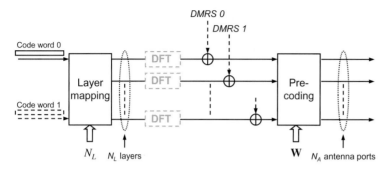

FIGURE 11.18

Precoder-based multi-antenna transmission for LTE uplink.

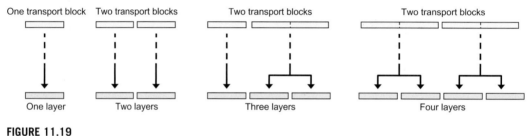

FIGURE 11.19

Uplink transport-channel-to-layer mapping (initial transmission).

As can be seen in Figure 11.18, the DFT precoding is actually taking place after layer mapping – that is, each layer is separately DFT precoded. To simplify the description this was not really visible in Figure 11.1, outlining the overall physical-layer transport-channel processing.

It can also be noted that, in contrast to Figure 10.19 in Section 10.3.3, the precoder in Figure 11.18 is not shaded. As discussed in Section 10.3.3, for downlink non-codebook-based precoding, the precoder part of the antenna mapping is not visible in the specification and the network can, in essence, apply an arbitrary precoding for the downlink transmission. Due to the use of precoded demodulation reference signals, the terminal can recover the different layers without knowledge of exactly what precoding has been applied at the transmitted side.

The same is also true for the uplink – that is, the presence of precoded demodulation reference signals would allow for the base station to demodulate the uplink multi-antenna transmission and recover the different layers without knowledge of the precoding taking place at the transmitter side. However, for LTE the uplink precoder matrix is selected by the network and conveyed to the terminal as part of the scheduling grant. The terminal should then follow the precoder matrix selected by the network. Thus, in the uplink, the precoder is visible in the specification and, in order to limit the downlink signaling, there is a limited set of precoder matrices specified for each transmission rank.

More specifically, for each combination of transmission rank N_L and number of antennas ports N_A, a set of precoder matrices of size $N_A \times N_L$ is defined, as illustrated in Tables 11.1 and 11.2 for

Table 11.1 Uplink Precoder Matrices for Two Antenna Ports

Transmission Rank	Codebook Index					
	0	**1**	**2**	**3**	**4**	**5**
1	$\begin{bmatrix} +1 \\ +1 \end{bmatrix}$	$\begin{bmatrix} +1 \\ -1 \end{bmatrix}$	$\begin{bmatrix} +1 \\ +1 \end{bmatrix}$	$\begin{bmatrix} +1 \\ -1 \end{bmatrix}$	$\begin{bmatrix} +1 \\ +1 \end{bmatrix}$	$\begin{bmatrix} +1 \\ -1 \end{bmatrix}$

Full-rank (2 × 2) identity matrix is not shown in the table.

Table 11.2 Subset of Uplink Precoder Matrices for Four Antenna Ports and Different Transmission Ranks

Transmission Rank	Codebook Index				
	0	**1**	**2**	**3**	...
1	$\begin{bmatrix} +1 \\ +1 \\ +1 \\ -1 \end{bmatrix}$	$\begin{bmatrix} +1 \\ +1 \\ +j \\ +j \end{bmatrix}$	$\begin{bmatrix} +1 \\ +1 \\ -1 \\ +1 \end{bmatrix}$	$\begin{bmatrix} +1 \\ +1 \\ -j \\ -j \end{bmatrix}$	...
2	$\begin{bmatrix} +1 & 0 \\ +1 & 0 \\ 0 & +1 \\ 0 & -j \end{bmatrix}$	$\begin{bmatrix} +1 & 0 \\ +1 & 0 \\ 0 & +1 \\ 0 & +j \end{bmatrix}$	$\begin{bmatrix} +1 & 0 \\ -j & 0 \\ 0 & +1 \\ 0 & +1 \end{bmatrix}$	$\begin{bmatrix} +1 & 0 \\ -j & 0 \\ 0 & 1 \\ 0 & -1 \end{bmatrix}$	
3	$\begin{bmatrix} +1 & 0 & 0 \\ +1 & 0 & 0 \\ 0 & +1 & 0 \\ 0 & 0 & +1 \end{bmatrix}$	$\begin{bmatrix} +1 & 0 & 0 \\ -1 & 0 & 0 \\ 0 & +1 & 0 \\ 0 & 0 & +1 \end{bmatrix}$	$\begin{bmatrix} +1 & 0 & 0 \\ 0 & +1 & 0 \\ +1 & 0 & 0 \\ 0 & 0 & +1 \end{bmatrix}$	$\begin{bmatrix} +1 & 0 & 0 \\ 0 & +1 & 0 \\ -1 & 0 & 0 \\ 0 & 0 & +1 \end{bmatrix}$	

Full-rank (4 × 4) identity matrix is not shown in the table.

two and four antenna ports respectively. For full-rank transmission – that is, when the transmission rank or number of layers equals the number of transmit antennas – only a single precoder matrix is defined, namely the identity matrix of size $N_A \times N_A$ (not shown in the tables). Note that, for the case of four antenna ports, only a subset of the defined matrices is shown. In total there are 24 rank-1 matrices, 16 rank-2 matrices, and 12 rank-3 matrices defined for four antenna ports, in addition to the single rank-4 matrix.

As can be seen, all the precoder matrices in Table 11.1 contain one and only one non-zero element in each row, and this is generally true for all precoder matrices defined for the uplink. As a consequence, the signal transmitted on a certain antenna port (corresponding to a certain row of the

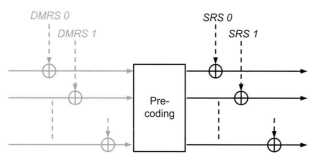

FIGURE 11.20

Illustration of SRS transmitted *after* uplink antenna precoding.

precoder matrix) always depends on one and only one specific layer (corresponding to a specific column of the precoder matrix). Expressed alternatively, the precoder matrix maps the layers to the antenna ports *with at most one layer being mapped to each antenna port*. Due to this, the good cubic-metric properties of the transmitted signal are also preserved for each antenna port when antenna precoding is applied. The precoder matrices of Tables 11.1 and 11.2 are therefore also referred to as *cubic-metric-preserving precoder matrices*.

In order to select a suitable precoder, the network needs information about the uplink channel. Such information can, for example, be based on measurements on the uplink sounding reference signals (Section 11.2.2). As indicated in Figure 11.20, sounding reference signals (SRS) are transmitted non-precoded – that is, directly on the different antenna ports. The received SRS thus reflect the channel of each antenna port, not including any precoding. Based on the received SRS, the network can thus decide on a suitable uplink transmission rank and corresponding uplink precoder matrix and provide information about the selected rank and precoder matrix as part of the scheduling grant.

The previous paragraph assumed the same number of antenna ports for PUSCH as for SRS. Clearly, this is a relevant situation and the SRS is, in this case, used to aid the selection of the precoding matrix, as discussed above. However, there are also situations when SRS and PUSCH use *different* numbers of antenna ports. One example is uplink transmission of two layers (two antenna ports), where the eNodeB would like to use SRS to probe the channel for potential four-layer transmission. In this case the SRS is transmitted on a *different* set of antenna ports than the PUSCH to aid the eNodeB in assessing the benefits, if any, of switching to four-layer transmission.

11.3.2 Uplink Multi-User MIMO

As described in Section 10.3.5, downlink multi-user MIMO (MU-MIMO) implies downlink transmission to different terminals using *the same time–frequency resource* and relying on the availability of multiple antennas, at least on the network side, to suppress interference between the transmissions. The term MU-MIMO originated from the resemblance to SU-MIMO (spatial multiplexing).

Uplink MU-MIMO is essentially the same thing but for the uplink transmission direction – that is, uplink MU-MIMO implies uplink transmissions from multiple terminals using *the same uplink time–frequency resource* and relying on the availability of multiple receive antennas at the base station

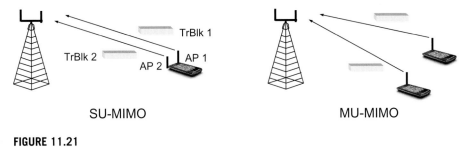

FIGURE 11.21

Relation between uplink SU-MIMO (spatial multiplexing) and MU-MIMO (SDMA).

to separate the two or more transmissions. Thus, MU-MIMO is really just another term for uplink *Space-Division Multiple Access* (SDMA).

Actually, on the uplink, the relation between MU-MIMO and SU-MIMO (spatial multiplexing) is even closer. Uplink spatial multiplexing, for example with two antenna ports and two layers, implies that the terminal transmits two transport blocks with one transport block transmitted on each layer and thus on each antenna port,[12] as illustrated in the left part of Figure 11.21. As illustrated in the right part of the figure, MU-MIMO is essentially equivalent to separating the two antennas into two different terminals and transmitting one transport block from each terminal. The base-station processing to separate the two transmissions could essentially be identical to the processing used to separate the two layers in the case of spatial multiplexing. It should be noted though that the separation of the two transmissions at the receiver side could be simplified, or at least the possible means to achieve this separation are extended, if the two terminals are well separated in space, something which is, for obvious reasons, not the case for two antennas attached to the same terminal. As an example, for sufficiently separated terminals, classical beam-forming relying on correlated receiver antennas can be used to separate the uplink transmissions. Alternatively, uncorrelated receiver antennas can be used, and the separation means are then essentially the same as for SU-MIMO.

One important benefit of uplink MU-MIMO is that one can get similar gains in *system throughput* as SU-MIMO (spatial multiplexing) without the need for multiple transmit antennas at the terminal side, allowing for less complex terminal implementation. It should be noted though that spatial multiplexing could still provide substantial gains in terms of *user throughput* and peak data rates that can be provided from a single terminal. The potential system gains of uplink MU-MIMO also rely on more than one terminal being actually available for transmission. The process of "pairing" terminals that should share the time–frequency resources is non-trivial and requires suitable radio-channel conditions.

Essentially, support for uplink MU-MIMO only requires the possibility to explicitly assign a specific orthogonal reference signal for the uplink transmission, thereby ensuring orthogonality between reference-signal transmissions from the different terminals involved in the MU-MIMO transmission. The orthogonal reference signals used for MU-MIMO are the same as for spatial multiplexing described in the previous section – that is, based on a combination of OCC (orthogonal cover codes)

---

[12]Note that the 2 × 2 precoder matrix is the identity matrix (see Table 11.1).

and different phase rotations. To support uplink MU-MIMO, each scheduling assignment includes a three-bit reference-signal indicator denoting a certain combination of OCC and phase rotation.

Also note that, similar to downlink MU-MIMO, nothing prevents the combination of spatial multiplexing and MU-MIMO – that is, multiplexing several terminals by means of MU-MIMO where one or several of the terminals are transmitting more than one layer. The eventual limitation is the number of orthogonal reference signals that can be assigned to the different users.

11.3.3 PUCCH Transmit Diversity

Precoder-based multi-layer transmission is only used for the uplink data transmission on PUSCH. However, in the case of a terminal with multiple transmit antennas, one obviously wants to use the full set of terminal antennas and corresponding terminal power amplifiers also for the L1/L2 control signaling on PUCCH in order to be able to utilize the full power resource and achieve maximum diversity. To achieve additional diversity, LTE release 10 also introduced the possibility for two-antenna *transmit diversity* for PUCCH. More specifically, the transmit diversity supported for PUCCH is referred to as Spatial Orthogonal-Resource Transmit Diversity (SORTD).

The basic principle of SORTD is simply to transmit the uplink control signaling using different resources (time, frequency, and/or code) on the different antennas. In essence, the PUCCH transmissions from the two antennas will be identical to PUCCH transmissions from two different terminals using different resources. Thus, SORTD creates additional diversity but achieves this by using twice as many PUCCH resources, compared to non-SORTD transmission.

For four physical antennas at the terminal, implementation-specific *antenna virtualization* is used. In essence, a transparent scheme is used to map the two-antenna-port signal to four physical antennas.

11.4 UPLINK L1/L2 CONTROL SIGNALING

Similar to the LTE downlink, there is also a need for uplink L1/L2 control signaling to support the transmission of downlink and uplink transport channels. Uplink L1/L2 control signaling consists of:

- hybrid-ARQ acknowledgements for received DL-SCH transport blocks;
- channel-state reports related to the downlink channel conditions, used to assist downlink scheduling; and
- scheduling requests, indicating that a terminal needs uplink resources for UL-SCH transmissions.

There is no information indicating the UL-SCH transport format signaled on the uplink. As mentioned in Chapter 8, the eNodeB is in complete control of the uplink UL-SCH transmissions and the terminal always follows the scheduling grants received from the network, including the UL-SCH transport format specified in those grants. Thus, the network knows the transport format used for the UL-SCH transmission in advance and there is no need for any explicit transport-format signaling on the uplink.

Uplink L1/L2 control signaling needs to be transmitted on the uplink regardless of whether or not the terminal has any uplink transport-channel (UL-SCH) data to transmit and thus regardless of whether or not the terminal has been assigned any uplink resources for UL-SCH transmission. Hence, two different methods are supported for the transmission of the uplink L1/L2 control signaling, depending on whether or not the terminal has been assigned an uplink resource for UL-SCH transmission:

- *No simultaneous transmission of UL-SCH and L1/L2 control.* If the terminal does not have a valid scheduling grant – that is, no resources have been assigned for the UL-SCH in the current subframe – a separate physical channel, the *Physical Uplink Control Channel* (PUCCH), is used for transmission of uplink L1/L2 control signaling. In the case of uplink carrier aggregation, the PUCCH can be transmitted on the primary component carrier only. A consequence of this is that transmissions on multiple component carriers in the downlink may need to be acknowledged on a single uplink component carrier.
- *Simultaneous transmission of UL-SCH and L1/L2 control.* If the terminal has a valid scheduling grant – that is, resources have been assigned for the UL-SCH in the current subframe – the uplink L1/L2 control signaling is time multiplexed with the coded UL-SCH on to the PUSCH prior to DFT precoding and OFDM modulation. Obviously, as the terminal has been assigned UL-SCH resources, there is no need to support transmission of the scheduling request in this case. Instead, scheduling information can be included in the MAC headers, as described in Chapter 12. In the case of carrier aggregation, all control signaling is transmitted on one of the uplink component carriers – that is, uplink control signaling cannot be spread across multiple UL-SCHs.

The reason to differentiate between the two cases above is to minimize the cubic metric for the uplink power amplifier in order to maximize coverage. However, in situations when there is sufficient power available in the terminal, simultaneous transmission of PUSCH and PUCCH can be used with no impact on the coverage. The possibility for simultaneous PUSCH and PUCCH transmission was therefore introduced in release 10 as one part of several features[13] adding flexibility at the cost of a somewhat higher cubic metric. In situations where this cost is not acceptable, simultaneous PUSCH and PUCCH can always be avoided by using the basic mechanism introduced in the first version of LTE.

These two cases, control on PUCCH and PUSCH respectively, are described in more detail in the following.

11.4.1 Uplink L1/L2 Control Signaling on PUCCH

If the terminal has not been assigned an uplink resource for UL-SCH transmission, the L1/L2 control information (channel-state reports, hybrid-ARQ acknowledgements, and scheduling requests) is transmitted on uplink resources (resource blocks) specifically assigned for uplink L1/L2 control on PUCCH. As mentioned in the previous section, the PUCCH can be transmitted on the primary component carrier only. However, as the primary component carrier is specified on a per-terminal basis, there may, from a network perspective, be PUCCH resources used on multiple component carriers.

Figure 11.22 illustrates the location of the PUCCH resources on one component carrier at the edges of the bandwidth allocated to that component carrier. Each such resource consists of 12 subcarriers (one resource block) within each of the two slots of an uplink subframe. To provide frequency diversity, the frequency resource is frequency hopping on the slot boundary – that is, one "frequency resource" consists of 12 subcarriers at the upper part of the spectrum within the first slot of a subframe

---

[13] Other examples of such features are simultaneous transmission on multiple uplink component carriers and uplink multi-cluster transmission.

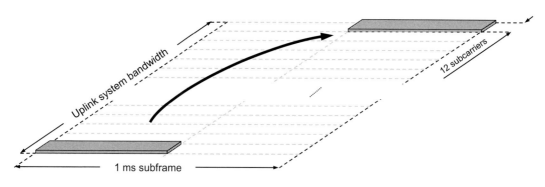

FIGURE 11.22

Uplink L1/L2 control signaling transmission on PUCCH.

and an equally sized resource at the lower part of the spectrum during the second slot of the subframe or vice versa. If more resources are needed for the uplink L1/L2 control signaling, for example in the case of very large overall transmission bandwidth supporting a large number of users, then additional resource blocks can be assigned next to the previously assigned resource blocks.

The reasons for locating the PUCCH resources at the edges of the overall available spectrum are twofold:

• Together with the frequency hopping described above, this maximizes the frequency diversity experienced by the control signaling.
• Assigning uplink resources for the PUCCH at other positions within the spectrum – that is, not at the edges – would have fragmented the uplink spectrum, making it impossible to assign very wide transmission bandwidths to a single terminal and still preserve the low-cubic-metric properties of the uplink transmission.

The bandwidth of one resource block during one subframe is too large for the control signaling needs of a single terminal. Therefore, to efficiently exploit the resources set aside for control signaling, multiple terminals can share the same resource block. This is done by assigning the different terminals to different orthogonal phase rotations of a cell-specific length-12 frequency-domain sequence, where the sequence is identical to a length-12 reference-signal sequence. Furthermore, as described in conjunction with the reference signals in Section 11.2, a linear phase rotation in the frequency domain is equivalent to applying a cyclic shift in the time domain. Thus, although the term "phase rotation" is used herein, the term cyclic shift is sometimes used with an implicit reference to the time domain.

The resource used by a PUCCH is therefore not only specified in the time–frequency domain by the resource-block pair, but also by the phase rotation applied. Similarly to the case of reference signals, there are up to 12 different phase rotations specified, providing up to 12 different orthogonal sequences from each cell-specific sequence. However, in the case of frequency-selective channels, not all the 12 phase rotations can be used if orthogonality is to be retained. Typically, up to six rotations are considered usable in a cell from a radio-propagation perspective, although inter-cell interference may result in a smaller number being useful from an overall system perspective. Higher-layer signaling is used to configure the number of rotations that is used in a cell.

As mentioned earlier, uplink L1/L2 control signaling includes hybrid-ARQ acknowledgements, channel-state reports, and scheduling requests. Different combinations of these types of messages are possible as described later, but to explain the structure for these cases it is beneficial to discuss separate transmission of each of the types first, starting with the hybrid-ARQ acknowledgement and the scheduling request. There are three formats defined for PUCCH, formats 1, 2, and 3, each capable of carrying a different number of bits.

11.4.1.1 PUCCH Format 1

Hybrid-ARQ acknowledgements are used to acknowledge the reception of one (or two in the case of spatial multiplexing) transport blocks on the DL-SCH. The hybrid-ARQ acknowledgement is only transmitted when the terminal correctly received control signaling related to DL-SCH transmission intended for this terminal on one of the PDCCHs. In the case of downlink carrier aggregation, there can be multiple simultaneous DL-SCHs for a single terminal, one per downlink component carrier and, consequently, multiple acknowledgement bits need to be conveyed in the uplink. This is described further below once the non-carrier-aggregated case has been treated.

If no valid DL-SCH-related control signaling is detected, then nothing is transmitted on the PUCCH (i.e. DTX). Apart from not unnecessarily occupying PUCCH resources that can be used for other purposes, this allows the eNodeB to perform three-state detection, ACK, NAK, or DTX, on the PUCCH received. Three-state detection is useful as NAK and DTX may need to be treated differently. In the case of NAK, retransmission of additional parity bits is useful for incremental redundancy, while for DTX the terminal has most likely missed the initial transmission of systematic bits and a better alternative than transmitting additional parity bits is to retransmit the systematic bits.

Scheduling requests are used to request resources for uplink data transmission. Obviously, a scheduling request should only be transmitted when the terminal is requesting resources, otherwise the terminal should be silent to save battery resources and not create unnecessary interference. Hence, unlike hybrid-ARQ acknowledgements, no explicit information bit is transmitted by the scheduling request; instead the information is conveyed by the presence (or absence) of energy on the corresponding PUCCH. However, the scheduling request, although used for a completely different purpose, shares the same PUCCH format as the hybrid-ARQ acknowledgement. This format is referred to as *PUCCH format 1* in the specifications.[14]

PUCCH format 1 uses the same structure in the two slots of a subframe, as illustrated in Figure 11.23. For transmission of a hybrid-ARQ acknowledgement, the single hybrid-ARQ acknowledgement bit related to one downlink component carrier is used to generate a BPSK symbol (in the case of downlink spatial multiplexing the two acknowledgement bits are used to generate a QPSK symbol). For a scheduling request, the same constellation point as for a negative acknowledgement is used. The modulation symbol is then used to generate the signal to be transmitted in each of the two PUCCH slots.

There are seven OFDM symbols per slot for a normal cyclic prefix (six in the case of an extended cyclic prefix). In each of those seven OFDM symbols, a length-12 sequence, obtained by phase rotation of the cell-specific sequence as described earlier, is transmitted. Three of the symbols are used

---

[14]There are actually three variants in the LTE specifications, formats 1, 1A, and 1B, used for transmission of scheduling requests and one or two hybrid-ARQ acknowledgements respectively. However, for simplicity, they are all referred to as format 1 herein.

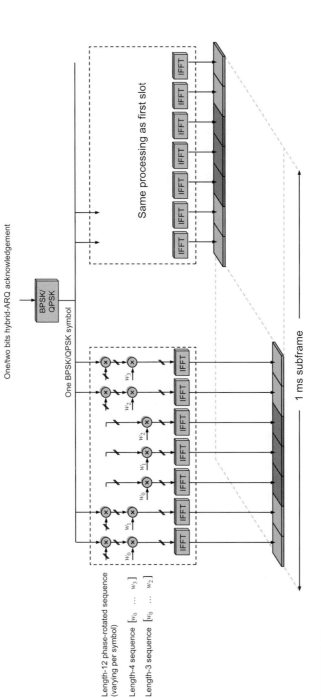

FIGURE 11.23

PUCCH format 1 (normal cyclic prefix).

as reference signals to enable channel estimation by the eNodeB and the remaining four[15] are modulated by the BPSK/QPSK symbols described earlier. In principle, the BPSK/QPSK modulation symbol could directly modulate the rotated length-12 sequence used to differentiate terminals transmitting on the same time–frequency resource. However, this would result in unnecessarily low capacity on the PUCCH. Therefore, the BPSK/QPSK symbol is multiplied by a length-4 orthogonal cover sequence.[16] Multiple terminals may transmit on the same time–frequency resource using the same phase-rotated sequence and be separated through different orthogonal covers. To be able to estimate the channels for the respective terminals, the reference signals also employ an orthogonal cover sequence, with the only difference being the length of the sequence – three for the case of a normal cyclic prefix. Thus, since each cell-specific sequence can be used for up to $3 \cdot 12 = 36$ different terminals (assuming all 12 rotations are available; typically at most six of them are used), there is a threefold improvement in the PUCCH capacity compared to the case of no cover sequence. The cover sequences are three Walsh sequences of length 4 for the data part and three DFT sequences of length 3 for the reference signals (for an extended cyclic prefix, the reference-signal cover sequence is of length 2).

A PUCCH format 1 resource, used for either a hybrid-ARQ acknowledgement or a scheduling request, is represented by a single scalar resource index. From the index, the phase rotation and the orthogonal cover sequence are derived.

The use of a phase rotation of a cell-specific sequence together with orthogonal sequences as described earlier provides orthogonality between different terminals in the same cell transmitting PUCCH on the same set of resource blocks. Hence, in the ideal case, there will be no intra-cell interference, which helps improve the performance. However, there will typically be inter-cell interference for the PUCCH as the different sequences used in neighboring cells are non-orthogonal. To randomize the inter-cell interference, the phase rotation of the sequence used in a cell varies on a symbol-by-symbol basis in a slot according to a hopping pattern derived from the physical-layer cell identity. On top of this, slot-level hopping is applied to the orthogonal cover and phase rotation to further randomize the interference. This is exemplified in Figure 11.24 assuming normal cyclic prefix and six of 12 rotations used for each cover sequence. To the phase rotation given by the cell-specific hopping a slot-specific offset is added. In cell A, a terminal is transmitting on PUCCH resource number 3, which in this example corresponds to using the (phase rotation, cover sequence) combination (6, 0) in the first slot and (11, 1) in the second slot of this particular subframe. PUCCH resource number 11, used by another terminal in cell A transmitting in the same subframe, corresponds to (11, 1) and (8, 2) in the first and second slots respectively of the subframe. In another cell the PUCCH resource numbers are mapped to different sets (rotation, cover sequence) in the slots. This helps to randomize the inter-cell interference.

For an extended cyclic prefix, the same structure as in Figure 11.23 is used, with the difference being the number of reference symbols in each slot. In this case, the six OFDM symbols in each slot are divided such that the two middle symbols are used for reference signals and the remaining four symbols used for the information. Thus, the length of the orthogonal sequence used to spread the

---

[15]The number of symbols used for reference signals and the acknowledgement is a trade-off between channel-estimation accuracy and energy in the information part; three symbols for reference symbols and four symbols for the acknowledgement have been found to be a good compromise.

[16]In the case of simultaneous SRS and PUCCH transmissions in the same subframe, a length-3 sequence is used, thereby making the last OFDM symbol in the subframe available for the sounding reference signal.

Phase rotation [multiples of $2\pi/12$]	Number of cover sequence					
	Even-numbered slot			Odd-numbered slot		
	0	1	2	0	1	2
0	0		12	12		16
1		6			14	
2	1		13	6		10
3		7			8	
4	2		14	0		4
5		8			2	
6	(3)		15	13		17
7		9			15	
8	4		16	7		(11)
9		10			9	
10	5		17	1		5
11		(11)				(3)

Cell A

Phase rotation [multiples of $2\pi/12$]	Number of cover sequence					
	Even-numbered slot			Odd-numbered slot		
	0	1	2	0	1	2
0		(11)			(3)	
1	0		12	12		16
2		6			14	
3	1		13	6		10
4		7			8	
5	2		14	0		4
6		8			2	
7	(3)		15	13		17
8		9			15	
9	4		16	7		(11)
10		10			9	
11	5		17	1		5

Cell B

FIGURE 11.24

Example of phase rotation and cover hopping for two PUCCH resource indices in two different cells.

reference symbols is reduced from 3 to 2 and the multiplexing capacity is lower. However, the general principles described above still apply.

As mentioned earlier, a PUCCH resource can be represented by an index. For a hybrid-ARQ acknowledgement, the resource index to use is given as a function of the first CCE in the PDCCH used to schedule the downlink transmission to the terminal. In this way, there is no need to explicitly include information about the PUCCH resources in the downlink scheduling assignment, which of course reduces overhead. Furthermore, as described in Chapter 12, hybrid-ARQ acknowledgements are transmitted a fixed time after the reception of a DL-SCH transport block and when to expect a hybrid ARQ on the PUCCH is therefore known to the eNodeB.

In addition to dynamic scheduling by using the PDCCH, there is also, as described in Chapter 13, the possibility to semi-persistently schedule a terminal according to a specific pattern. In this case there is no PDCCH to derive the PUCCH resource index from. Instead, the configuration of the semi-persistent scheduling pattern includes information on the PUCCH index to use for the hybrid-ARQ acknowledgement. In either of these two cases, a terminal is using PUCCH resources only when it has been scheduled in the downlink. Thus, the amount of PUCCH resources required for hybrid-ARQ acknowledgements does not necessarily increase with an increasing number of terminals in the cell, but, for dynamic scheduling, is rather related to the number of CCEs in the downlink control signaling.

The description above was mainly concerned with the case of downlink carrier aggregation not being used. For downlink carrier aggregation, there can be multiple simultaneous DL-SCHs scheduled for a single terminal, one per downlink component carrier and, consequently, multiple acknowledgement bits need to be conveyed in the uplink (one, or two in the case of spatial multiplexing, per

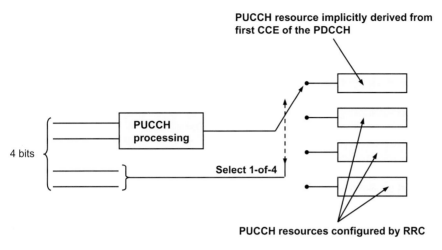

FIGURE 11.25

Resource selection for carrier aggregation.

downlink component carrier). PUCCH format 1 can be used to support more than two bits in the uplink by using resource selection. Assume four bits are to be transmitted in the uplink. With resource selection, two bits indicate which PUCCH resource to use while the remaining two bits are transmitted using the normal PUCCH structure but on the resource pointed to by the first two bits. This is illustrated in Figure 11.25. In total, four PUCCH resources are needed. One resource is derived from the first CCE using the same rule as in the absence of carrier aggregation (assuming that the scheduling assignment is transmitted on, and relating to, the primary component carrier). The remaining resources are semi-statically configured by RRC signaling.

For more than four bits, resource selection is less efficient and PUCCH format 3, described in a subsequent section, was therefore introduced in LTE release 10 in conjunction with carrier aggregation.

Unlike the hybrid-ARQ acknowledgements, whose occurrence is known to the eNodeB from the downlink scheduling decisions, the need for uplink resources for a certain terminal is in principle unpredictable by the eNodeB. One way to handle this would be to have a contention-based mechanism for requesting uplink resources. The random-access mechanism is based on this principle and can, to some extent, also be used for scheduling requests, as discussed in Chapter 13. Contention-based mechanisms typically work well for low intensities, but for higher scheduling-request intensities, the collision rate between different terminals simultaneously requesting resources becomes too large. Therefore, LTE provides a contention-free scheduling-request mechanism on the PUCCH, where each terminal in the cell is given a reserved resource on which it can transmit a request for uplink resources. The contention-free resource is represented by a PUCCH resource index as described earlier, occurring at every nth subframe. The more frequently these time instants occur, the lower the scheduling-request delay at the cost of higher PUCCH resource consumption. As the eNodeB configures all the terminals in the cell, when and on which resources a terminal can request resources is known to the eNodeB. A single scheduling request resource is also sufficient for the case

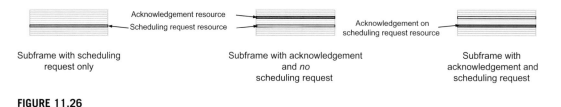

FIGURE 11.26

Multiplexing of scheduling request and hybrid-ARQ acknowledgement from a single terminal.

of carrier aggregation, as it only represents a request for uplink resources, which is independent of whether carrier aggregation is used or not.

The discussion earlier has considered transmission of *either* a hybrid-ARQ acknowledgement *or* a scheduling request. However, there are situations when the terminal needs to transmit *both* of them. In such a situation, the hybrid-ARQ acknowledgement is transmitted on the scheduling-request resource (see Figure 11.26). This is possible as the same PUCCH structure is used for both of them and the scheduling request carries no explicit information. By comparing the amount of energy detected on the acknowledgement resource and the scheduling-request resource for a specific termi-nal, the eNodeB can determine whether the terminal is requesting resources or not. Once the resource used for transmission of the acknowledgement is detected, the hybrid-ARQ acknowledgement can be decoded. Other, more advanced methods jointly decoding hybrid ARQ and scheduling request can also be envisioned.

11.4.1.2 *PUCCH Format 2*

Channel-state reports, the contents of which are discussed in Chapter 13, are used to provide the eNodeB with an estimate of the channel properties as seen from the terminal to aid channel-dependent scheduling. A channel-state report consists of multiple bits transmitted in one subframe. PUCCH format 1, which is capable of at most two bits of information per subframe, obviously cannot be used for this purpose. Transmission of channel-state reports on the PUCCH is instead handled by PUCCH format 2, which is capable of multiple information bits per subframe.[17]

PUCCH format 2, illustrated for a normal cyclic prefix in Figure 11.27, is based on a phase rota-tion of the same cell-specific sequence as format 1. After block coding using a punctured Reed–Müller code and QPSK modulation of the channel-state information, there are 10 QPSK symbols to transmit in the subframe: the first five symbols are transmitted in the first slot and the remaining five in the last slot.

Assuming a normal cyclic prefix, there are seven OFDM symbols per slot. Of the seven OFDM symbols in each slot, two[18] are used for reference-signal transmission to allow coherent demodula-tion at the eNodeB. In the remaining five, the respective QPSK symbol to be transmitted is multiplied

---

[17]There are actually three variants in the LTE specifications, formats 2, 2A, and 2B, where the last two formats are used for simultaneous transmission of channel-state reports and hybrid-ARQ acknowledgements, as discussed later in this section. However, for simplicity, they are all referred to as format 2 herein.

[18]Similarly to format 1, the number of symbols used for reference signals and the coded channel-quality information is a trade-off between channel-estimation accuracy and energy in the information part. Two symbols for reference symbols and five symbols for the coded information part in each slot were found to be the best compromise.

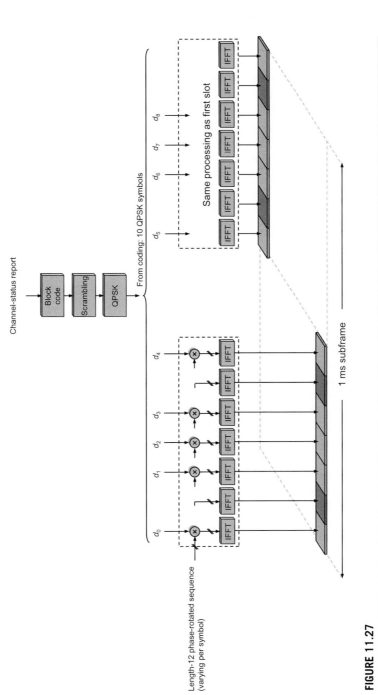

FIGURE 11.27

PUCCH format 2 (normal cyclic prefix).

by a phase-rotated length-12 cell-specific sequence and the result is transmitted in the corresponding OFDM symbol. For an extended cyclic prefix, where there are six OFDM symbols per slot, the same structure is used but with one reference-signal symbol per slot instead of two.

Basing the format 2 structure on phase rotations of the same cell-specific sequence as format 1 is beneficial as it allows the two formats to be transmitted in the same resource block. As phase-rotated sequences are orthogonal, one rotated sequence in the cell can be used either for one PUCCH instance using format 2 or three PUCCH instances using format 1. Thus, the "resource consumption" of one channel-state report is equivalent to three hybrid-ARQ acknowledgements (assuming normal cyclic prefix). Note that no orthogonal cover sequences are used for format 2.

The rotation angles to use in the different symbols for PUCCH format 2 are hopping in a similar way as for format 1, motivated by interference randomization. Resources for PUCCH format 2 can, similar to format 1, be represented by a scalar index, which can be seen as a "channel number" and higher-layer signaling is used to configure each terminal with a resource to transmit its channel-state report on, as well as when those reports should be transmitted. Hence, the eNodeB has full knowledge of when and on which resources each of the terminals will transmit channel-state reports on PUCCH. The channel-state reports on PUCCH are also known as *periodic* reports; as will be discussed later, there are also *aperiodic* reports but those reports can only be transmitted on PUSCH.

Periodic CSI reports in combination with carrier aggregation are provided for by transmitting multiple periodic reports, one per component carrier, offset in time such that reports for two different component carriers do not collide.

11.4.1.3 *PUCCH Format 3*

For downlink carrier aggregation, multiple hybrid-ARQ acknowledgement bits need to be fed back in the case of simultaneous transmission on multiple component carriers. As discussed above, PUCCH format 1 in combination with resource selection can be used, but for more than four bits this is not an efficient solution from a performance perspective. To enable the possibility of transmitting more than four bits in an efficient way, PUCCH format 3 can be used. A terminal capable of more than two downlink component carriers – that is, capable of more than fours bits for hybrid-ARQ acknowledgements – needs to support PUCCH format 3. For such a terminal, PUCCH format 3 can also be used for less than four bits of feedback relating to simultaneous transmission on multiple component carriers if configured by higher-layer signaling not to use PUCCH format 1 with resource selection.

The basis for PUCCH format 3, illustrated in Figure 11.28, is DFT-precoded OFDM – that is, the same transmission scheme as used for UL-SCH. The acknowledgement bits, one or two per downlink component carrier depending on the transmission mode configured for that particular component carrier, are concatenated with a scheduling request bit (if present) into a sequence of bits where bits corresponding to unscheduled transport blocks are set to zero. Block coding is applied,[19] followed by scrambling using a cell-specific scrambling sequence to randomize inter-cell interference. The resulting 48 bits are QPSK-modulated and divided into two groups, one per slot, of 12 QPSK symbols each.

Assuming a normal cyclic prefix, there are seven OFDM symbols per slot. Similarly to PUCCH format 2, two OFDM symbols (one in the case of an extended cyclic prefix) in each slot are used for reference signal transmission, leaving five symbols for data transmission. In each slot, the block of 12 DFT-precoded QPSK symbols is transmitted in the five available DFTS-OFDM symbols. To further

---

[19]A (32, k) Reed–Müller code is used, but for 12 or more bits in TDD two Reed–Müller codes are used in combination.

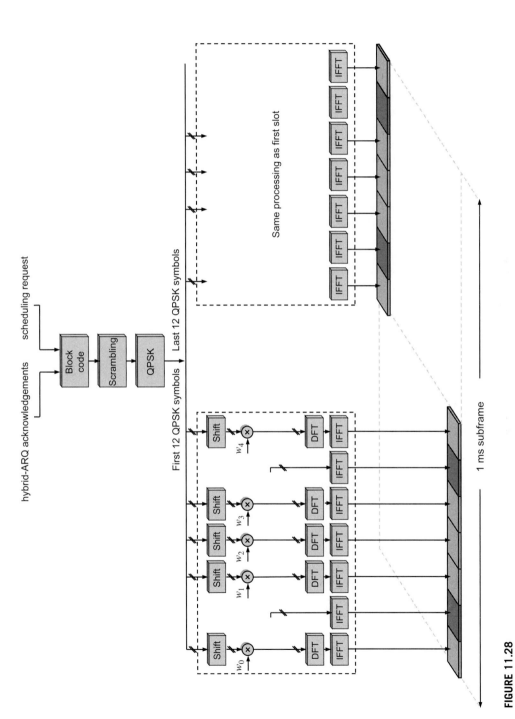

FIGURE 11.28

PUCCH format 3 (normal cyclic prefix).

randomize the inter-cell interference, a cyclic shift of the 12 inputs to the DFT, varying between OFDM symbols in a cell-specific manner, is applied to the block of 12 QPSK symbols prior to DFT precoding.

To increase the multiplexing capacity, a length-5 orthogonal sequence is used with each of the five OFDM symbol carrying data in a slot being multiplied by one element of the sequence. Thus, up to five terminals may share the same resource-block pair for PUCCH format 3. Different length-5 sequences are used in the two slots to improve the performance in high-Doppler scenarios. To facilitate channel estimation for the different transmissions sharing the same resource block, different reference-signal sequences are used.

The length-5 orthogonal cover sequences are obtained as five DFT sequences. There is also the possibility to use a length-4 Walsh sequence for the second slot in order to leave the last OFDM symbol unused for the case when sounding is configured in the subframe.

In the same manner as for the other two PUCCH formats, a resource can be represented by a single index from which the orthogonal sequence and the resource-block number can be derived. A terminal can be configured with four different resources for PUCCH format 3. In the scheduling grant for a secondary carrier, two bits[20] inform the terminal which of the four resources to use. In this was, the scheduler can avoid PUCCH collisions between different terminals by assigning them to different resources.

Note that, due to the differences in the underlying structure of PUCCH format 3 compared to the other two formats, resource blocks cannot be shared between format 3 and the other two formats.

11.4.1.4 *Simultaneous Transmission of Multiple Feedback Reports*

The above description has focused on transmission of hybrid-ARQ acknowledgements or channel-state reports alone. However, there is also a need to handle the case when hybrid ARQ or scheduling requests need to be transmitted in the same subframe. Transmitting multiple PUCCHs simultaneously from the same terminal would increase the cubic metric. Therefore, a single PUCCH structure supporting simultaneous transmission of multiple feedback signals is used.

In principle, the following four situations requiring simultaneous transmission of multiple feedback signals from a single terminal can be envisioned:

1. *Hybrid-ARQ acknowledgement and scheduling request.* For PUCCH format 1, this is supported by transmitting the hybrid-ARQ acknowledgement on the scheduling-request resource as described earlier. For PUCCH format 3, the transmission structure supports inclusion of a scheduling-request bit.
2. *Scheduling request and channel-state report.* The eNodeB is in control of when a terminal may transmit a scheduling request and when the terminal should report the channel state. Hence, this situation can be avoided by proper configuration, but if this is not done the terminal drops the channel-state report and transmits the scheduling request only. Missing a channel-state report is not detrimental and only incurs some degradation in the scheduling and rate-adaptation accuracy, whereas the scheduling request is critical for uplink transmissions.

---

[20]The power control bits in an assignment relating to a secondary component carrier are reinterpreted as an *acknowledgement resource indicator* (ARI) – see Chapter 10.

3. *Scheduling request, hybrid-ARQ acknowledgement, and channel-state report.* This is similar to the previous situation; the channel-state report is dropped and multiplexing of hybrid ARQ and scheduling request is handled as described in point 1 above.
4. *Hybrid-ARQ acknowledgement and channel-state report.* Simultaneous transmission of acknowledgement and channel-state report from the same terminal is possible, as described below. There is also the possibility to configure the terminal to drop the channel-state report and only transmit the acknowledgement. In the case of carrier aggregation, the channel-state report is always dropped if it happens to collide with PUCCH format 3.

Transmission of data in the downlink implies transmission of hybrid-ARQ acknowledgements in the uplink. At the same time, since data is transmitted in the downlink, up-to-date channel-state reports are beneficial to optimize the downlink transmissions. Hence, simultaneous transmission of hybrid-ARQ acknowledgements and channel-state reports is supported by LTE. The basis is PUCCH format 2 as described earlier for both normal and extended cyclic prefixes, although the detailed solution differs between the two.

For a normal cyclic prefix, each slot in PUCCH format 2 has two OFDM symbols used for reference signals. When transmitting a hybrid-ARQ acknowledgement at the same time as a channel-state report, the second reference signal in each slot is modulated by the acknowledgement, as illustrated in Figure 11.29a. Either BPSK or QPSK is used, depending on whether one or two acknowledgement bits are to be fed back. The fact that the acknowledgement is superimposed on the reference signal needs to be accounted for at the eNodeB. One possibility is to decode the acknowledgement bit(s) modulated on to the second reference symbol using the first reference symbol for channel estimation. Once the acknowledgement bit(s) have been decoded, the modulation imposed on the second reference symbol can be removed and channel estimation and decoding of the channel-state report can be handled in the same way as in the absence of simultaneous hybrid-ARQ acknowledgement. This two-step approach works well for low to medium Doppler frequencies; for higher Doppler frequencies the acknowledgement and channel-state reports are preferably decoded jointly.

For an extended cyclic prefix, there is only a single reference symbol per slot. Hence, it is not possible to overlay the hybrid-ARQ acknowledgement on the reference symbol. Instead, the acknowledgement bit(s) are jointly coded with the channel-state report prior to transmission using PUCCH format 2, as illustrated in Figure 11.29b.

The time instances when to expect channel-state reports and hybrid-ARQ acknowledgements are known to the eNodeB, which therefore knows whether to expect a hybrid-ARQ acknowledgement along with the channel-state report or not. If the PDCCH assignment is missed by the terminal, then only the channel-state report will be transmitted as the terminal is not aware that it has been scheduled. In the absence of a simultaneous channel-state report, the eNodeB can employ DTX detection to discriminate between a missed assignment and a failed decoding of downlink data. However, one consequence of the structures above is that DTX detection is cumbersome, if not impossible. This implies that incremental redundancy needs to be operated with some care if the eNodeB has scheduled data such that the acknowledgement occurs at the same time as a channel-state report. As the terminal may have missed the original transmission attempt in the downlink, it may be preferable for the eNodeB to select the redundancy version of the retransmission such that systematic bits are also included in the retransmission.

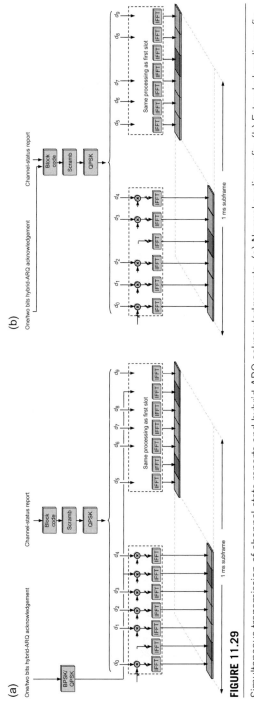

FIGURE 11.29

Simultaneous transmission of channel-state reports and hybrid-ARQ acknowledgements. (a) Normal cyclic prefix. (b) Extended cyclic prefix.

One possibility to circumvent this is to configure the terminal to drop the channel-state report in the case of simultaneous transmission of a hybrid-ARQ acknowledgement. In this case, the eNodeB can detect DTX as the acknowledgement can be transmitted using PUCCH format 1, as described earlier. Obviously, there will be no channel-state report sent in this case, which needs to be taken into account in the scheduling process. Dropping the channel-state report can also be useful for handling coverage-limited situations when the available transmission power is insufficient for a channel-state report in addition to the acknowledgement.

11.4.1.5 Resource-Block Mapping for PUCCH

The signals described above for all three of the PUCCH formats are, as already explained, transmitted on a resource-block pair with one resource block in each slot. The resource-block pair to use is determined from the PUCCH resource index. Multiple resource-block pairs can be used to increase the control-signaling capacity; when one resource-block pair is full, the next PUCCH resource index is mapped to the next resource-block pair in sequence. The mapping is in principle done such that PUCCH format 2 (channel-state reports) is transmitted closest to the edges of the uplink cell bandwidth with PUCCH format 1 (hybrid-ARQ acknowledgements, scheduling requests) next, as illustrated in Figure 11.30. The location of PUCCH format 3 is configurable and it can, for example, be located between formats 1 and 2. A semi-static parameter, provided as part of the system information, controls on which resource-block pair the mapping of PUCCH format 1 starts. Furthermore, the semi-statically configured scheduling requests are located at the outermost parts of the format 1 resources. As the amount of resources necessary for hybrid-ARQ acknowledgements varies dynamically, this maximizes the amount of contiguous spectrum available for PUSCH.

In many scenarios, the configuration of the PUCCH resources can be done such that the three PUCCH formats are transmitted on separate sets of resource blocks. However, for the smallest cell bandwidths, this would result in too high an overhead. Therefore, it is possible to mix the first two PUCCH formats in one of the resource-block pairs – for example, in Figure 11.30 this is the case for the resource with index 2. Although this mixture is primarily motivated by the smaller cell bandwidths, it can equally well be used for the larger cell bandwidths. In the resource-block pair where PUCCH formats 1 and 2 are mixed, the rotation angles are divided between the two formats. Furthermore,

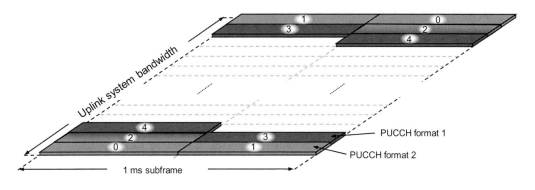

FIGURE 11.30

Allocation of resource blocks for PUCCH.

some of the phase rotations are reserved as "guard"; hence the efficiency of such a mixed resource-block pair is slightly lower than a resource-block pair carrying only one of the first two PUCCH formats.

11.4.2 Uplink L1/L2 Control Signaling on PUSCH

If the terminal is transmitting data on PUSCH – that is, has a valid scheduling grant in the subframe – control signaling is time multiplexed[21] with data on the PUSCH instead of using the PUCCH (in release 10, simultaneous PUSCH and PUCCH can be used). Only hybrid-ARQ acknowledgements and channel-state reports are transmitted on the PUSCH. Obviously there is no need to request a scheduling grant when the terminal is already scheduled; instead, in-band buffer-status reports are sent as part of the MAC headers, as described in Chapter 12.

Time multiplexing of channel-state reports and hybrid-ARQ acknowledgements is illustrated in Figure 11.31. However, although they both use time multiplexing there are some differences in the details for the two types of uplink L1/L2 control signaling motivated by their different properties.

The hybrid-ARQ acknowledgement is important for proper operation of the downlink. For one and two acknowledgements, robust QPSK modulation is used, regardless of the modulation scheme used for the data, while for a larger number of bits the same modulation scheme as for the data is used. Furthermore, the hybrid-ARQ acknowledgement is transmitted near to the reference symbols as the channel estimates are of better quality close to the reference symbols. This is especially important at high Doppler frequencies, where the channel may vary during a slot. Unlike the data part, the hybrid-ARQ acknowledgement cannot rely on retransmissions and strong channel coding to handle these variations.

In principle, the eNodeB knows when to expect a hybrid-ARQ acknowledgement from the terminal and can therefore perform the appropriate demultiplexing of the acknowledgement and the data part. However, there is a certain probability that the terminal has missed the scheduling assignment on the PDCCH, in which case the eNodeB will expect a hybrid-ARQ acknowledgement while the terminal will not transmit one. If the rate-matching pattern were to depend on whether an acknowledgement is transmitted or not, all the coded bits transmitted in the data part could be affected by a missed PDCCH assignment, which is likely to cause the UL-SCH decoding to fail. To avoid this error, the hybrid-ARQ acknowledgements are therefore punctured into the coded UL-SCH bit stream. Thus, the non-punctured bits are not affected by the presence/absence of hybrid-ARQ acknowledgements and the problem of a mismatch between the rate matching in the terminal and the eNodeB is avoided.

The contents of the channel-state reports is described in Chapter 12; at this stage it suffices to note that a channel-state report consists of *Channel-Quality Indicator* (CQI), *Precoding Matrix Indicator* (PMI), and *Rank Indicator* (RI). The CQI and PMI are time multiplexed with the coded data bits from PUSCH and transmitted using the same modulation as the data part. Channel-state reports are mainly useful for low-to-medium Doppler frequencies for which the radio channel is relatively constant, hence the need for special mapping is less pronounced. The RI, however, is mapped differently than the CQI and PMI; as illustrated in Figure 11.31, the RI is located near to the reference symbols using a similar mapping as the hybrid-ARQ acknowledgements. The more robust mapping of

---

[21] In the case of spatial multiplexing, the CQI/PMI is time multiplexed with one of the codewords, implying that it is spatially multiplexed with the other codeword.

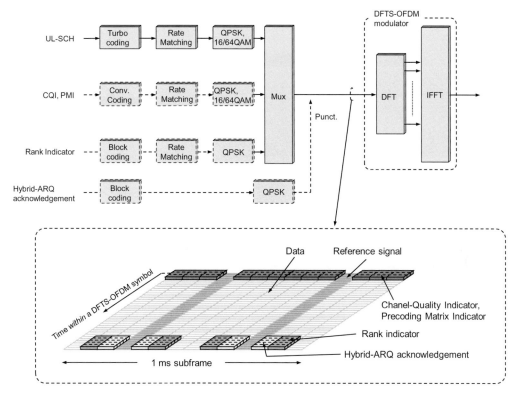

FIGURE 11.31

Multiplexing of control and data on to PUSCH. Note that there are *two* time dimensions and no frequency dimension in the "grid" at the bottom, the reason being that the signal prior to DFT precoding is illustrated. The acknowledgement (and the RI) will thus be located near but not as close as possible to the reference signal as possible as the "grid" is read column by column.

the RI is motivated by the fact that the RI is required in order to correctly interpret the CQI/PMI. The CQI/PMI, on the other hand, is simply mapped across the full subframe duration. Modulation-wise, the RI is handled in the same way as the hybrid-ARQ acknowledgements.

For uplink spatial multiplexing, in which case two transport blocks are transmitted simultaneously on the PUSCH, the CQI and PMI are multiplexed with the coded transport block using the highest modulation-and-coding scheme, followed by applying the multiplexing scheme described above per layer (Figure 11.32). The intention behind this approach is to transmit the CQI and PMI on the (one or two) layers with the best quality.[22]

The hybrid-ARQ acknowledgements and the rank indicator are replicated across all transmission layers and multiplexed with the coded data in each layer in the same way as the single layer case

---

[22] Assuming the MCS follows the channel quality, this holds for one, two, and four layers but not necessarily for three layers.

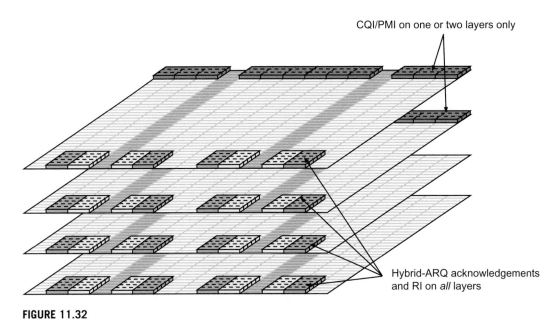

FIGURE 11.32

Multiplexing of CQI/PMI, RI, and hybrid-ARQ acknowledgements in the case of uplink spatial multiplexing.

described above. The bits may, though, have been scrambled differently on the different layers. In essence, as the same information is transmitted on multiple layers with different scrambling, this provides diversity.

The basis for channel-state reports on the PUSCH is *aperiodic reports*, where the eNodeB requests a channel-state report from the terminal by setting the CSI request bit in the scheduling grant, as mentioned in Chapter 10. UL-SCH rate matching takes the presence of the channel-state reports into account; by using a higher code rate a suitable number of resource elements is made available for transmission of the channel-state report. Since the reports are explicitly requested by the eNodeB, their presence is known and the appropriate rate de-matching can be done at the receiver. If one of the configured transmission instances for a periodic report coincides with the terminal being scheduled on the PUSCH, the periodic report is "rerouted" and transmitted on the PUSCH resources. Also, in this case there is no risk of mismatch in rate matching; the transmission instants for periodic reports are configured by robust RRC signaling and the eNodeB knows in which subframes such reports will be transmitted.

The channel coding of the channel-state reports depends on the report size. For the smaller sizes such as a periodic report that otherwise would have been transmitted on the PUCCH, the same block coding as used for the PUCCH reports is used. For the larger reports, a tail-biting convolutional code is used for CQI/PMI, whereas the RI uses a (3, 2) block code for a single component carrier.

Unlike the data part, which relies on rate adaptation to handle different radio conditions, this cannot be used for the L1/L2 control-signaling part. Power control could, in principle, be used as an alternative, but this would imply rapid power variations in the time domain, which negatively impact the RF properties. Therefore, the transmission power is kept constant over the subframe and the amount of resource elements allocated to L1/L2 control signaling – that is, the code rate of the control signaling – is varied according to the scheduling decision for the data of the data part. High data rates are typically scheduled when the radio conditions are advantageous and hence a smaller amount of resource needs to be used by the L1/L2 control signaling compared to the case of poor radio conditions. To account for different hybrid-ARQ operating points, an offset between the code rate for the control-signaling part and the modulation-and-coding scheme used for the data part can be configured via higher-layer signaling.

For carrier aggregation, control signaling is time multiplexed on one uplink component carrier only – that is, uplink control information cannot be split across multiple uplink component carriers. Apart from the aperiodic CSI reports, which are transmitted upon the component carrier that triggered the report, the primary component carrier is used for uplink control signaling if scheduled in the same subframe, otherwise one of the secondary component carriers is used.

11.5 UPLINK TIMING ALIGNMENT

The LTE uplink allows for uplink intra-cell orthogonality, implying that uplink transmissions received from different terminals within a cell do not cause interference to each other. A requirement for this *uplink orthogonality* to hold is that the signals transmitted from different terminals within the same subframe but within different frequency resources (different resource blocks) arrive approximately time aligned at the base station. More specifically, any timing misalignment between received signals received should fall within the cyclic prefix. To ensure such receiver-side time alignment, LTE includes a mechanism for *transmit-timing advance*.

In essence, timing advance is a negative offset, at the terminal, between the start of a received downlink subframe and a transmitted uplink subframe. By controlling the offset appropriately for each terminal, the network can control the timing of the signals received at the base station from the terminals. Terminals far from the base station encounter a larger propagation delay and therefore need to start their uplink transmissions somewhat in advance, compared to terminals closer to the base station, as illustrated in Figure 11.33. In this specific example, the first terminal is located close to the base station and experiences a small propagation delay, $T_{P,1}$. Thus, for this terminal, a small value of the timing advance offset $T_{A,1}$ is sufficient to compensate for the propagation delay and to ensure the correct timing at the base station. However, a larger value of the timing advance is required for the second terminal, which is located at a larger distance from the base station and thus experiences a larger propagation delay.

The timing-advance value for each terminal is determined by the network based on measurements on the respective uplink transmissions. Hence, as long as a terminal carries out uplink data transmission, this can be used by the receiving base station to estimate the uplink receive timing and thus be a source for the timing-advance commands. Sounding reference signals can be used as a regular signal to measure upon, but in principle the base station can use any signal transmitted from the terminals.

Based on the uplink measurements, the network determines the required timing correction for each terminal. If the timing of a specific terminal needs correction, the network issues a timing-advance command for this specific terminal, instructing it to retard or advance its timing relative

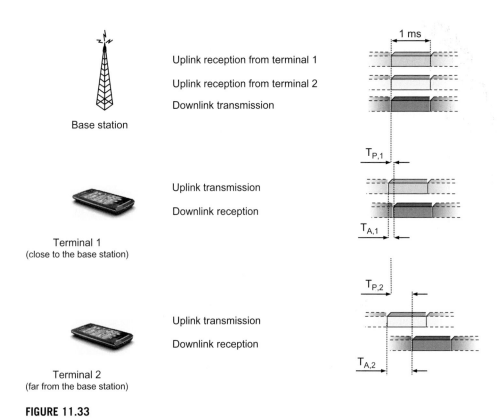

FIGURE 11.33

Uplink timing advance.

to the current uplink timing. The user-specific timing-advance command is transmitted as a MAC control element (see Chapter 8 for a description of MAC control elements) on the DL-SCH. The maximum value possible for timing advance is 0.67 ms, corresponding to a terminal-to-base-station distance of slightly more than 100 km. This is also the value assumed when determining the processing time for decoding, as discussed in Section 12.1. Typically, timing-advance commands to a terminal are transmitted relatively infrequently – for example, one or a few times per second.

If the terminal has not received a timing-advance command during a (configurable) period, the terminal assumes it has lost the uplink synchronization. In this case, the terminal must re-establish uplink timing using the random-access procedure prior to any PUSCH or PUCCH transmission in the uplink.

For carrier aggregation, there may be multiple component carriers transmitted from a single terminal. In principle, different timing advance commands for different component carriers could be envisioned. One motivation for this could be inter-band carrier aggregation, where the different component carriers are received at different geographical locations, for example by using remote radio heads for some of the bands but not others. However, such deployments are not common and in the interest of simplicity LTE is using a single timing-advance command valid for all uplink component carriers.

Retransmission Protocols

The LTE downlink and uplink transmission schemes have been described from a physical layer perspective in Chapters 10 and 11 respectively. In this chapter, the operation of the two mechanisms responsible for retransmission handling in LTE, the MAC and RLC sublayers, will be described.

Retransmissions of missing or erroneous data units are handled primarily by the hybrid-ARQ mechanism in the MAC layer, complemented by the retransmission functionality of the RLC protocol. The reasons for having a two-level retransmission structure can be found in the trade-off between fast and reliable feedback of the status reports. The hybrid-ARQ mechanism targets very fast retransmissions and, consequently, feedback on success or failure of the decoding attempt is provided to the transmitter after each received transport block. Although it is in principle possible to attain an arbitrarily low error probability of the hybrid-ARQ feedback, it comes at a cost in transmission power. Keeping the cost reasonable typically results in a feedback error rate of around 1%, which results in a hybrid-ARQ residual error rate of a similar order. Such an error rate is in many cases far too high; high data rates with TCP may require virtually error-free delivery of packets to the TCP protocol layer. As an example, for sustainable data rates exceeding 100 Mbit/s, a packet-loss probability less than 10^{-5} is required [72]. The reason is that TCP assumes packet errors to be due to congestion in the network. Any packet error therefore triggers the TCP congestion-avoidance mechanism with a corresponding decrease in data rate.

Compared to the hybrid-ARQ acknowledgements, the RLC status reports are transmitted relatively infrequently and thus the cost of obtaining a reliability of 10^{-5} or lower is relatively small. Hence, the combination of hybrid-ARQ and RLC attains a good combination of small round-trip time and a modest feedback overhead where the two components complement each other – fast retransmissions due to the hybrid-ARQ mechanism and reliable packet delivery due to the RLC. As the MAC and RLC protocol layers are located in the same network node, a tight interaction between the two protocols is possible. Hence, to some extent, the combination of the two can be viewed as *one* retransmission mechanism with *two* feedback channels. However, note that, as discussed in Chapter 8 and illustrated in Figure 12.1, the RLC operates per logical channel, while the hybrid-ARQ operates per transport channel (that is, per component carrier). One hybrid-ARQ entity may therefore retransmit data belonging to multiple logical channels. Similarly, one RLC entity may handle data transmitted across multiple component carriers in the presence of carrier aggregation. Finally, for carrier aggregation, hybrid-ARQ retransmissions must occur on the same component carrier as the original transmission, as each component carrier has its own independent hybrid-ARQ entity. RLC retransmissions, on the other hand, are not tied to a specific component carrier as, in essence, carrier aggregation is invisible above the MAC layer.

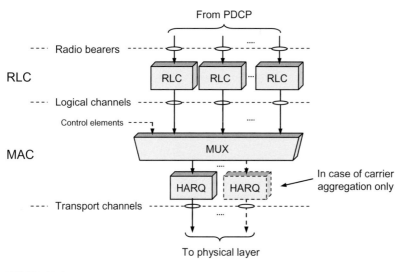

From PDCP

Radio bearers

RLC

Logical channels

Control elements

MAC

MUX

In case of carrier
aggregation only

Transport channels

To physical layer

FIGURE 12.1

RLC and hybrid-ARQ retransmission mechanisms in LTE.

In the following section, the principles behind the hybrid-ARQ and RLC protocols will be discussed in more detail.

12.1 HYBRID ARQ WITH SOFT COMBINING

The hybrid-ARQ functionality spans both the physical layer and the MAC layer; generation of different redundancy versions at the transmitter as well as the soft combining at the receiver are handled by the physical layer, while the hybrid-ARQ protocol is part of the MAC layer. In the presence of carrier aggregation, there is, as already stated, one independent hybrid-ARQ entity per component carrier and terminal. Unless otherwise noted, the description below holds for one component carrier – that is, the description is on a per-component carrier basis.

The basis for the LTE hybrid-ARQ mechanism is a structure with multiple stop-and-wait protocols, each operating on a single transport block. In a stop-and-wait protocol, the transmitter stops and waits for an acknowledgement after each transmitted transport block. This is a simple scheme; the only feedback required is a single bit indicating positive or negative acknowledgement of the transport block. However, since the transmitter stops after each transmission, the throughput is also low. LTE therefore applies *multiple* stop-and-wait processes operating in parallel such that, while waiting for acknowledgement from one process, the transmitter can transmit data to another hybrid-ARQ process. This is illustrated in Figure 12.2; while processing the data received in the first hybrid-ARQ process the receiver can continue to receive using the second process and so on. This structure, multiple hybrid-ARQ processes operating in parallel to form one hybrid-ARQ entity, combines the simplicity of a stop-and-wait protocol while still allowing continuous transmission of data.

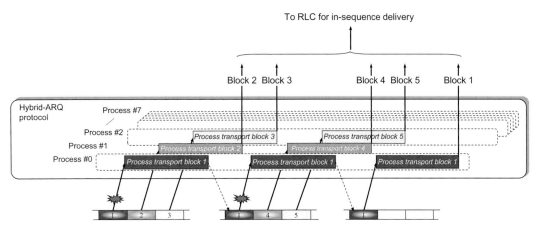

FIGURE 12.2

Multiple parallel hybrid-ARQ processes forming one hybrid-ARQ entity.

There is one hybrid-ARQ entity per terminal (and per component carrier in the case of carrier aggregation; see Figure 12.1). Spatial multiplexing, where two transport blocks can be transmitted in parallel on the same transport channel as described in Chapter 10, is supported by one hybrid-ARQ entity having two sets of hybrid-ARQ processes with independent hybrid-ARQ acknowledgements. The details for the physical-layer transmission of the downlink and uplink hybrid-ARQ acknowledgements were described in Chapters 10 and 11.

Upon receiving a transport block for a certain hybrid-ARQ process, the receiver makes an attempt to decode the transport block and informs the transmitter about the outcome through a hybrid-ARQ acknowledgement, indicating whether the transport block was correctly decoded or not. The time from reception of data until transmission of the hybrid-ARQ acknowledgement is fixed, hence the transmitter knows from the timing relation which hybrid-ARQ process a received acknowledgement relates to. This is beneficial from an overhead perspective as there is no need to signal the process number along with the acknowledgement.

An important part of the hybrid-ARQ mechanism is the use of *soft combining*, which implies that the receiver combines the received signal from multiple transmission attempts. The principles of hybrid-ARQ with soft combining were outlined in Chapter 6, with incremental redundancy being the basic scheme in LTE. Clearly, the receiver needs to know when to perform soft combining prior to decoding and when to clear the soft buffer – that is, the receiver needs to differentiate between the reception of an initial transmission (prior to which the soft buffer should be cleared) and the reception of a retransmission. Similarly, the transmitter must know whether to retransmit erroneously received data or to transmit new data. Therefore, an explicit *new-data indicator* is included for each of the one or two scheduled transport blocks along with other downlink scheduling information on the PDCCH. The new-data indicator is present in both downlink assignments and uplink grants, although the meaning is slightly different for the two.

For downlink data transmission, the new-data indicator is toggled for a new transport block – that is, it is essentially a single-bit sequence number. Upon reception of a downlink scheduling assignment, the terminal checks the new-data indicator to determine whether the current transmission should be soft combined with the received data currently in the soft buffer for the hybrid-ARQ process in question, or if the soft buffer should be cleared.

For uplink data transmission, there is also a new-data indicator transmitted on the downlink PDCCH. In this case, toggling the new-data indicator requests transmission of a new transport block, otherwise the previous transport block for this hybrid-ARQ process should be retransmitted (in which case the eNodeB should perform soft combining).

The use of multiple parallel hybrid-ARQ processes operating in parallel can result in data being delivered from the hybrid-ARQ mechanism out of sequence. For example, transport block 5 in Figure 12.2 was successfully decoded before transport block 1, which required two retransmissions. Out-of-sequence delivery can also occur in the case of carrier aggregation, where transmission of a transport block on one component carrier could be successful while a retransmission is required on another component carrier. To handle out-of-sequence delivery from the hybrid-ARQ protocol, the RLC protocol includes an in-sequence-delivery mechanism, as described in Section 12.2.

Hybrid-ARQ protocols can be characterized as synchronous vs. asynchronous, related to the flexibility in the time domain, as well as adaptive vs. non-adaptive, related to the flexibility in the frequency domain:

- An *asynchronous* hybrid-ARQ protocol implies that retransmissions can occur at any time, whereas a *synchronous* protocol implies that retransmissions occur at a fixed time after the previous transmission (see Figure 12.3). The benefit of a synchronous protocol is that there is no need to explicitly signal the hybrid-ARQ process number as this information can be derived from the subframe number. On the other hand, an asynchronous protocol allows for more flexibility in the scheduling of retransmissions.
- An *adaptive* hybrid-ARQ protocol implies that the frequency location and possibly also the more detailed transmission format can be changed between retransmissions. A *non-adaptive protocol,*

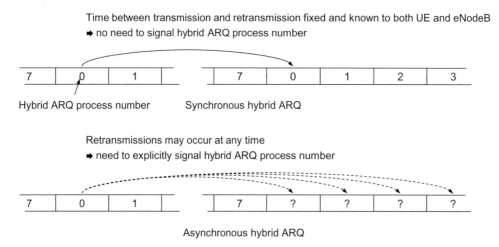

FIGURE 12.3

Synchronous and asynchronous hybrid ARQ.

in contrast, implies that the retransmission must occur at the same frequency resources and with the same transmission format as the initial transmission.

In the case of LTE, asynchronous adaptive hybrid ARQ is used for the downlink. For the uplink, synchronous hybrid ARQ is used. Typically, the retransmissions are non-adaptive, but there is also the possibility to use adaptive retransmissions as a complement.

12.1.1 Downlink Hybrid ARQ

In the downlink, retransmissions are scheduled in the same way as new data – that is, they may occur at any time and at an arbitrary frequency location within the downlink cell bandwidth. Hence, the downlink hybrid-ARQ protocol is *asynchronous* and *adaptive*. The support for asynchronous and adaptive hybrid ARQ for the LTE downlink is motivated by the need to avoid collisions with, for example, transmission of system information and MBSFN subframes. Instead of dropping a retransmission that otherwise would collide with MBSFN subframes or transmission of system information, the eNodeB can move the retransmission in time and/or frequency to avoid the overlap in resources.

Support for soft combining is, as described in the introduction, provided through an explicit new-data indicator, toggled for each new transport block. In addition to the new-data indicator, hybrid-ARQ-related downlink control signaling consists of the hybrid-ARQ process number (three bits for FDD, four bits for TDD) and the redundancy version (two bits), both explicitly signaled in the scheduling assignment for each downlink transmission.

Downlink spatial multiplexing implies, as already mentioned, transmission of two transport blocks in parallel on a component carrier. To provide the possibility to retransmit only one of the transport blocks, which is beneficial as error events for the two transport blocks can be fairly uncorrelated, each transport block has its own separate new-data indicator and redundancy-version indication. However, there is no need to signal the process number separately as each process consists of two subprocesses in the case of spatial multiplexing or, expressed differently, once the process number for the first transport block is known, the process number for the second transport block is given implicitly.

Transmissions on downlink component carriers are acknowledged independently. On each component carrier, transport blocks are acknowledged by transmitting one or two bits on the uplink, as described in Chapter 11. In the absence of spatial multiplexing, there is only a single transport block within a TTI and consequently only a single acknowledgement bit is required in response. However, if the downlink transmission used spatial multiplexing, there are two transport blocks per TTI, each requiring its own hybrid-ARQ acknowledgement bit. The total number of bits required for hybrid-ARQ acknowledgements thus depends on the number of component carriers and the transmission mode for each of the component carriers. As each downlink component carrier is scheduled separately from its own PDCCH, the hybrid-ARQ process numbers are signaled independently for each component carrier.

Obviously, the terminal should not transmit a hybrid-ARQ acknowledgement in response to reception of system information, paging messages and other broadcast traffic. Hence, hybrid-ARQ acknowledgements are only sent in the uplink for "normal" unicast transmission.

12.1.2 Uplink Hybrid ARQ

Shifting the focus to the uplink, a difference compared to the downlink case is the use of synchronous, non-adaptive operation as the basic principle of the hybrid-ARQ protocol, motivated by the lower overhead compared to an asynchronous, adaptive structure. Hence, uplink retransmissions always occur at an *a priori* known subframe; in the case of FDD operation uplink retransmissions occur eight subframes after the prior transmission attempt for the same hybrid-ARQ process. The set of resource blocks used for the retransmission on a component carrier is identical to the initial transmission. Thus, the only control signaling required in the downlink for a retransmission is a hybrid-ARQ acknowledgement, transmitted on the PHICH as described in Chapter 10. In the case of a negative acknowledgement on the PHICH, the data is retransmitted.

Spatial multiplexing is handled in the same way as in the downlink – two transport blocks transmitted in parallel on the uplink, each with its own modulation-and-coding scheme and new-data indicator but sharing the same hybrid-ARQ process number. The two transport blocks are individually acknowledged and hence two bits of information are needed in the downlink to acknowledge an uplink transmission using spatial multiplexing. Uplink carrier aggregation is another example of a situation where multiple acknowledgements are needed – one or two acknowledgement bits per uplink component carrier. TDD is a third example, as will be discussed further below, where uplink transmissions in different subframes may need to be acknowledged in the same downlink subframe. Thus, multiple PHICHs may be needed as each PHICH is capable of transmitting a single bit only. Each of the PHICHs is transmitted on the same downlink component carrier that was used to schedule the initial uplink transmission.

Despite the fact that the basic mode of operation for the uplink is synchronous, non-adaptive hybrid ARQ, there is also the possibility to operate the uplink hybrid ARQ in a synchronous, *adaptive* manner, where the resource-block set and modulation-and-coding scheme for the retransmissions is changed. Although non-adaptive retransmissions are typically used due to the very low overhead in terms of downlink control signaling, adaptive retransmissions are sometimes useful to avoid fragmenting the uplink frequency resource or to avoid collisions with random-access resources. This is illustrated in Figure 12.4. A terminal is scheduled for an initial transmission in subframe n; a transmission that is not correctly received and consequently a retransmission is required in subframe $n + 8$ (assuming FDD; for TDD the timing obviously depends on the downlink–uplink allocation, as discussed further below). With non-adaptive hybrid ARQ, the retransmissions occupy the same part of the uplink spectrum as the initial transmission. Hence, in this example the spectrum is fragmented, which limits the bandwidth available to another terminal (unless the other terminal is capable of multi-cluster transmission). In subframe $n + 16$, an example of an adaptive retransmission is found; to make room for another terminal to be granted a large part of the uplink spectrum, the retransmission is moved in the frequency domain. It should be noted that the uplink hybrid-ARQ protocol is still synchronous – that is, a retransmission should always occur eight subframes after the previous transmission.

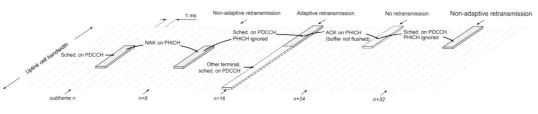

FIGURE 12.4

Non-adaptive and adaptive hybrid-ARQ operation.

The support for both adaptive and non-adaptive hybrid ARQ is realized by *not* flushing the transmission buffer when receiving a positive hybrid-ARQ acknowledgement on PHICH for a given hybrid-ARQ process. Instead, the actual control of whether data should be retransmitted or not is done by the new-data indicator included in the uplink scheduling grant sent on the PDCCH. The new-data indicator is toggled for each new transport block. If the new-data indicator is toggled, the terminal flushes the transmission buffer and transmits a new data packet. However, if the new-data indicator does not request transmission of a new transport block, the previous transport block is retransmitted. Hence, clearing of the transmission buffer is not handled by the PHICH but by the PDCCH as part of the uplink grant. The negative hybrid-ARQ acknowledgement on the PHICH could instead be seen as a single-bit scheduling grant for retransmissions where the set of bits to transmit and all the resource information are known from the previous transmission attempt. In Figure 12.4 an example of postponing a transmission is seen in subframe $n + 24$. The terminal has received a positive acknowledgement and therefore does *not* retransmit the data. However, the transmission buffer is not flushed, which later is exploited by an uplink grant requesting retransmission in subframe $n + 32$.

A consequence of the above method of supporting both adaptive and non-adaptive hybrid ARQ is that the PHICH and PDCCH related to the same uplink subframe have the same timing. If this were not the case, the complexity would increase as the terminal would not know whether to obey the PHICH or wait for a PDCCH overriding the PHICH.

As explained earlier, the new-data indicator is explicitly transmitted in the uplink grant. However, unlike the downlink case, the redundancy version is *not* explicitly signaled for each retransmission. With a single-bit acknowledgement on the PHICH, this is not possible. Instead, as the uplink hybrid-ARQ protocol is synchronous, the redundancy version follows a predefined pattern, starting with zero when the initial transmission is scheduled by the PDCCH. Whenever a retransmission is requested by a negative acknowledgement on the PHICH, the next redundancy version in the sequence is used. However, if a retransmission is explicitly scheduled by the PDCCH overriding the PHICH, there is the potential to affect the redundancy version to use. Grants for retransmissions use the same format as ordinary grants (for initial transmissions). One of the information fields in an uplink grant is, as described in Chapter 10, the modulation-and-coding scheme. Of the 32 different combinations this five-bit field can take, three of them are reserved. Those three combinations represent different redundancy versions; hence, if one of these combinations is signaled as part of an uplink grant indicating a retransmission, the corresponding redundancy version is used for the transmission. The transport-block size is already known from the initial transmission as it, by definition, cannot change between retransmission attempts. In Figure 12.4, the initial transmission in subframe n uses the first redundancy version in sequence as the transport-block size must be indicated for the initial transmission. The retransmission in subframe $n + 8$ uses the next redundancy version in the sequence, while the explicitly scheduled retransmission in subframe $n + 16$ can use any redundancy scheme as indicated on the PDCCH.

Whether to exploit the possibility to signal an arbitrary redundancy version in a retransmission scheduled by the PDCCH is a trade-off between incremental-redundancy gain and robustness. From an incremental-redundancy perspective, changing the redundancy value between retransmissions is typically beneficial to fully exploit the gain from incremental redundancy. However, as the modulation-and-coding scheme is normally not indicated for uplink retransmissions, as either the single-bit PHICH is used or the modulation-and-coding field is used to explicitly indicate a new redundancy version, it is implicitly assumed that the terminal did not miss the initial scheduling grant. If this is the case, it is necessary to explicitly indicate the modulation-and-coding scheme, which also implies that the first redundancy version in the sequence is used.

In order to minimize the overhead and not introduce any additional signaling in the uplink grants, the PHICH that the terminal will expect the hybrid-ARQ acknowledgement upon is derived from the number of the first resource block upon which the corresponding uplink PUSCH transmission occurred. This principle is also compatible with semi-persistently scheduled transmission (see Chapter 13) as well as retransmissions. In addition, the resources used for a particular PHICH further depend on the reference-signal phase rotation signaled as part of the uplink grant. In this way, multiple terminals scheduled on the same set of resources using multi-user MIMO will use different PHICH resources as their reference signals are assigned different phase rotations through the corresponding field in the uplink grant. For spatial multiplexing, where two PHICH resources are needed, the second PHICH uses the same principle as the first, but to ensure that different PHICHs are used for the two transport blocks, the resource for the second PHICH is derived not from the first but from the second resource block upon which the PUSCH was transmitted.[1]

The above association is also used in the case of carrier aggregation. For the case when no cross-carrier scheduling is used and each uplink component carrier is scheduled on its corresponding downlink component carrier, this is straightforward and there is no risk of multiple uplink component carriers being associated with the same PHICH resource. With cross-carrier scheduling, on the other hand, transmissions on multiple uplink component carriers may need to be acknowledged on a single downlink component carrier, as illustrated in Figure 12.5. Avoiding PHICH collisions in this case is up to the scheduler by ensuring that different reference-signal phase rotations or different resource-block starting positions are used for the different uplink component carriers. For semi-persistent scheduling the reference-signal phase rotation is always set to zero, but since semi-persistent scheduling is supported on the primary component carrier only, there is no risk of collisions between component carriers.

In the case of multiple component carriers being used for the uplink data transmission in a subframe, there are one (in the absence of spatial multiplexing) or two (in the presence of spatial multiplexing)

No cross-carrier scheduling

UL grant 1   UL grant 2
(RB_1, CS_1) (RB_2, CS_2)

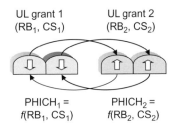

$PHICH_1 =$ $PHICH_2 =$
$f(RB_1, CS_1)$ $f(RB_2, CS_2)$

Cross-carrier scheduling

UL grant 1   UL grant 2
(RB_1, CS_1) (RB_2, CS_2)

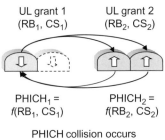

$PHICH_1 =$ $PHICH_2 =$
$f(RB_1, CS_1)$ $f(RB_2, CS_2)$

PHICH collision occurs
if $(RB_1, CS_1) = (RB_2, CS_2)$

FIGURE 12.5

PHICH association.

[1] In essence, this implies that uplink spatial multiplexing must use at least two resource blocks in the frequency domain.

PHICHs transmitted per uplink component carrier. As a general principle, LTE transmits the PHICH on the same component carrier that was used for the grant scheduling the corresponding uplink data transmission. Not only is this principle general in the sense that it can handle symmetric as well as asymmetric carrier aggregation scenarios, it is also beneficial from a terminal power consumption perspective as the terminal only need to monitor the component carriers it monitors for uplink scheduling grants (especially as the PDCCH may override the PHICH to support adaptive retransmissions as discussed above).

12.1.3 Hybrid-ARQ Timing

Clearly, the receiver must know to which hybrid-ARQ process a received acknowledgement is associated. This is handled by the timing of the hybrid-ARQ acknowledgement being used to associate the acknowledgement with a certain hybrid-ARQ process; the timing relation between the reception of data in the downlink and transmission of the hybrid-ARQ acknowledgement in the uplink (and vice versa) is fixed. From a latency perspective, the time between the reception of downlink data at the terminal and transmission of the hybrid-ARQ acknowledgement in the uplink should be as short as possible. At the same time, an unnecessarily short time would increase the demand on the terminal processing capacity and a trade-off between latency and implementation complexity is required. The situation is similar for uplink data transmissions. For LTE, this trade-off led to the decision to have eight hybrid-ARQ processes in both uplink and downlink for FDD. For TDD, the number of processes depends on the downlink–uplink allocation, as discussed below.

Starting with the FDD case, transmission of acknowledgements in the uplink in response to downlink data transmission is illustrated in Figure 12.6. Downlink data on the DL-SCH is transmitted to the terminal in subframe n and received by the terminal, after the propagation delay T_p, in subframe n. The terminal attempts to decode the received signal, possibly after soft combining with a previous transmission attempt, and transmits the hybrid-ARQ acknowledgement in uplink subframe $n + 4$ (note that the start of an uplink subframe at the terminal is offset by T_{TA} relative to the start of the corresponding downlink subframe at the terminal as a result of the timing-advance procedure described

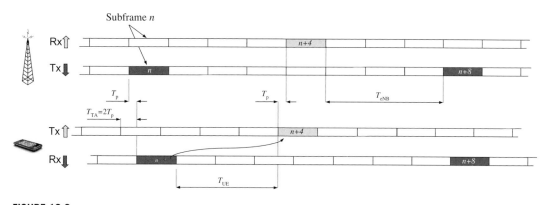

FIGURE 12.6

Timing relation between downlink data in subframe n and uplink hybrid-ARQ acknowledgement in subframe $n + 4$ for FDD.

in Section 11.5). Upon reception of the hybrid-ARQ acknowledgement, the eNodeB can, if needed, retransmit the downlink data in subframe $n + 8$. Thus, eight hybrid-ARQ processes are used – that is, the hybrid-ARQ round-trip time is 8 ms.

The description in the previous paragraph, as well as the illustration in Figure 12.6, describe the timing for downlink data transmission – that is, data on DL-SCH and acknowledgements on PUCCH (or PUSCH). However, the timing of uplink data transmission – that is, data on PUSCH and acknowledgements on PHICH – is identical, namely uplink data transmission in subframe n results in a PHICH transmission in subframe $n + 4$. This is also the same timing relation as for uplink scheduling grants in general (as described in Section 13.2.2) and allows a scheduling grant on the PDCCH to override the PHICH, as shown in Figure 12.4.

In Figure 12.6 it is seen that the processing time available to the terminal, T_{UE}, depends on the value of the timing advance or, equivalently, on the terminal-to-base-station distance. As a terminal must operate at any distance up to the maximum size supported by the specifications, the terminal must be designed such that it can handle the worst-case scenario. LTE is designed to handle at least 100 km, corresponding to a maximum timing advance of 0.67 ms. Hence, there is approximately 2.3 ms left for the terminal processing, which is considered a reasonable trade-off between the processing requirements imposed on a terminal and the associated delays.

For the eNodeB, the processing time available, denoted as T_{eNB}, is 3 ms and thus of the same order as for the terminal. In the case of downlink data transmission, the eNodeB performs scheduling of any retransmissions during this time and, for uplink data transmission, the time is used for decoding of the received signal. The timing budget is thus similar for the terminal and the eNodeB, which is motivated by the fact that, although the eNodeB typically has more processing power than a terminal, it also has to serve multiple terminals and to perform the scheduling operation.

Having the same number of hybrid-ARQ processes in uplink and downlink is beneficial for half-duplex FDD operation, discussed in Chapter 13. By proper scheduling, the uplink hybrid-ARQ transmission from the terminal will coincide with transmission of uplink data and the acknowledgements related to the reception of uplink data will be transmitted in the same subframe as the downlink data. Thus, using the same number of hybrid-ARQ processes in uplink and downlink results in a 50:50 split between transmission and reception for a half-duplex terminal.

Shifting to TDD operation, the time relation between the reception of data in a certain hybrid-ARQ process and the transmission of the hybrid-ARQ acknowledgement depends on the downlink–uplink allocation. Obviously, an uplink hybrid-ARQ acknowledgement can only be transmitted in an uplink subframe and a downlink acknowledgement only in a downlink subframe. The amount of processing time required in the terminal and the eNodeB remains the same though, as the same turbo decoders are used and the scheduling decisions are similar. Therefore, for TDD the acknowledgement of a transport block in subframe n is transmitted in subframe $n + k$, where $k \geq 4$ and is selected such that $n + k$ is an uplink subframe when the acknowledgement is to be transmitted from the terminal (on PUCCH or PUSCH) and a downlink subframe when the acknowledgement is transmitted from the eNodeB (on PHICH). The value of k depends on the downlink–uplink configuration, as shown in Table 12.1 for both downlink and uplink transmissions. From the table it is seen that, as a consequence of the downlink–uplink configuration, the value of k is sometimes larger than the FDD value $k = 4$. For example, assuming configuration 2, an uplink transmission on PUSCH received in subframe 2 should be acknowledged on PHICH in subframe $2 + 6 = 8$. Similarly, for the same configuration, downlink transmission on PDSCH in subframe 0 should be acknowledged on PUCCH (or PUSCH) in subframe $0 + 7 = 7$.

Table 12.1 Number of Hybrid-ARQ Processes and Acknowledgement Timing k for Different TDD Configurations

Configuration (DL:UL)	Downlink Proc.	PDSCH reception in subframe n										Uplink Proc.	PUSCH reception in subframe n									
		0	1	2	3	4	5	6	7	8	9		0	1	2	3	4	5	6	7	8	9
0 (2:3)	4	4	6	–	–	–	4	6	–	–	–	6	–	–	4	7	6	–	–	4	7	6
1 (3:2)	7	7	6	–	–	4	7	6	–	–	4	4	–	–	4	6	–	–	–	4	6	–
2 (4:1)	10	**7**	6	–	**4**	8	7	6	–	4	8	2	–	–	6	–	–	–	–	6	–	–
3 (7:3)	9	4	11	–	–	–	7	6	6	5	5	3	–	–	6	6	6	–	–	–	–	–
4 (8:2)	12	12	11	–	–	8	7	7	6	5	4	2	–	–	6	6	–	–	–	–	–	–
5 (9:1)	15	12	11	–	9	8	7	6	5	4	13	1	–	–	6	–	–	–	–	–	–	–
6 (5:5)	6	7	7	–	–	–	7	7	–	–	5	6	–	–	4	6	6	–	–	4	7	–

From the table it is also seen that the number of hybrid-ARQ processes used for TDD depends on the downlink–uplink configuration, implying that the hybrid-ARQ round-trip time is configuration dependent for TDD (actually, it may even vary between subframes, as seen in Figure 12.7). For the downlink-heavy configurations 2, 3, 4, and 5, the number of downlink hybrid-ARQ processes is larger than for FDD. The reason is the limited number of uplink subframes available, resulting in k-values well beyond 4 for some subframes (Table 12.1).

The PHICH timing in TDD is identical to the timing when receiving an uplink grant, as described in Section 13.2.2. The reason is the same as in FDD, namely to allow a PDCCH uplink grant to override the PHICH in order to implement adaptive retransmissions, as illustrated in Figure 12.4.

The downlink–uplink allocation for TDD has implications on the number of transport blocks to acknowledge in a single subframe. For FDD, in the absence of carrier aggregation, there is always a one-to-one relation between uplink and downlink subframes. Hence, a subframe only needs to carry acknowledgements for one subframe in the other direction. For TDD, in contrast, there is not necessarily a one-to-one relation between uplink and downlink subframes. This can be seen in the possible downlink–uplink allocations described in Chapter 9.

For uplink transmissions using UL-SCH, each uplink transport block is acknowledged individually by using the PHICH. Hence, in uplink-heavy asymmetries (downlink–uplink configuration 0), the terminal may need to receive two hybrid-ARQ acknowledgements in downlink subframes 0 and 5 in the absence of uplink spatial multiplexing; for spatial multiplexing up to four acknowledgements per component carrier are needed. There are also some subframes where no PHICH will be transmitted and therefore the amount of PHICH groups may vary between subframes for TDD.

For downlink transmissions, there are some configurations where DL-SCH receipt in multiple downlink subframes needs to be acknowledged in a single uplink subframe, as illustrated in Figure 12.7 (the illustration corresponds to the two entries shown in bold in Table 12.1). Two different mechanisms to handle this are provided in TDD: multiplexing and bundling.

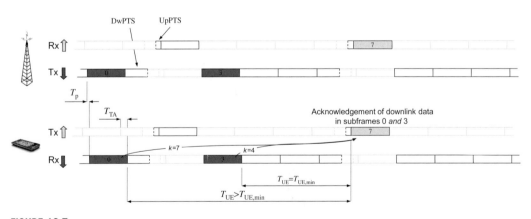

FIGURE 12.7

Example of timing relation between downlink data and uplink hybrid-ARQ acknowledgement for TDD (configuration 2).

Multiplexing implies that independent acknowledgements for each of the received transport blocks are fed back to the eNodeB. This allows independent retransmission of erroneous transport blocks. However, it also implies that multiple bits need to be transmitted from the terminal, which may limit the uplink coverage. This is the motivation for the bundling mechanism.

Bundling of acknowledgements implies that the outcome of the decoding of downlink transport blocks from multiple downlink subframes can be combined into a single hybrid-ARQ acknowledgement transmitted in the uplink. Only if both of the downlink transmissions in subframes 0 and 3 in the example in Figure 12.7 are correctly decoded will a positive acknowledgement be transmitted in uplink subframe 7.

Combining acknowledgements related to multiple downlink transmissions into a single uplink message assumes that the terminal has not missed any of the scheduling assignments upon which the acknowledgement is based. Assume, as an example, that the eNodeB scheduled the terminal in two (subsequent) subframes, but the terminal missed the PDCCH transmission in the first of the two subframes and successfully decoded the data transmitted in the second subframe. Obviously, without any additional mechanism, the terminal will transmit an acknowledgement based on the assumption that it was scheduled in the second subframe only, while the eNodeB will interpret acknowledgement as the terminal successfully received both transmissions. To avoid such errors, the *downlink assignment index* (see Section 10.4.4) in the scheduling assignment on the PDCCH is used. The downlink assignment index in essence informs the terminal about the number of transmissions it should base the combined acknowledgement upon. If there is a mismatch between the assignment index and the number of transmissions the terminal received, the terminal concludes at least one assignment was missed and transmits no hybrid-ARQ acknowledgement, thereby avoiding acknowledging transmissions not received.

TDD is not the only example when transmissions in multiple downlink subframes must be acknowledged in a single uplink subframe. Carrier aggregation in FDD with different numbers of uplink and downlink component carriers[2] in essence leads to the same problem. In LTE, there is at most one PUCCH and it is always transmitted on the primary component carrier. Hence, even for a symmetric carrier aggregation configuration, the possibility to transmit more than two hybrid-ARQ acknowledgement bits in the uplink must be supported. As discussed in Chapter 10, this is handled either by resource selection or by using PUCCH format 3. Acknowledgements in response to UL-SCH reception are transmitted using multiple PHICHs, where the PHICH is transmitted on the same downlink component carrier as the uplink scheduling grant initiating the uplink transmission.

12.2 RADIO-LINK CONTROL

The *radio-link control* (RLC) protocol takes data in the form of RLC SDUs from PDCP and delivers them to the corresponding RLC entity in the receiver by using functionality in MAC and physical layers. The relation between RLC and MAC, including multiplexing of multiple logical channels into a single transport channel, is illustrated in Figure 12.8. Multiplexing of several logical channels into a single transport channel is mainly used for priority handling, as described in Section 13.2 in conjunction with downlink and uplink scheduling.

---

[2]In LTE release 10, the number of downlink component carriers is always at least as large as the number of uplink component carriers (see Chapter 9).

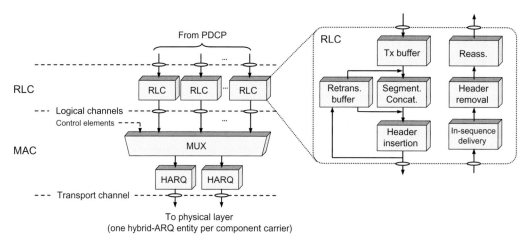

FIGURE 12.8

MAC and RLC structure (single-terminal view).

There is one RLC entity per logical channel configured for a terminal, where each RLC entity is responsible for:

- segmentation, concatenation, and reassembly of RLC SDUs;
- RLC retransmission; and
- in-sequence delivery and duplicate detection for the corresponding logical channel.

Other noteworthy features of the RLC are: (1) the handling of varying PDU sizes; and (2) the possibility for close interaction between the hybrid-ARQ and RLC protocols. Finally, the fact that there is one RLC entity per logical channel and one hybrid-ARQ entity per component carrier implies that one RLC entity may interact with multiple hybrid-ARQ entities in the case of carrier aggregation.

12.2.1 Segmentation, Concatenation, and Reassembly of RLC SDUs

The purpose of the segmentation and concatenation mechanism is to generate RLC PDUs of appropriate size from the incoming RLC SDUs. One possibility would be to define a fixed PDU size, a size that would result in a compromise. If the size were too large, it would not be possible to support the lowest data rates. Also, excessive padding would be required in some scenarios. A single small PDU size, however, would result in a high overhead from the header included with each PDU. To avoid these drawbacks, which is especially important given the very large dynamic range of data rates supported by LTE, the RLC PDU size varies dynamically.

Segmentation and concatenation of RLC SDUs into RLC PDUs are illustrated in Figure 12.9. The header includes, among other fields, a sequence number, which is used by the reordering and retransmission mechanisms. The reassembly function at the receiver side performs the reverse operation to reassemble the SDUs from the received PDUs.

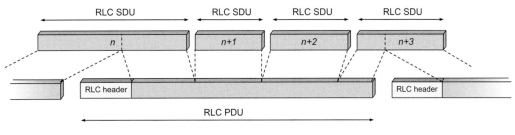

FIGURE 12.9

Generation of RLC PDUs from RLC SDUs.

12.2.2 RLC Retransmission

Retransmission of missing PDUs is one of the main functionalities of the RLC. Although most of the errors can be handled by the hybrid-ARQ protocol, there are, as discussed at the beginning of the chapter, benefits of having a second-level retransmission mechanism as a complement. By inspecting the sequence numbers of the received PDUs, missing PDUs can be detected and a retransmission requested from the transmitting side.

Different services have different requirements; for some services (for example, transfer of a large file), error-free delivery of data is important, whereas for other applications (for example, streaming services), a small amount of missing packets is not a problem. The RLC can therefore operate in three different modes, depending on the requirements from the application:

- *Transparent mode* (TM), where the RLC is completely transparent and is essentially bypassed. No retransmissions, no segmentation/reassembly, and no in-sequence delivery take place. This configuration is used for control-plane broadcast channels such as BCCH, CCCH, and PCCH, where the information should reach multiple users. The size of these messages are selected such that all intended terminals are reached with a high probability and hence there is neither need for segmentation to handle varying channel conditions, nor retransmissions to provide error-free data transmission. Furthermore, retransmissions are not possible for these channels as there is no possibility for the terminal to feed back status reports as no uplink has been established.
- *Unacknowledged mode* (UM) supports segmentation/reassembly and in-sequence delivery, but not retransmissions. This mode is used when error-free delivery is not required, for example voice-over IP, or when retransmissions cannot be requested, for example broadcast transmissions on MTCH and MCCH using MBSFN.
- *Acknowledged mode* (AM) is the main mode of operation for TCP/IP packet data transmission on the DL-SCH. Segmentation/reassembly, in-sequence delivery, and retransmissions of erroneous data are all supported.

In the following, the acknowledged mode is described. The unacknowledged mode is similar, with the exception that no retransmissions are done and that each RLC entity is unidirectional.

In acknowledged mode, the RLC entity is bidirectional – that is, data may flow in both directions between the two peer entities. This is obviously needed as the reception of PDUs needs to be acknowledged back to the entity that transmitted those PDUs. Information about missing PDUs is provided by the receiving end to the transmitting end in the form of so-called *status reports*. Status reports can either be transmitted autonomously by the receiver or requested by the transmitter. To keep track of the PDUs in transit, the transmitter attaches an RLC header to each PDU, including, among other fields, a sequence number.

Both RLC entities maintain two windows, the transmission and reception windows respectively. Only PDUs in the transmission window are eligible for transmission; PDUs with sequence number below the start of the window have already been acknowledged by the receiving RLC. Similarly, the receiver only accepts PDUs with sequence numbers within the reception window. The receiver also discards any duplicate PDUs as each PDU should be assembled into an SDU only once.

12.2.3 In-Sequence Delivery

In-sequence delivery implies that data blocks are delivered by the receiver in the same order as they were transmitted. This is an essential part of RLC; the hybrid-ARQ processes operate independently and transport blocks may therefore be delivered out of sequence, as seen in Figure 12.2. In-sequence delivery implies that SDU n should be delivered prior to SDU $n + 1$. This is an important aspect as several applications require the data to be received in the same order as it was transmitted. TCP can, to some extent, handle IP packets arriving out of sequence, although with some performance impact, while for some streaming applications in-sequence delivery is essential. The basic idea behind in-sequence delivery is to store the received PDUs in a buffer until all PDUs with lower sequence number have been delivered. Only when all PDUs with lower sequence number have been used for assembling SDUs is the next PDU used. RLC retransmission, provided when operating in acknowledged mode only, operates on the same buffer as the in-sequence delivery mechanism.

12.2.4 RLC Operation

The operation of the RLC with respect to retransmissions and in-sequence delivery is perhaps best understood by the simple example in Figure 12.10, where two RLC entities are illustrated, one in the transmitting node and one in the receiving node. When operating in acknowledged mode, as assumed below, each RLC entity has both transmitter and receiver functionality, but in this example only one

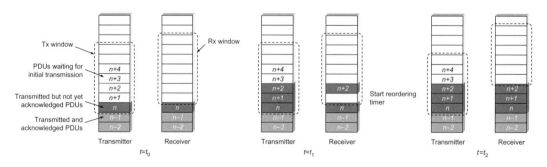

FIGURE 12.10

In-sequence delivery.

of the directions is discussed as the other direction is identical. In the example, PDUs numbered from n to $n + 4$ are awaiting transmission in the transmission buffer. At time t_0, PDUs with sequence number up to and including n have been transmitted and correctly received, but only PDUs up to and including $n - 1$ have been acknowledged by the receiver. As seen in the figure, the transmission window starts from n, the first not-yet-acknowledged PDU, while the reception window starts from $n + 1$, the next PDU expected to be received. Upon reception of PDU n, the PDU is forwarded to the SDU reassembly functionality for further processing.

The transmission of PDUs continues and, at time t_1, PDUs $n + 1$ and $n + 2$ have been transmitted but, at the receiving end, only PDU $n + 2$ has arrived. One reason for this could be that the missing PDU, $n + 1$, is being retransmitted by the hybrid-ARQ protocol and therefore has not yet been delivered from the hybrid ARQ to the RLC. The transmission window remains unchanged compared to the previous figure, as none of the PDUs n and higher have been acknowledged by the receiver. Hence, any of these PDUs may need to be retransmitted as the transmitter is not aware of whether they have been received correctly or not. The reception window is not updated when PDU $n + 2$ arrives. The reason is that PDU $n + 2$ cannot be forwarded to SDU assembly as PDU $n + 1$ is missing. Instead, the receiver waits for the missing PDU $n + 1$. Clearly, waiting for the missing PDU for an infinite time would stall the queue. Hence, the receiver starts a timer, the *reordering timer*, for the missing PDU. If the PDU is not received before the timer expires, a retransmission is requested. Fortunately, in this example, the missing PDU arrives from the hybrid-ARQ protocol at time t_2, before the timer expires. The reception window is advanced and the reordering timer is stopped as the missing PDU has arrived. PDUs $n + 1$ and $n + 2$ are delivered for reassembly into SDUs.

Duplicate detection is also the responsibility of the RLC, using the same sequence number as used for reordering. If PDU $n + 2$ arrives again (and is within the reception window), despite it having already been received, it is discarded.

The example above illustrates the basic principle behind in-sequence delivery, which is supported by both acknowledged and unacknowledged modes. However, the acknowledged mode of operation also provides retransmission functionality. To illustrate the principles behind this, consider Figure 12.11, which is a continuation of the example above. At time t_3, PDUs up to $n + 5$ have been transmitted. Only PDU $n + 5$ has arrived and PDUs $n + 3$ and $n + 4$ are missing. Similar to the case

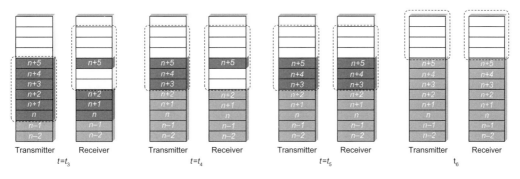

FIGURE 12.11

Retransmission of missing PDUs.

above, this causes the reordering timer to start. However, in this example no PDUs arrive prior to the expiration of the timer. The expiration of the timer at time t_4 triggers the receiver to send a control PDU containing a status report, indicating the missing PDUs, to its peer entity. Control PDUs have higher priority than data PDUs to avoid the status reports being unnecessarily delayed and negatively impact the retransmission delay. Upon reception of the status report at time t_5, the transmitter knows that PDUs up to $n + 2$ have been received correctly and the transmission window is advanced. The missing PDUs $n + 3$ and $n + 4$ are retransmitted and, this time, correctly received.

The retransmission was triggered by the reception of a status report in this example. However, as the hybrid-ARQ and RLC protocols are located in the same node, tight interaction between the two is possible. The hybrid-ARQ protocol at the transmitting end could therefore inform the RLC at the transmitting end in case the transport block(s) containing PDUs $n + 3$ and $n + 4$ have failed. The RLC can use this to trigger retransmission of missing PDUs without waiting for an explicit RLC status report, thereby reducing the delays associated with RLC retransmissions.

Finally, at time t_6, all PDUs, including the retransmissions, have been delivered by the transmitter and successfully received. As $n + 5$ was the last PDU in the transmission buffer, the transmitter requests a status report from the receiver by setting a flag in the header of the last RLC data PDU. Upon reception of the PDU with the flag set, the receiver will respond by transmitting the requested status report, acknowledging all PDUs up to and including $n + 5$. Reception of the status report by the transmitter causes all the PDUs to be declared as correctly received and the transmission window is advanced.

Status reports can, as mentioned earlier, be triggered for multiple reasons. However, to control the amount of status reports and to avoid flooding the return link with an excessive number of status reports, it is possible to use a status prohibit timer. With such a timer, status reports cannot be transmitted more often than once per time interval as determined by the timer.

For the initial transmission, it is relatively straightforward to rely on a dynamic PDU size as a means to handle the varying data rates. However, the channel conditions and the amount of resources may also change between RLC retransmissions. To handle these variations, already transmitted PDUs can be (re)segmented for retransmissions. The reordering and retransmission mechanisms described above still apply; a PDU is assumed to be received when all the segments have been received. Status reports and retransmissions operate on individual segments; only the missing segment of a PDU needs to be retransmitted.

Power Control, Scheduling, and Interference Handling

13

This chapter deals with some radio-resource-management issues, including uplink power control, downlink and uplink scheduling, and different means for inter-cell interference coordination to handle/avoid severe interference between different cells including cells of different layers in so-called *heterogeneous network deployments*.

13.1 UPLINK POWER CONTROL

Uplink power control for LTE is the set of algorithms and tools by which the transmit power for different uplink physical channels and signals are controlled to ensure that they, if possible, are received at the cell site with appropriate power. This means that the transmission should be received with sufficient power to allow for proper demodulation of the corresponding information. At the same time, the transmit power should not be unnecessarily high as that would cause unnecessary interference to other cells.[1] The transmit power will thus depend on the channel properties, including the channel attenuation and the noise and interference level at the receiver side. Furthermore, in the case of DL-SCH transmission on PDSCH, if the received power is too low one can either increase the transmit power or reduce the data rate by use of rate control. Thus, in this case there is an intimate relation between power control and rate control.

How to set the transmit power for random access will be discussed in Chapter 14. Here we will mainly discuss the power-control mechanism for the PUCCH and PUSCH physical channels. We will also briefly discuss the power setting for sounding reference signals. Uplink *demodulation* reference signals are always transmitted together and time-multiplexed with PUSCH or PUCCH. The demodulation reference signals are then transmitted with the same power as the corresponding physical channel. This is also true in the case of uplink spatial multiplexing if the reference signal power is defined as the total power of all demodulation reference signals transmitted by the terminal. Expressed differently, the power of a single demodulation reference signal is equal to the corresponding *per-layer* PDSCH power.

Fundamentally, LTE uplink power control is a combination of an *open-loop* mechanism, implying that the terminal transmit power depends on estimates of the downlink path loss, and a *closed-loop* mechanism, implying that the network can, in addition, directly adjust the terminal transmit power by means of explicit *power-control commands* transmitted on the downlink. In practice, these power-control commands are determined based on prior network measurements of the received uplink power, thus the term "*closed loop*".

[1] In the case of uplink MU-MIMO there may also be interference to the cell itself.

4G LTE/LTE-Advanced for Mobile Broadband.
© 2011 Erik Dahlman, Stefan Parkvall & Johan Sköld. Published by Elsevier Ltd. All rights reserved.

13.1.1 Uplink Power Control – Some Basic Rules

Before going into the details of the power control algorithms for PUSCH and PUCCH, some basic rules for the power assignment to different physical channels will be discussed. These rules mainly deal with the presence of different transmit-power limitations and how these limitations impact the transmit-power setting for different physical channels. This is especially of interest in the case of the simultaneous transmission of multiple physical channels from the same terminal, a situation that may occur for LTE release 10 and beyond:

- With LTE release 10 there is the possibility for *carrier aggregation*, implying that multiple PUSCH may be transmitted in parallel on different component carriers.
- With LTE release 10 there is also the possibility for *simultaneous PUSCH/PUCCH transmission* on the same or different component carriers.

In principle, each physical channel is separately and independently power controlled. However, in the case of multiple physical channels to be transmitted in parallel from the same terminal, the total power to be transmitted for all physical channels may, in some cases, exceed the maximum terminal output power P_{TMAX} corresponding to the terminal power class. As will be seen below, the basic strategy is then to first ensure that transmission of any L1/L2 control signaling is assigned the power assumed to be needed for reliable transmission. The remaining available power is then assigned to the remaining physical channels.

For each uplink component carrier configured for a terminal there is also an associated and explicitly configured *maximum per-carrier transmit power* $P_{CMAX,c}$, which may be different for different component carriers (indicated by the index c). Furthermore, although it obviously does not make sense for $P_{CMAX,c}$ to exceed the maximum terminal output power P_{TMAX}, the sum of $P_{CMAX,c}$ for all configured component carriers may very well, and typically will, exceed P_{TMAX}. The reason is that, in many cases, the terminal will not be scheduled for uplink transmission on all its configured component carriers and the terminal should also in that case be able to transmit with its maximum output power.

As will be seen in the next sections, the power control of each physical channel explicitly ensures that the total transmit power for a given component carrier does not exceed $P_{CMAX,c}$ for that carrier. However, the separate power-control algorithms do not ensure that the total transmit power for all component carriers to be transmitted by the terminal does not exceed the maximum terminal output power P_{TMAX}. Rather, this is ensured by a subsequent *power scaling* applied to the physical channels to be transmitted. This power scaling is carried out in such a way that any L1/L2 control signaling has higher priority, compared to data (UL-SCH) transmission.

If PUCCH is to be transmitted in the subframe it is first assigned the power determined by its corresponding power-control algorithm, before any power is assigned to any PUSCH to be transmitted in parallel PUCCH. This ensures that L1/L2 control signaling on PUCCH is assigned the power assumed to be needed for reliable transmission before any power is assigned for data transmission.

If PUCCH is not transmitted in the subframe but L1/L2 control signaling is multiplexed on to PUSCH, the PUSCH carrying the L1/L2 control signaling is first assigned the power determined by its corresponding power-control algorithm, before any power is assigned to any other PUSCH to be transmitted in parallel. Once again, this ensures that L1/L2 control signaling is assigned the power assumed to be needed before any power is assigned for other PUSCH transmissions only carrying UL-SCH. Note that, in the case of transmission of multiple PUSCH in parallel (carrier aggregation), at most one PUSCH may include L1/L2 control signaling. Also, there cannot be PUCCH transmission and L1/L2 control signaling multiplexed on to PUSCH in the same subframe. Thus, there will never be any conflict between the above rules.

If the remaining available transmit power is not sufficient to fulfill the power requirements of any remaining PUSCH to be transmitted, the powers of these remaining physical channels, which only carry UL-SCH, are scaled so that the total power for all physical channels to be transmitted does not exceed the maximum terminal output power.

Overall, the PUSCH power scaling, including the priority for PUSCH with L1/L2 control signaling, can thus be expressed as:

$$\sum_i w_c \cdot P_{\text{PUSCH},c} \leq P_{\text{TMAX}} - P_{\text{PUCCH}}, \tag{13.1}$$

where $P_{\text{PUSCH},c}$ is the transmit power for PUSCH on carrier c as determined by the power-control algorithm (before power scaling but including the per-carrier limitation $P_{\text{CMAX},c}$), P_{PUCCH} is the transmit power for PUCCH (which is zero if there is no PUCCH transmission in the subframe), and w_c is the power-scaling factor for PUSCH on carrier c ($w_c \leq 1$). For any PUSCH carrying L1/L2 control signaling the scaling factor w_c should be set to 1. For the remaining PUSCH, some scaling factors may be set to zero by decision of the terminal, in practice implying that the PUSCH, as well as the corresponding UL-SCH mapped to the PUSCH, are not transmitted. For the remaining PUSCH the scaling factors w_c are set to the same value less than or equal to 1 to ensure that the above inequality is fulfilled. Thus, all PUSCH that are actually transmitted are power scaled by the same factor.

After this overview of some general rules for the power setting of different terminals, especially for the case of multiple physical channels transmitted in parallel from the same terminal, the power control carried out separately for each physical channel will be described in more detail.

13.1.2 **Power Control for PUCCH**

For PUCCH, the appropriate received power is simply the power needed to achieve a desired – that is, a sufficiently low – error rate in the decoding of the L1/L2 control information transmitted on the PUCCH. However, it is then important to bear the following in mind:

- In general, decoding performance is not determined by the *received signal strength* but rather by the *received signal-to-interference-plus-noise ratio* (SINR). What is an appropriate received power thus depends on the interference level at the receiver side, an interference level that may differ between different deployments and which may also vary in time as, for example, the load of the network varies.
- As described in Chapter 11, there are different PUCCH formats which are used to carry different types of uplink L1/L2 control information (hybrid-ARQ acknowledgements, scheduling requests, channel-state reports, or combinations thereof). The different PUCCH formats thus carry different numbers of information bits per subframe and the information they carry may also have different error-rate requirements. The required received SINR may therefore differ between the different PUCCH formats, something that needs to be taken into account when setting the PUCCH transmit power in a given subframe.

Overall, power-control for PUCCH can be described by the following expression:

$$P_{\text{PUCCH}} = \min\left\{P_{\text{CMAX},c}, P_{0,\text{PUCCH}} + PL_{\text{DL}} + \Delta_{\text{Format}} + \delta\right\}. \qquad (13.2)$$

In the expression above, P_{PUCCH} is the PUCCH transmit power to use in a given subframe and PL_{DL} is the downlink path loss as estimated by the terminal. The "min $\{P_{\text{CMAX},c}, \ldots\}$" term ensures that the PUCCH transmit power as determined by the power control will not exceed the per-carrier maximum power $P_{\text{CMAX},c}$.

The parameter $P_{0,\text{PUCCH}}$ in expression (13.2) is a cell-specific parameter that is broadcast as part of the cell system information. Considering only the part $P_{0,\text{PUCCH}} + PL_{\text{DL}}$ in the PUCCH power-control expression and assuming that the (estimated) downlink path loss accurately reflects the true uplink path loss, it is obvious that $P_{0,\text{PUCCH}}$ can be seen as the *desired* or *target* received power. As discussed earlier, the required received power will depend on the uplink noise/interference level. From this point of view, the value of $P_{0,\text{PUCCH}}$ should take the interference level into account and thus vary in time as the interference level varies. However, in practice it is not feasible to have $P_{0,\text{PUCCH}}$ varying with the instantaneous interference level. One simple reason is that the terminal does not read the system information continuously and thus the terminal would anyway not have access to a fully up-to-date $P_{0,\text{PUCCH}}$ value. Another reason is that the uplink path-loss estimates derived from downlink measurements will anyway not be fully accurate, for example due to differences between the instantaneous downlink and uplink path loss, as well as due to measurement inaccuracies.

Thus, in practice, $P_{0,\text{PUCCH}}$ may reflect the average interference level, or perhaps only the relatively constant noise level. More rapid interference variations can then be taken care of by closed-loop power control, see below.

For the transmit power to reflect the typically different SINR requirements for different PUCCH formats, the PUCCH power-control expression includes the term Δ_{Format}, which adds a format-dependent power offset to the transmit power. The power offsets are defined such that a baseline PUCCH format, more exactly the format corresponding to the transmission of a single hybrid-ARQ acknowledgement (format 1 with BPSK modulation, as described in Section 11.4.1.1), has an offset equal to 0 dB, while the offsets for the remaining formats can be explicitly configured by the network. For example, PUCCH format 1 with QPSK modulation, carrying two simultaneous acknowledgements and used in the case of downlink spatial multiplexing, should have a power offset of roughly 3 dB, reflecting the fact that twice as much power is needed to communicate two acknowledgements instead of just a single acknowledgement.

Finally, it is possible for the network to directly adjust the PUCCH transmit power by providing the terminal with explicit power-control commands that adjust the term δ in the power-control expression above. These power-control commands are *accumulative* – that is, each received power-control command increases or decreases the term δ by a certain amount. The power-control commands for PUCCH can be provided to the terminal by two different means:

- As mentioned in Section 10.4, a power-control command is included in each downlink scheduling assignment – that is, the terminal receives a power-control command every time it is explicitly scheduled on the downlink. One reason for uplink PUCCH transmissions is the transmission of hybrid-ARQ acknowledgements as a response to downlink DL-SCH transmissions. Such downlink transmissions are typically associated with downlink scheduling assignments on PDCCH and the corresponding power-control commands could thus be used to adjust the PUCCH transmit power prior to the transmission of the hybrid-ARQ acknowledgements.

- Power-control commands can also be provided on a special PDCCH that simultaneously provides power-control commands to multiple terminals (PDCCH using DCI format 3/3A; see Section 10.4.7). In practice, such power-control commands are then typically transmitted on a regular basis and can be used to adjust the PUCCH transmit power, for example prior to (periodic) uplink channel-state reports. They can also be used in the case of semi-persistent scheduling (see Section 13.2.3), in which case there may be uplink transmission of both PUSCH (UL-SCH) and PUCCH (L1/L2 control) without any explicit scheduling assignments/grants.

The power-control command carried within the uplink scheduling grant consists of two bits, corresponding to the four different update steps $-1, 0, +1$, or $+3$ dB. The same is true for the power-control command carried on the special PDCCH assigned for power control when this is configured to DCI format 3A. On the other hand, when the PDCCH is configured to use DCI format 3, each power-control command consists of a single bit, corresponding to the update steps -1 and $+1$ dB. In the latter case, twice as many terminals can be power controlled by a single PDCCH. One reason for including the possibility for 0 dB (no change of power) as one power-control step is that a power-control command is included in *every* downlink scheduling assignment and it is desirable not to have to update the PUCCH transmit power for each assignment.

13.1.3 **Power Control for PUSCH**

Power-control for PUSCH transmission can be described by the following expression:

$$P_{\text{PUSCH},c} = \min\left\{P_{\text{CMAX},c} - P_{\text{PUCCH}}, P_{0,\text{PUSCH}} + \alpha \cdot PL_{\text{DL}} + 10 \cdot \log_{10}(M) + \Delta_{\text{MCS}} + \delta\right\}, \qquad (13.3)$$

where M indicates the instantaneous PUSCH bandwidth measured in number of resource blocks and the term Δ_{MCS} is similar to the term Δ_{Format} in the expression for PUCCH power control – that is, it reflects the fact that different SINR is required for different modulation schemes and coding rates used for the PUSCH transmission.

The above expression is clearly similar to the power-control expression for PUCCH transmission, with some key differences:

- The use of "$P_{\text{CMAX},c} - P_{\text{PUCCH}}$" reflects the fact that the transmit power available for PUSCH on a carrier is the maximum allowed per-carrier transmit power *after power has been assigned to any PUCCH transmission* on that carrier. This ensures priority of L1/L2 signaling on PUCCH over data transmission on PUSCH in the power assignment, as described in Section 13.1.1.
- The term $10 \cdot \log_{10}(M)$ reflects the fact that what is fundamentally controlled by the parameter $P_{0,\text{PUSCH}}$ is the power *per resource block*. For a larger resource assignment, a correspondingly higher received power and thus a correspondingly higher transmit power is needed.[2]

---

[2]One could also have included a corresponding term in the expression for PUCCH power control. However, as the PUCCH bandwidth always corresponds to one resource block, the term would always equal zero.

- The parameter α, which can take a value smaller than or equal to 1, allows for so-called *partial path-loss compensation*, as described below.

In general, the parameters $P_{0,PUSCH}$, α, and Δ_{MCS} can be different for the different component carriers configured for a terminal.

In the case of PUSCH transmission, the explicit power-control commands controlling the term δ above are included in the uplink scheduling grants, rather than in the downlink scheduling assignments. This makes sense as PUSCH transmissions are preceded by an uplink scheduling grant except for the case of semi-persistent scheduling. Similar to the power-control commands for PUCCH in the downlink scheduling assignment, the power-control commands for PUSCH are multi-level. Furthermore, also in the same way as for PUCCH power control, explicit power-control commands for PUSCH can be provided on the special PDCCH that simultaneously provides power-control commands to multiple terminals. These power-control commands can, for example, be used for the case of PUSCH transmission using semi-persistent scheduling.

Assuming α equal to 1, also referred to as *full path-loss compensation*, the PUSCH power-control expression becomes very similar to the corresponding expression for PUCCH. Thus, the network can select a modulation-and-coding scheme (MCS) and the power-control mechanism, including the term Δ_{MCS}, will ensure that the received SINR will match the SINR required for that modulation-and-coding scheme, *assuming that the terminal transmit power does not reach its maximum value.*

In the case of PUSCH transmission, it is also possible to "turn off" the Δ_{MCS} function by setting all Δ_{MCS} values to zero. In that case, the PUSCH received power will be matched to a certain MCS given by the selected value of $P_{0,PUSCH}$.

With the parameter α less than 1, the PUSCH power control operates with so-called *partial path-loss compensation* – that is, an increased path loss is not fully compensated for by a corresponding increase in the uplink transmit power. In that case, the received power, and thus the received SINR per resource block, will vary with the path loss and, consequently, the scheduled modulation-and-coding scheme should vary accordingly. Clearly, in the case of fractional path-loss compensation, the Δ_{MCS} function should be disabled. Otherwise, the terminal transmit power would be further reduced when the modulation-and-coding scheme is reduced to match the partial path-loss compensation.

Figure 13.1 illustrates the differences between full path-loss compensation ($\alpha = 1$) and partial path-loss compensation ($\alpha < 1$). As can be seen, with partial path-loss compensation, the terminal transmit power increases more slowly than the increase in path loss (left-hand figure) and, consequently, the received power, and thus also the received SINR, is reduced as the path loss increases (right-hand figure). To compensate for this, the modulation-and-coding scheme – that is, the PUSCH data rate – should be reduced as the path loss increases.

The potential benefit of partial path-loss compensation is a relatively lower transmit power for terminals closer to the cell border, implying less interference to other cells. At the same time, this also leads to a reduced data rate for these terminals. It should also be noted that a similar effect can be achieved with full path-loss compensation by having the scheduled modulation-and-coding scheme depend on the estimated downlink path loss, which can be derived from the power headroom report, and rely on Δ_{MCS} to reduce the relative terminal transmit power for terminals with higher path loss. However, an even better approach would then be to not only base the modulation-and-coding scheme selection on the path loss to the current cell, but also on the path loss to the neighboring interfered cells.

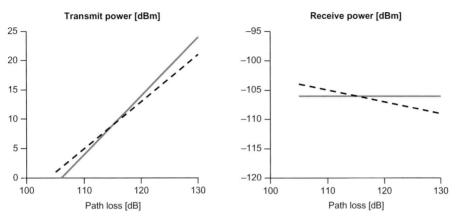

FIGURE 13.1

Full vs. partial path-loss compensation. Solid curve: full compensation ($\alpha = 1$). Dashed curve: partial compensation ($\alpha = 0.8$ in this example).

13.1.4 Power Control for SRS

The SRS transmit power basically follows that of the PUSCH, compensating for the exact bandwidth of the SRS transmission and with an additional power offset. Thus, the power control for SRS transmission can be described according to:

$$P_{SRS} = \min\left\{P_{CMAX,c}, P_{0,PUSCH} + \alpha \cdot PL_{DL} + 10 \cdot \log_{10}\left(M_{SRS}\right) + \delta + P_{SRS}\right\}, \quad (13.4)$$

where the parameters $P_{0,PUSCH}$, α, and δ are the same as for PUSCH power control, as discussed in Section 13.1.3. Furthermore, M_{SRS} is the bandwidth, expressed as number of resource blocks, of the SRS transmission and P_{SRS} is a configurable offset.

13.1.5 Power Headroom

To assist the scheduler in the selection of a combination of modulation-and-coding scheme and resource size M that does not lead to the terminal being power limited, the terminal can be configured to provide regular *power headroom* reports on its power usage (see also Section 13.2.2.2). According to Section 13.1.1 there is a separate transmit-power limitation for each component carrier. Thus, power headroom should be measured and reported separately for each component carrier.

There are two different types of power-headroom reports defined for LTE release 10, *Type 1* and *Type 2*. Type 1 reporting reflects the power headroom assuming PUSCH-only transmission on the carrier, while the Type-2 report assumes combined PUSCH and PUCCH transmission.

The Type-1 power headroom valid for a certain subframe, assuming that the terminal was really scheduled for PUSCH transmission in that subframe, is given by the following expression:

$$\text{Power Headroom} = P_{CMAX,c} - \left(P_{0,PUSCH} + \alpha \cdot PL_{DL} + 10 \cdot \log_{10}\left(M\right) + \Delta_{MCS} + \delta\right), \quad (13.5)$$

where the values for M and Δ_{MCS} correspond to the resource assignment and modulation-and-coding scheme used in the subframe to which the power-headroom report corresponds. It can be noted that the power headroom is not a measure of the difference between the maximum per-carrier transmit power and the actual carrier transmit power. Rather, comparing with expression (13.3) it can be seen that the power headroom is a measure of the difference between $P_{\text{CMAX},c}$ and the transmit power that would have been used *assuming that there would have been no upper limit on the transmit power.* Thus, the power headroom can very well be negative. More exactly, a negative power headroom indicates that the per-carrier transmit power was limited by $P_{\text{CMAX},c}$ at the time of the power headroom reporting. As the network knows what modulation-and-coding scheme and resource size the terminal used for transmission in the subframe to which the power-headroom report corresponds, it can determine what are the valid combinations of modulation-and-coding scheme and resource size M, assuming that the downlink path loss PL_{DL} and the term δ have not changed substantially.

Type-1 power headroom can also be reported for subframes where there is no actual PUSCH transmission. In such cases, $10 \cdot \log_{10}(M)$ and Δ_{MCS} in the expression above are set to zero:

$$\text{Power Headroom} = P_{\text{CMAX},c} - \left(P_{0,\text{PUSCH}} + \alpha \cdot PL_{\text{DL}} + \delta\right). \tag{13.6}$$

This can be seen as the power headroom assuming a default transmission configuration corresponding to the minimum possible resource assignment ($M = 1$) and the modulation-and-coding scheme associated with $\Delta_{\text{MCS}} = 0\,\text{dB}$.

Similarly, Type-2 power headroom reporting is defined as the difference between the maximum per-carrier transmit power and the sum of the PUSCH and PUCCH transmit power given by their corresponding power-control expressions (13.2) and (13.1) respectively, once again not taking into account any maximum per-carrier power when calculating the PUSCH and PUCCH transmit power.

Similar to Type-1 power headroom reporting, the Type-2 power headroom can also be reported for subframes in which no PUSCH and/or PUCCH is transmitted. In that case a virtual PUSCH and or PUCCH transmit power is calculated, assuming the smallest possible resource assignment ($M = 1$) and $\Delta_{\text{MCS}} = 0\,\text{dB}$ for PUSCH and $\Delta_{\text{Format}} = 0$ for PUCCH.

13.2 SCHEDULING AND RATE ADAPTATION

The purpose of the scheduler is to determine to/from which terminal(s) to transmit data and on which set of resource blocks. The scheduler is a key element and to a large degree determines the overall behavior of the system. The basic operation is so-called *dynamic* scheduling, where the eNodeB in each 1 ms TTI transmits scheduling information to the selected set of terminals, controlling the uplink and downlink transmission activity. The scheduling decisions are transmitted on the PDCCHs as described in Chapter 10. To reduce the control signaling overhead, there is also the possibility of *semi-persistent scheduling*. Semi-persistent scheduling will be described further in Section 13.2.3.

For carrier aggregation, each component carrier is independently scheduled with individual scheduling assignments/grants and one DL-SCH/UL-SCH per scheduled component carrier. Semi-persistent scheduling is only supported on the primary component carriers, motivated by the fact that the main usage is for small payloads not requiring multiple component carriers.

The *downlink scheduler* is responsible for dynamically controlling the terminal(s) to transmit to and, for each of these terminals, the set of resource blocks upon which the terminal's DL-SCH (or

DL-SCHs in the case of carrier aggregation) is transmitted. Transport-format selection (selection of transport-bock size, modulation-and-coding scheme, resource-block allocation, and antenna mapping) for each component carrier and logical channel multiplexing for downlink transmissions are controlled by the eNodeB, as illustrated in the left part of Figure 13.2.

The *uplink scheduler* serves a similar purpose, namely to dynamically control which terminals are to transmit on their UL-SCH (or UL-SCHs in the case of carrier aggregation) and on which uplink resources. The uplink scheduler is in complete control of the transport format the terminal will use, whereas the logical-channel multiplexing is controlled by the terminal according to a set of rules. Thus, uplink scheduling is *per terminal* and not per radio bearer. This is illustrated in the right part of Figure 13.2, where the scheduler controls the transport format and the terminal controls the logical-channel multiplexing.

In the following, dynamic downlink and uplink scheduling will be described, as well as related functionality such as uplink priority handling, scheduling request and buffer status reporting, semi-persistent scheduling, half-duplex FDD operation, channel-state reporting, and DRX functionality.

13.2.1 Downlink Scheduling

The task of the downlink scheduler is to dynamically determine the terminal(s) to transmit to and, for each of these terminals, the set of resource blocks upon which the terminal's DL-SCH should be transmitted. In most cases, a single terminal cannot use the full capacity of the cell, for example due to lack of data. Also, as the channel properties may vary in the frequency domain, it is useful to be able to transmit to different terminals on different parts of the spectrum. Therefore, multiple terminals can be scheduled in parallel in a subframe, in which case there is one DL-SCH per scheduled terminal and component carrier, each dynamically mapped to a (unique) set of frequency resources.

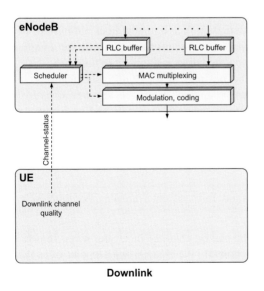
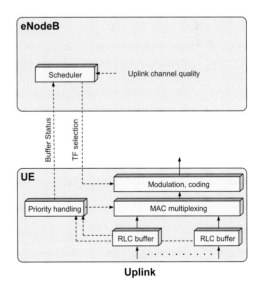

Downlink   **Uplink**

FIGURE 13.2

Transport-format selection in downlink (left) and uplink (right).

The scheduler is in control of the instantaneous data rate used, and the RLC segmentation and MAC multiplexing will therefore be affected by the scheduling decision. Although formally part of the MAC layer but to some extent better viewed as a separate entity, the scheduler is thus controlling most of the functions in the eNodeB associated with downlink data transmission:

- *RLC*. Segmentation/concatenation of RLC SDUs is directly related to the instantaneous data rate. For low data rates, it may only be possible to deliver a part of an RLC SDU in a TTI, in which case segmentation is needed. Similarly, for high data rates, multiple RLC SDUs may need to be concatenated to form a sufficiently large transport block.
- *MAC*. Multiplexing of logical channels depends on the priorities between different streams. For example, radio resource control signaling, such as handover commands, typically has a higher priority than streaming data, which in turn has higher priority than a background file transfer. Thus, depending on the data rate and the amount of traffic of different priorities, the multiplexing of different logical channels is affected. Hybrid-ARQ retransmissions also need to be accounted for.
- *L1*. Coding, modulation and, if applicable, the number of transmission layers and the associated precoding matrix are obviously affected by the scheduling decision. The choices of these parameters are mainly determined by the radio conditions and the selected data rate – that is, the transport block size.

The scheduling decision is communicated to each of the scheduled terminals through the downlink L1/L2 control signaling as described in Chapter 10, using one PDCCH per downlink assignment.

Each terminal monitors a set of PDCCHs as described in Chapter 10 for downlink scheduling assignments. A scheduling assignment is transmitted in the same subframe as the data. If a valid assignment matching the identity of the terminal is found, then the terminal receives and processes the transmitted signal as indicated in the assignment. Once the transport block is successfully decoded, the terminal will demultiplex the received data into the appropriate logical channels.

In the case of carrier aggregation, there is one PDCCH per component carrier. Furthermore, if cross-carrier scheduling is configured (see Chapter 10), the downlink assignment does not have to be transmitted on the same component carrier as the associated data, as information about the component carrier containing the associated data is included in the scheduling assignment in this case.

The scheduling strategy is implementation specific and not part of the 3GPP specifications. In principle, any of the schedulers described in Chapter 6 can be applied. However, the overall goal of most schedulers is to take advantage of the channel variations between terminals and preferably to schedule transmissions to a terminal when the channel conditions are advantageous. Most scheduling strategies therefore need information about:

- channel conditions at the terminal;
- buffer status and priorities of the different data flows;
- the interference situation in neighboring cells (if some form of interference coordination is implemented).

Information about the channel conditions at the terminal can be obtained in several ways. In principle, the eNodeB can use any information available, but typically the channel-state reports from the terminal, further described in Section 13.2.5, are used. However, additional sources of channel knowledge, for example exploiting channel reciprocity to estimate the downlink quality from uplink channel estimates in the case of TDD, can also be exploited by a particular scheduler implementation.

In addition to the channel-state information, the scheduler should take buffer status and priority levels into account. Obviously it does not make sense to schedule a terminal with empty transmission buffers. Priorities of the different types of traffic may also vary; RRC signaling may be prioritized over user data. Furthermore, RLC and hybrid-ARQ retransmissions, which are in no way different from other types of data from a scheduler perspective, are typically also given priority over initial transmissions.

Downlink inter-cell interference coordination is also part of the implementation-specific scheduler strategy. A cell may signal to its neighboring cells the intention to transmit with a lower transmission power in the downlink on a set of resource blocks. This information can then be exploited by neighboring cells as a region of low interference where it is advantageous to schedule terminals at the cell edge, terminals that otherwise could not attain high data rates due to the interference level. Inter-cell interference handling is further discussed in Section 13.3.

13.2.2 Uplink Scheduling

The basic function of the *uplink scheduler* is similar to its downlink counterpart, namely to dynamically determine, for each 1 ms interval, which terminals are to transmit and on which uplink resources. As discussed before, the LTE uplink is primarily based on maintaining orthogonality between different uplink transmissions and the shared resource controlled by the eNodeB scheduler is time–frequency resource units. In addition to assigning the time–frequency resources to the terminal, the eNodeB scheduler is also responsible for controlling the transport format the terminal will use for each of the uplink component carriers. As the scheduler knows the transport format the terminal will use when it is transmitting, there is no need for outband control signaling from the terminal to the eNodeB. This is beneficial from a coverage perspective, taking into account that the cost per bit of transmitting outband control information can be significantly higher than the cost of data transmission, as the control signaling needs to be received with higher reliability. It also allows the scheduler to tightly control the uplink activity to maximize the resource usage compared to schemes where the terminal autonomously selects the data rate, as autonomous schemes typically require some margin in the scheduling decisions. A consequence of the scheduler being responsible for selection of the transport format is that accurate and detailed knowledge about the terminal situation with respect to buffer status and power availability is more accentuated in LTE compared to systems where the terminal autonomously controls the transmission parameters.

The basis for uplink scheduling is *scheduling grants*, containing the scheduling decision and providing the terminal information about the resources and the associated transport format to use for transmission of the UL-SCH on one component carrier. Only if the terminal has a valid grant is it allowed to transmit on the corresponding UL-SCH; autonomous transmissions are not possible without a corresponding grant. Dynamic grants are valid for one subframe – that is, for each subframe in which the terminal is to transmit on the UL-SCH, the scheduler issues a new grant. Uplink component carriers are scheduled independently; if the terminal is to transmit simultaneously on multiple component carriers, multiple scheduling grants are needed.

The terminal monitors a set of PDCCHs as described in Chapter 10 for uplink scheduling grants. Upon detection of a valid uplink grant, the terminal will transmit its UL-SCH according to the information in the grant. Obviously, the grant cannot relate to the same subframe it was received in as the uplink subframe has already started when the terminal has decoded the grant. The terminal also needs

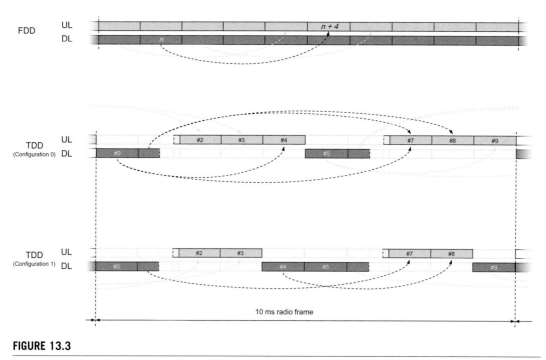

FIGURE 13.3

Timing relation for uplink grants in FDD and TDD configurations 0 and 1.

some time to prepare the data to transmit. Therefore, a grant received in subframe n affects the uplink transmission in a later subframe.

For FDD, the grant timing is straightforward. An uplink grant received in a subframe n triggers an uplink transmission in subframe $n + 4$, as illustrated in Figure 13.3. This is the same timing relation as used for uplink retransmission triggered by the PHICH, motivated by the possibility to override the PHICH by a dynamic scheduling grant, as described in Chapter 12.

For TDD, the situation is slightly more complicated as subframe $n + 4$ may not be an uplink subframe. Hence, for TDD configurations 1–6 the timing relation is modified such that the uplink transmission occurs in subframe $n + k$, where k is the smallest value larger than or equal to 4 such that subframe $n + k$ is an uplink subframe. This provides at least the same processing time for the terminal as in the FDD case while minimizing the delay from receipt of the uplink grant to the actual transmission. Note that this implies that the time between grant receipt and uplink transmission may differ between different subframes. Furthermore, for the downlink-heavy configurations 1–5, another property is that uplink scheduling grants can only be received in some of the downlink subframes.

For TDD configuration 0 there are more uplink subframes than downlink subframes, which calls for the possibility to schedule transmissions in multiple uplink subframes from a single downlink subframe. The same timing relation as for the other TDD configurations is used but with slight modifications. Recall from Chapter 10 that the grant transmitted in the downlink contains an uplink index consisting of two bits. For downlink–uplink configuration 0, the index field specifies which uplink subframe(s) a grant received in a downlink subframe applies to. For example, as illustrated in Figure 13.3, an uplink scheduling grant received in downlink subframe 0 applies to one or both of the uplink subframes 4 and 7, depending on which of the bits in the uplink index are set.

Similarly to the downlink case, the uplink scheduler can exploit information about channel conditions, buffer status, and priorities of the different data flows, and, if some form of interference coordination is employed, the interference situation in neighboring cells. Channel-dependent scheduling, which typically is used for the downlink, can be used for the uplink as well. In the uplink, estimates of the channel quality can be obtained from the use of uplink channel sounding, as described in Chapter 11. For scenarios where the overhead from channel sounding is too costly, or when the variations in the channel are too rapid to be tracked, for example at high terminal speeds, uplink diversity can be used instead. The use of frequency hopping as discussed in Chapter 11 is one example of obtaining diversity in the uplink.

Inter-cell interference coordination can be used in the uplink for similar reasons as in the downlink by exchanging information between neighboring cells, as discussed in Section 13.3.

13.2.2.1 *Uplink Priority Handling*

Multiple logical channels of different priorities can be multiplexed into the same transport block using the same MAC multiplexing functionality as in the downlink (described in Chapter 8). However, unlike the downlink case, where the prioritization is under control of the scheduler and up to the implementation, the uplink multiplexing is done according to a set of well-defined rules in the terminal as a scheduling grant applies to a specific uplink carrier of a *terminal*, not to a specific radio bearer within the terminal. Using radio-bearer-specific scheduling grants would increase the control signaling overhead in the downlink and hence per-terminal scheduling is used in LTE.

The simplest multiplexing rule would be to serve logical channels in strict priority order. However, this may result in starvation of lower-priority channels; all resources would be given to the high-priority channel until its transmission buffer is empty. Typically, an operator would instead like to provide at least some throughput for low-priority services as well. Therefore, for each logical channel in an LTE terminal, a *prioritized data rate* is configured in addition to the priority value. The logical channels are then served in decreasing priority order up to their prioritized data rate, which avoids starvation as long as the scheduled data rate is at least as large as the sum of the prioritized data rates. Beyond the prioritized data rates, channels are served in strict priority order until the grant is fully exploited or the buffer is empty. This is illustrated in Figure 13.4.

13.2.2.2 *Scheduling Requests and Buffer Status Reports*

The scheduler needs knowledge about the amount of data awaiting transmission from the terminals to assign the proper amount of uplink resources. Obviously, there is no need to provide uplink resources to a terminal with no data to transmit as this would only result in the terminal performing padding to

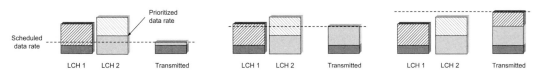

FIGURE 13.4

Prioritization of two logical channels for three different uplink grants.

fill up the granted resources. Hence, as a minimum, the scheduler needs to know whether the terminal has data to transmit and should be given a grant. This is known as a *scheduling request*.

A scheduling request is a simple flag, raised by the terminal to request uplink resources from the uplink scheduler. Since the terminal requesting resources by definition has no PUSCH resource, the scheduling request is transmitted on the PUCCH. Each terminal can be assigned a dedicated PUCCH scheduling request resource, occurring every nth subframe, as described in Chapter 11. With a dedicated scheduling-request mechanism, there is no need to provide the identity of the terminal requesting to be scheduled as the identity of the terminal is implicitly known from the resources upon which the request is transmitted. When data with higher priority than already existing in the transmit buffers arrives at the terminal and the terminal has no grant and hence cannot transmit the data, the terminal transmits a scheduling request at the next possible instant, as illustrated in Figure 13.5. Upon reception of the request, the scheduler can assign a grant to the terminal. If the terminal does not receive a scheduling grant until the next possible scheduling-request instant, then the scheduling request is repeated. There is only a single scheduling-request bit, irrespective of the number of uplink component carriers the terminal is capable of. In the case of carrier aggregation, the scheduling request is transmitted on the primary component carrier, in line with the general principle of PUCCH transmission on the primary component carrier only.

The use of a single bit for the scheduling request is motivated by the desire to keep the uplink overhead small, as a multi-bit scheduling request would come at a higher cost. A consequence of the single-bit scheduling request is the limited knowledge at the eNodeB about the buffer situation at the terminal when receiving such a request. Different scheduler implementations handle this differently. One possibility is to assign a small amount of resources to ensure that the terminal can exploit them efficiently without becoming power limited. Once the terminal has started to transmit on the UL-SCH, more detailed information about the buffer status and power headroom can be provided through the inband MAC control message, as discussed below. Knowledge of the service type may also be used – for example, in the case of voice the uplink resource to grant is preferably the size of a typical voice-over-IP package. The scheduler may also exploit, for example, path-loss measurements used for mobility and handover decisions to estimate the amount of resources the terminal may efficiently utilize.

An alternative to a dedicated scheduling-request mechanism would be a contention-based design. In such a design, multiple terminals share a common resource and provide their identity as part of the request. This is similar to the design of the random access. The number of bits transmitted from a terminal as part of a request would in this case be larger, with the correspondingly larger need for resources. In contrast, the resources are shared by multiple users. Basically, contention-based designs

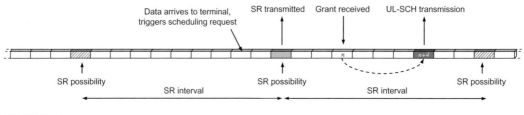

FIGURE 13.5

Scheduling-request transmission.

are suitable for a situation where there are a large number of terminals in the cell and the traffic intensity, and hence the scheduling intensity, is low. In situations with higher intensities, the collision rate between different terminals simultaneously requesting resources would be too high and lead to an inefficient design.

Although the scheduling-request design for LTE relies on dedicated resources, a terminal that has not been allocated such resources obviously cannot transmit a scheduling request. Instead, terminals without scheduling-request resources configured rely on the random-access mechanism described in Chapter 14. In principle, an LTE terminal can therefore be configured to rely on a contention-based mechanism if this is advantageous in a specific deployment.

Terminals that already have a valid grant obviously do not need to request uplink resources. However, to allow the scheduler to determine the amount of resources to grant to each terminal in future subframes, information about the buffer situation and the power availability is useful, as discussed above. This information is provided to the scheduler as part of the uplink transmission through MAC control elements (see Chapter 8 for a discussion on MAC control elements and the general structure of a MAC header). The LCID field in one of the MAC subheaders is set to a reserved value indicating the presence of a buffer status report, as illustrated in Figure 13.6.

From a scheduling perspective, buffer information for each logical channel is beneficial, although this could result in a significant overhead. Logical channels are therefore grouped into logical-channel groups and the reporting is done per group. The buffer-size field in a buffer-status report indicates the amount of data awaiting transmission across all logical channels in a logical-channel group. A buffer-status report represents one or all four logical-channel groups and can be triggered for the following reasons:

- Arrival of data with higher priority than currently in the transmission buffer – that is, data in a logical-channel group with higher priority than the one currently being transmitted – as this may impact the scheduling decision.
- Change of serving cell, in which case a buffer-status report is useful to provide the new serving cell with information about the situation in the terminal.
- Periodically as controlled by a timer.

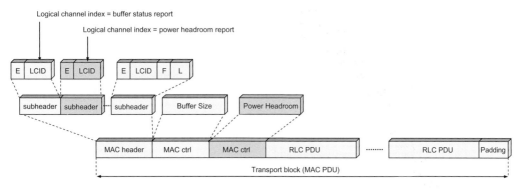

FIGURE 13.6

Signaling of buffer status and power-headroom reports.

- Instead of padding. If the amount of padding required to match the scheduled transport block size is larger than a buffer-status report, a buffer-status report is inserted. Clearly it is better to exploit the available payload for useful scheduling information instead of padding if possible.

In addition to buffer status, the amount of transmission power available in each terminal is also relevant for the uplink scheduler. Obviously, there is little reason to schedule a higher data rate than the available transmission power can support. In the downlink, the available power is immediately known to the scheduler as the power amplifier is located in the same node as the scheduler. For the uplink, the power availability, or power headroom (as discussed in Section 13.1.5), is defined as the difference between the nominal maximum output power and the estimated output power for UL-SCH transmission. This quantity can be positive as well as negative (on a dB scale), where a negative value would indicate that the network has scheduled a higher data rate than the terminal can support given its current power availability. The power headroom depends on the power-control mechanism and thereby indirectly on factors such as the interference in the system and the distance to the base stations.

Information about the power headroom is fed back from the terminals to the eNodeB in a similar way as the buffer-status reports – that is, only when the terminal is scheduled to transmit on the UL-SCH. Type-1 reports are provided for all component carriers simultaneously, while Type-2 reports are provided for the primary component carrier only.

A power headroom report can be triggered for the following reasons:

- Periodically as controlled by a timer.
- Change in path loss, since the last power headroom report is larger than a (configurable) threshold.
- Instead of padding (for the same reason as buffer-status reports).

It is also possible to configure a prohibit timer to control the minimum time between two power-headroom reports and thereby the signaling load on the uplink.

13.2.3 Semi-Persistent Scheduling

The basis for uplink and downlinks scheduling is dynamic scheduling, as described in Sections 13.2.1 and 13.2.2. dynamic scheduling with a new scheduling decision taken in each subframe allows for full flexibility in terms of the resources used and can handle large variations in the amount of data to transmit at the cost of the scheduling decision being sent on a PDCCH in each subframe. In many situations, the overhead in terms of control signaling on the PDCCH is well motivated and relatively small compared to the payload on DL-SCH/UL-SCH. However, some services, most notably voice-over IP, are characterized by regularly occurring transmission of relatively small payloads. To reduce the control signaling overhead for those services, LTE provides semi-persistent scheduling in addition to dynamic scheduling.

With semi-persistent scheduling, the terminal is provided with the scheduling decision on the PDCCH, together with an indication that this applies to every nth subframe until further notice. Hence, control signaling is only used once and the overhead is reduced, as illustrated in Figure 13.7. The periodicity of semi-persistently scheduled transmissions – that is, the value of n – is configured by RRC signaling in advance, while activation (and deactivation) is done using the PDCCH using the semi-persistent C-RNTI.[3] For example, for voice-over IP the scheduler can configure a periodicity

---

[3] Each terminal has two identities, the "normal" C-RNTI for dynamic scheduling and the semi-persistent C-RNTI for semi-persistent scheduling.

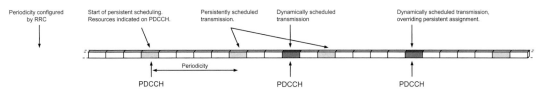

FIGURE 13.7

Example of semi-persistent scheduling.

of 20 ms for semi-persistent scheduling and, once a talk spurt starts, the semi-persistent pattern is triggered by the PDCCH.

After enabling semi-persistent scheduling, the terminal continues to monitor the PDCCH for uplink and downlink scheduling commands. When a dynamic scheduling command is detected, it takes precedence over the semi-persistent scheduling in that particular subframe, which is useful if the semi-persistently allocated resources occasionally need to be increased. For example, for voice-over IP in parallel with web browsing it may be useful to override the semi-persistent resource allocation with a larger transport block when downloading the web page.

For the downlink, only initial transmissions use semi-persistent scheduling. Retransmissions are explicitly scheduled using a PDCCH assignment. This follows directly from the use of an asynchronous hybrid-ARQ protocol in the downlink. Uplink retransmissions, in contrast, can either follow the semi-persistently allocated subframes or be dynamically scheduled.

Semi-persistent scheduling is only supported on the primary component carrier and any transmission on a secondary component carrier must be dynamically scheduled. This is reasonable as semi-persistent scheduling is intended for low-rate services for which a single component carrier is sufficient.

13.2.4 Scheduling for Half-Duplex FDD

Half-duplex FDD implies that a single terminal cannot receive and transmit at the same time while the eNodeB still operates in full duplex. In LTE, half-duplex FDD is implemented as a scheduler constraint, implying it is up to the scheduler to ensure that a single terminal is not scheduled simultaneous in uplink and downlink. Hence, from a terminal perspective, subframes are dynamically used for uplink or downlink. Briefly, the basic principle for half-duplex FDD is that a terminal is receiving in the downlink unless it has been explicitly instructed to transmit in the uplink (either UL-SCH transmission or hybrid-ARQ acknowledgements triggered by a downlink transmission). The timing and structure for control signaling are identical between half- and-full duplex FDD terminals.

An alternative design approach would be to base half-duplex FDD on the TDD control signaling structure and timing, with a semi-static configuration of subframes to either downlink or uplink. However, this would complicate supporting a mixture of half- and-full duplex terminals in the same cell as the timing of the control signaling would differ. It would also imply a waste of uplink spectrum resources. All terminals need to be able to receive subframes 0 and 5, as those subframes are used for system information and synchronization signals. Hence, if a fixed uplink–downlink allocation were to be used, no uplink transmissions could take place in those two subframes, resulting in a

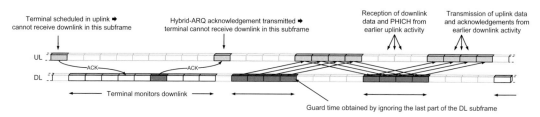

FIGURE 13.8

Example of half-duplex FDD terminal operation.

loss in uplink spectral efficiency of 20%. Clearly this is not attractive and led to the choice of implementing half-duplex FDD as a scheduling strategy instead.

An example of half-duplex operation as seen from a terminal perspective is shown in Figure 13.8. In the leftmost part of the figure, the terminal is explicitly scheduled in the uplink and, consequently, cannot receive data in the downlink in the same subframe. The uplink transmission implies the receipt of an acknowledgement on the PHICH four subframes later, as mentioned in Chapter 12, and therefore the terminal cannot be scheduled in the uplink in this subframe. Similarly, when the terminal is scheduled to receive data in the downlink in subframe n, the corresponding hybrid-ARQ acknowledgement needs to be transmitted in the uplink subframe $n + 4$, preventing downlink reception in subframe $n + 4$. The scheduler can exploit this by scheduling downlink data in four consecutive subframes and uplink transmission in the four next subframes when the terminal needs to transmit hybrid-ARQ acknowledgements in the uplink anyway, and so on. Hence, at most half of the time can be used in the downlink and half in the uplink or, in other words, the asymmetry in half-duplex FDD is 4:4. Efficient support of half-duplex FDD is one of the reasons why the same number of hybrid-ARQ processes was selected in uplink and downlink.

Note that, as the eNodeB is operating in full duplex, regardless of the duplex capability of the terminals, the cell capacity is hardly affected by the presence of half-duplex terminals as, given a sufficient number of terminals with data to transmit/receive, the scheduler can with a high likelihood find a set of terminals to schedule in the uplink and another set to schedule in the downlink in a given subframe.

Similar to TDD, a half-duplex terminal needs some guard time for switching between uplink and downlink. For half-duplex FDD, guard time for the downlink-to-uplink switch is created by allowing the terminal to skip receipt of the last OFDM symbols in a downlink subframe immediately preceding an uplink subframe, as described in Chapter 9. Guard time for the uplink-to-downlink switch is handled by setting the appropriate amount of timing advance in the terminals.

13.2.5 Channel-State Reporting

As mentioned several times, the possibility for downlink channel-dependent scheduling – that is, selecting the downlink transmission configuration and related parameters depending on the instantaneous downlink channel conditions – is a key feature of LTE. An important part of the support for downlink channel-dependent scheduling is *channel-state reports* provided by terminals to the network, reports on which the latter can base its scheduling decisions.

The channel-state reports consist of one or several pieces of information:

- *Rank indication* (RI), providing a recommendation on the transmission rank to use or, expressed differently, the number of layers that should preferably be used for downlink transmission to the terminal. RI only needs to be reported by terminals that are configured to be in one of the spatial-multiplexing transmission modes. There is at most one RI reported, valid across the full bandwidth – that is, the RI is frequency non-selective. Frequency-dependent transmission rank would be impossible to utilize since all layers are transmitted on the same set of resource blocks in LTE.
- *Precoder matrix indication* (PMI), indicating which of the precoder matrices (see Chapter 10) should preferably be used for the downlink transmission. The reported precoder matrix is determined assuming the number of layers indicated by the RI. The precoder recommendation may be frequency selective, implying that the terminal may recommend different precoders for different parts of the downlink spectrum. Furthermore, the network can restrict the set of matrices from which the terminal should select the recommended precoder, so-called codebook subset restriction, to avoid reporting precoders that are not useful in the antenna setup used.
- *Channel-quality indication* (CQI), representing the highest modulation-and-coding scheme that, if used, would mean PDSCCH transmissions (using the recommended RI and PMI) were received with a block-error rate of at most 10%. The reason to use CQI as a feedback quantity instead of, for example, the signal-to-noise ratio, is to account for different receiver implementation in the terminal. Also, basing the feedback reports on CQI instead of signal-to-noise ratio also simplifies the testing of terminals; a terminal delivering data with more than 10% block-error probability when using the modulation-and-coding scheme indicated by the CQI would fail the test. As will be discussed further below, multiple CQI reports, each representing the channel quality in a certain part of the downlink spectrum, can be part of a channel-state report.

Together, a combination of the RI, PMI, and CQI forms a channel-state report. Exactly what is included in a channel-state report depends on the reporting mode the terminal is configured to be in. As mentioned earlier, RI and PMI do not need to be reported unless the terminal is in a spatial-multiplexing transmission mode. However, also given the transmission mode, there are different reporting modes that typically differ as to what set of resource blocks the report is valid for and whether precoding information is reported or not. The type of information useful to the network also depends on the particular implementation and antenna deployment.

Although referred to as channel-state reports, what a terminal delivers to the network are not explicit reports of the downlink channel state. Rather, what the terminal delivers are *recommendations* on the transmission rank and precoding matrix to use, together with an indication of the highest possible modulation-and-coding scheme that the network preferably should not exceed. Information about the actual modulation scheme and coding rate used for DL-SCH transmission as well as the set of resource blocks used for the transmission is always included in the downlink scheduling assignment. Hence, the eNodeB is free to follow the CSI report or to select transmission parameters on its own.

The modulation-and-coding scheme used for DL-SCH transmission can, and often will, differ from the reported CQI as the scheduler needs to account for additional information not available to the terminal when recommending a certain CQI. For example, the set of resource blocks used for the DL-SCH transmission also need to account for other users. Furthermore, the amount of data awaiting transmission in the eNodeB also needs to be accounted for. Obviously, there is no need to select a very high data rate, even if the channel conditions would permit this, if there is only a small amount of data to transmit and a sufficient number of resource blocks can be allocated to the terminal in question.

With regards to the precoder-related recommendations, the network has two choices:

- The network may follow the latest terminal recommendation, in which case the eNodeB only has to confirm (a one-bit indicator in the downlink scheduling assignment) that the precoder configuration recommended by the terminal is used for the downlink transmission. On receiving such a confirmation, the terminal will use its recommended configuration when demodulating and decoding the corresponding DL-SCH transmission. Since the PMI computed in the terminal can be frequency selective, an eNodeB following the precoding matrix recommended by the terminal may have to apply different precoding matrices for different (sets of) resource blocks.
- The network may select a different precoder, information about which then needs to be explicitly included in the downlink scheduling assignment. The terminal then uses this configuration when demodulating and decoding the DL-SCH. To reduce the amount of downlink signaling, only a single precoding matrix can be signaled in the scheduling assignment, implying that, if the network overrides the recommendation, then the precoding is frequency non-selective. The network may also choose to override the transmission rank only, in which case the terminal assumes that a subset of the columns in each of the recommended precoder matrices is used.

There are two types of channel-state reports in LTE, *aperiodic* and *periodic*, which are different in terms of how a report is triggered:

- Aperiodic channel-state reports are delivered when explicitly requested by the network by means of the channel-state-request flag included in uplink scheduling grants (see Section 10.4.5). An aperiodic channel-state report is always delivered using the PUSCH – that is, on a dynamically assigned resource.
- Periodic channel-state reports are configured by the network to be delivered with a certain periodicity, possibly as often as once every 2 ms, on a semi-statically configured PUCCH resource. However, similar to hybrid-ARQ acknowledgements normally delivered on PUCCH, channel-state reports are "re-routed" to the PUSCH[4] if the terminal has a valid uplink grant and is anyway to transmit on the PUSCH.

Aperiodic and periodic reports, despite both providing estimates on the channel conditions, are quite different in terms of their detailed contents and the usage. In general, aperiodic reports are larger and more detailed than their periodic counterparts. There are several reasons for this. First, the PUSCH, upon which the aperiodic report is transmitted, is capable of a larger payload, and hence a more detailed report, than the PUCCH used for the periodic reports. Furthermore, as aperiodic reports are transmitted on a per-need basis only, the overhead from these reports are less of an issue compared to periodic reports. Finally, if the network requests a report it is likely that it will transmit a large amount of data to the terminal, which makes the overhead from the report less of an issue compared to a periodic report that is transmitted irrespective of whether the terminal in question will be scheduled in the near future or not. Hence, as the structure and usage of aperiodic and periodic reports is different, they are described separately below, starting with aperiodic reports.

---

[4] In release 10, a terminal can be configured for simultaneous PUSCH and PUCCH transmission, in which case the periodic channel-state reports can remain on the PUCCH.

Three aperiodic reporting modes are supported in LTE, where each mode has several submodes depending on the configuration:

- Wideband reports, reflecting the average channel quality across the entire cell bandwidth with a single CQI value. Despite a single average CQI value being provided for the whole bandwidth, the PMI reporting is frequency selective. Frequency-selective reporting is obtained, for reporting purposes only, by dividing the overall downlink bandwidth (of each component carrier) into a number of equally sized *sub-bands*, where each sub-band consists of a set of consecutive resource blocks. The size of a sub-band, ranging from four to eight resource blocks, depends on the cell bandwidth. The PMI is then reported for each sub-band. For transmission modes supporting spatial multiplexing, the CQI and the PMI are calculated assuming the channel rank indicated by the RI, otherwise rank-1 is assumed. Wideband reports are smaller than their frequency-selective counterparts, but obviously do not provide any information about the frequency domain.
- UE-selected reports, where the terminal selects the best M sub-bands and reports, in addition to the indices of the selected sub-bands, one CQI reflecting the average channel quality over the selected M sub-bands together with one wideband CQI reflecting the channel quality across the full downlink carrier bandwidth. This type of report thus provides frequency-domain information about the channel conditions. The sub-band size, ranging from two to four resource blocks, and the value of M, ranging from 1 to 6, depends on the downlink carrier bandwidth. Depending on the transmission mode configured, the PMI and RI are also provided as part of this type of report.
- Configured reports, where the network configured the set of sub-bands the terminal should generate reports for. The terminal reports one wideband CQI reflecting the channel quality across the full downlink carrier bandwidth and one CQI per configured sub-band. The sub-band size depends on the downlink carrier bandwidth and is in the range of four to eight resource blocks. Depending on the transmission mode configured, the PMI and RI are also provided as part of this type of report.

The different aperiodic reporting modes are summarized in Table 13.1.

Periodic reports are configured by the network to be delivered with a certain periodicity. The limited, compared to PUSCH, payload supported on the PUCCH also implies that the different types of information in a periodic report may not be possible to transmit in a single subframe. Therefore, some of the reporting modes will transmit one or several of the wideband CQI, the wideband CQI including PMI, the RI, and the CQI for the UE-selected sub-bands at different time points. Furthermore, the RI can typically be reported less often, compared to the reporting of PMI and CQI, reflecting the fact that the suitable number of layers typically varies on a slower basis, compared to the channel variations that impact the choice of precoder matrix and modulation rate, and coding scheme.

Two periodic reporting modes are supported in LTE, again with different submodes possible:

- Wideband reports, reflecting the average channel quality across the entire cell bandwidth with a single CQI value. If PMI reporting is enabled, a single PMI valid across the full bandwidth is reported.
- UE-selected reports. Although named in the same way as for aperiodic reports, the principle for UE-selected periodic reports is different. The total bandwidth (of a component carrier) is divided into one to four *bandwidth parts*, with the number of bandwidth parts obtained form the cell bandwidth. For each bandwidth part, the terminal selects the best sub-band within that part. The

Table 13.1 Possible Aperiodic Reporting Modes for Different Transmission Modes

Transmission Mode	Reporting Mode							
	Wideband CQI		Frequency-Selective CQI					
			UE-Selected Sub-Bands			Conf. Sub-Bands		
	1-0: No PMI	1-1: Wideband PMI	1-2: Selective PMI	2-0: No PMI	2-1: Wideband PMI	2-2: Selective PMI	3-0: No PMI	3-1: Wideband PMI
1 Single antenna, CRS				•			•	
2 Transmit diversity				•			•	
3 Open-loop spatial mux.				•			•	
4 Closed-loop spatial mux.			•			•		•
5 Multi-user MIMO								•
6 Codebook-based beam-forming			•			•		•
7 Single-layer trans., DM-RS				•			•	
8 Dual-layer trans., DM-RS			•	•		•	•	•
9 Multi-layer trans., DM-RS			•	•		•	•	•

sub-band size ranges from four to eight resource blocks. Since the supported payload size of the PUCCH is limited, the reporting cycles through the bandwidth parts and in one subframe report the wideband CQI and PMI (if enabled) for that bandwidth part, as well as the best sub-band and the CQI for that sub-band. The RI (if enabled) is reported in a separate subframe.

The different aperiodic reporting modes are summarized in Table 13.2. Note that all PMI reporting, if enabled, is of wideband type. There is no support for frequency-selective PMI in periodic reporting, as the amount of bits would result in a too large overhead.

A typical use of periodic and aperiodic reporting could be to configure lightweight periodic CSI reporting PUCCH, for example to provide feedback of the wideband CQI and no PMI information (mode 1-0). Upon arrival of data to transmit in the downlink to a specific terminal, aperiodic reports could be requested as needed, for example with frequency-selective CQI and PMI (mode 3-1).

Channel-state reports, irrespective of whether they are aperiodic or periodic, need a known reference signal as input to the CSI computation. The cell-specific reference signals can be used for this purpose, and is the only possibility in release 8/9, but in release 10 channel-state reports can also be based on the CSI-RS (see Chapter 10 for a discussion on CSI-RS). Which reference signal to base the channel-state reports upon is linked to the transmission mode. For transmission modes already supported in release 8/9, the cell-specific reference signals are used, while for transmission mode 9, introduced in release 10, the CSI-RS is used.

The discussion above also holds for carrier aggregation, although with some modifications and enhancements as channel-state reports for multiple downlink component carriers are needed.

For aperiodic reporting, the two-bit[5] CSI request in the downlink control signaling allows for three different types of CSI reports to be requested (the fourth bit combination represents no CSI request). Of these three alternatives, one is used to trigger a CSI reports for the downlink component carrier associated with the uplink component carrier for which the scheduling grant relates to. The remaining alternatives point to one of two configurable combinations of component carriers for which the CSI report should be generated. Thus, as an example, for a terminal capable of two downlink component carriers, aperiodic reports can, with the proper configuration, be requested for the primary component carrier, the secondary component carrier, or both.

For periodic reporting, the basic principle in the case of carrier aggregation is to configure the reporting cycles such that the CSI reports for the different component carriers are not transmitted simultaneously.

13.2.6 Discontinuous Reception (DRX) and Component Carrier Deactivation

Packet-data traffic is often highly bursty, with occasional periods of transmission activity followed by longer periods of silence. Clearly, from a delay perspective, it is beneficial to monitor the downlink control signaling in each subframe to receive uplink grants or downlink data transmissions and instantaneously react on changes in the traffic behavior. At the same time this comes at a cost in terms of power consumption at the terminal; the receiver circuitry in a typical terminal represents a non-negligible amount of power consumption. To reduce the terminal power consumption, LTE includes mechanisms for *discontinuous reception* (DRX).

---

[5] Recall from Chapter 10 that the CSI request field expanded from one to two bits in the case of carrier aggregation.

Table 13.2 Possible Periodic Reporting Modes for Different Transmission Modes

Transmission Mode		Reporting Mode							
		Wideband CQI		Frequency-Selective CQI					
				UE-Selected Sub-Bands			Conf. Sub-Bands		
		1-0: No PMI	1-1: Wideband PMI	1-2: Selective PMI	2-0: No PMI	2-1: Wideband PMI	2-2: Selective PMI	3-0: No PMI	3-1: Wideband PMI
1	Single antenna, CRS	●			●				
2	Transmit diversity	●			●				
3	Open-loop spatial mux.	●			●				
4	Closed-loop spatial mux.		●			●			
5	Multi-user MIMO		●			●			
6	Codebook-based beam-forming		●			●			
7	Single-layer trans., DM-RS	●			●				
8	Dual-layer trans., DM-RS	●	●		●	●			
9	Multi-layer trans., DM-RS	●	●		●	●			

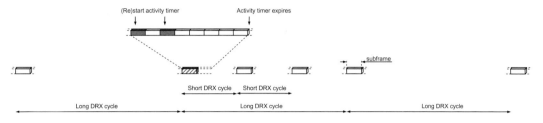

FIGURE 13.9

Illustration of DRX operation.

The basic mechanism for DRX is a configurable DRX cycle in the terminal. With a DRX cycle configured, the terminal monitors the downlink control signaling only in one subframe per DRX cycle, sleeping with the receiver circuitry switched off in the remaining subframes. This allows for a significant reduction in power consumption: the longer the cycle, the lower the power consumption. Naturally, this implies restrictions to the scheduler as the terminal can be addressed only in the active subframes.

In many situations, if the terminal has been scheduled and active with receiving or transmitting data in one subframe, it is highly likely it will be scheduled again in the near future. One reason could be that it was not possible to transmit all the data in the transmission buffer in one subframe and additional subframes are required. Waiting until the next active subframe according to the DRX cycle, although possible, would result in additional delays. Hence, to reduce the delays, the terminal remains in the active state for a certain configurable time after being scheduled. This is implemented by the terminal (re)starting an inactivity timer every time it is scheduled and remaining awake until the time expires, as illustrated at the top of Figure 13.9.

Retransmissions take place regardless of the DRX cycle. Thus, the terminal receives and transmits hybrid-ARQ acknowledgements as normal in response to data transmission. In the uplink, this also includes retransmissions in the subframes given by the synchronous hybrid-ARQ timing relation. In the downlink, where asynchronous hybrid ARQ is used, the retransmission time is not fixed in the specifications. To handle this, the terminal monitors the downlink for retransmissions in a configurable time window after the previous transmission.

The above mechanism, a (long) DRX cycle in combination with the terminal remaining awake for some period after being scheduled, is sufficient for most scenarios. However, some services, most notably voice-over IP, are characterized by periods of regular transmission, followed by periods of no or very little activity. To handle these services, a second short DRX cycle can optionally be used in addition to the long cycle described above. Normally, the terminal follows the long DRX cycle, but if it has recently been scheduled, it follows a shorter DRX cycle for some time. Handling voice-over IP in this scenario can be done by setting the short DRX cycle to 20 ms, as the voice codec typically delivers a voice-over-IP packet per 20 ms. The long DRX cycle is then used to handle longer periods of silence between talk spurts.

The DRX mechanism described above is common to all component carriers configured in the terminal. Hence, if the terminal is in DRX it is not receiving on any component carrier, but when it wakes up, all (activated) component carriers will be woken up. Although discontinuous reception greatly reduces the terminal power consumption, it is possible to go one step further in the case of carrier aggregation. Obviously, from a power-consumption perspective, it is beneficial to receive on as few component carriers as possible. LTE therefore supports deactivation of downlink component carriers. A deactivated component carrier maintains the configuration provided by RRC but cannot be used for reception, neither PDCCH nor PDSCH. When the need arises, a downlink component carrier can be activated rapidly and used for reception within a few subframes. A typical use would be to configure several component carriers but deactivate all component carriers except the primary one. When a data burst starts, the network could activate several component carriers to maximize the downlink data rate. Once the data burst is delivered, the component carriers could be deactivated again to reduce terminal power consumption.

Activation and deactivation of downlink component carriers are done through MAC control elements. A command to activate a component carrier takes effect eight subframes after receipt – that is, if the MAC control element is received in subframe n, then the additional component carriers are activated starting from subframe $n + 8$. There is also a timer-based mechanism for deactivation such that a terminal may, after a configurable time with no activity on a certain component carrier, deactivate that component carrier. The primary component carrier is always active as it must be possible for the network to communicate with the terminal.

In the uplink there is no explicit activation of uplink component carriers. However, whenever a downlink component carrier is activated or deactivated, the corresponding uplink component carrier is also activated or deactivated.

13.3 INTER-CELL INTERFERENCE COORDINATION

Like all modern mobile-communication technologies, LTE can be deployed with one-cell frequency reuse. Fundamentally, this implies that the LTE transmission structure has been designed so that reliable transmission is also possible for the low signal-to-interference ratios (SIR) that may occur in a reuse-one deployment when the same time–frequency resource issued in neighboring cells (SIR as low as −5 dB or even somewhat lower in the worst case[6]). This is especially true for transmission of critical information such as system information and L1/L2 control signaling. It should be noted that data transmission on DL-SCH can always be made sufficiently reliable by selecting a sufficiently low instantaneous data rate in combination with hybrid-ARQ retransmissions.

Still, a one-cell frequency reuse typically implies relatively large SIR variations over the cell area. As a consequence, the data rates that can be offered to the end-user may also vary substantially, with only relatively low data rates being available at the "cell edge".

If one was *only* interested in maximizing the data rates that could be offered to users at the cell edge – that is, maximizing the "worst-case-user" quality – a reuse larger than one could actually be preferred. For cell-edge users receiving high interference from neighboring cells, the negative impact of reduced bandwidth availability due to frequency reuse larger than one could be more than compensated for by the higher SIR and corresponding higher achievable data rate per MHz, leading to overall higher achievable data rates. However, the overall system efficiency would be degraded as the

---

[6]This assumes a relatively homogeneous deployment. In *heterogeneous network deployments* with large differences in the cell output powers, the SIR could be even lower and special means may need to be taken to ensure reliable transmission, as discussed in Section 13.4.

majority of users are not at the cell edge and thus have a relatively good SIR even with one-cell reuse. For such users, any further SIR improvement due to larger reuse would typically not be able to compensate for the reduced bandwidth availability for at least two reasons:

- The relative interference reduction and corresponding increase in SIR would not be as large as for users on the cell edge.
- The achievable data rates do not vary linearly with the SIR and, especially for high SIR, a further SIR improvement may give a relatively small increase in achievable data rate per MHz, as was discussed in Chapter 2.

Furthermore, the assumption of low SIR at the cell edge in the case of one-cell reuse is based on the assumption that there really are transmissions ongoing in the neighboring cells. There are always time instances when there is no or at least limited transmission ongoing in a cell, especially at low-load conditions. The cell-edge SIR may then be relatively good even with a one-cell reuse and a higher reuse with a corresponding reduction in the bandwidth available per cell would not be beneficial even from a cell-edge-user point of view.

Thus, the basic mode of operation should be one-cell reuse, giving each cell access to the overall available spectrum. However, system performance, and especially the quality for cell-edge users, could be further enhanced if one could at least partly coordinate the scheduling between neighboring cells. The basic principle of such *inter-cell interference coordination* (ICIC) would be to, if possible, avoid high-power transmission on time–frequency resources on which cell-edge users are scheduled in neighboring cells, users that would otherwise experience high interference and correspondingly low data rates. This kind of "selective" interference avoidance would benefit the cell-edge-user quality and could also enhance overall system performance.

The above discussion implicitly assumed downlink transmission with terminals at the cell edge being interfered by downlink transmissions from other cells. The concept of inter-cell interference coordination is equally applicable to the uplink, although the interference situation in this case is somewhat different.

For the uplink, the interference level experienced by a certain link does not depend on where the transmitting terminal is located, but rather on the location of the *interfering* terminals, with interfering terminals closer to the cell border causing more interference to neighboring cells. The location of the transmitting terminal is still important though, as a terminal closer to the cell site can raise its transmission power to compensate for high interference coming from terminals in neighboring cells, something that may not be possible for cell-edge terminals. Thus, the fundamental goal of uplink inter-cell interference is the same as for the downlink – that is, to coordinate the, in this case, uplink scheduling between cells to avoid simultaneous transmissions from terminals at the cell border in neighboring cells causing severe interference to each other.

In the case of scheduling located at a higher-level node above the eNodeB, coordinated scheduling between cells of different eNodeB would, at least conceptually, be straightforward. However, in LTE there is no higher-level node and scheduling is carried out locally at the eNodeB – that is, in practice at the cell site.[7] Thus, the best that can be done in the radio-access specifications is to introduce messages that convey information about the scheduling strategy between neighboring eNodeBs using the

---

[7] As discussed below, in case of so-called centralized RAN (C-RAN) this may change somewhat.

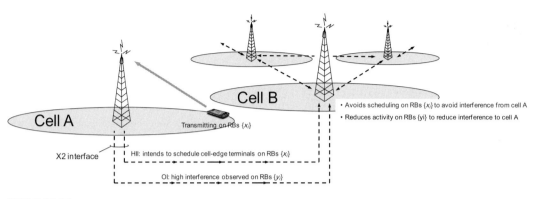

FIGURE 13.10

Illustration of uplink ICIC based on HII and OI X2 signaling.

X2 interface. An eNodeB can then use the information provided by a neighboring eNodeB as input to its own scheduling process. For LTE, a number of such ICIC-related X2 messages have been defined.

To assist uplink interference coordination, two messages are defined, the *High Interference Indicator* (HII) and the *Overload Indicator* (OI), see also Figure 13.10.

The *High Interference Indicator* provides information about the set of resource blocks within which an eNodeB is likely to schedule transmissions from cell-edge terminals – that is, resource blocks on which a neighboring cell can expect higher interference. Although nothing is explicitly specified on how an eNodeB should react to the HII (or any other ICIC-related X2 signaling) received from a neighboring eNodeB, a reasonable action for the receiving eNodeB would be to try to avoid scheduling its own cell-edge terminals on the same resource blocks, thereby reducing the uplink interference to cell-edge transmissions in its own cell as well as in the cell from which the HII was received. The HII can thus be seen as a *proactive* tool for ICIC, trying to prevent the occurrence of *too-low-SIR* situations.

In contrast to the HII, the *Overload Indicator* (OL) is a *reactive* ICIC tool, essentially indicating, at three levels (Low/Medium/High), the uplink interference experienced by a cell on its different resource blocks. A neighboring eNodeB receiving the OI could then change its scheduling behavior to improve the interference situation for the eNodeB issuing the OI.

For the downlink, *the Relative Narrowband Transmit Power* (RNTP) is defined to support ICIC operation (see Figure 13.11). The RNTP is similar to the HII in the sense that it provides information, for each resource block, whether or not the relative transmit power of that resource block is to exceed a certain level. Similar to the HII, a neighboring cell can use the information provided by the received RNTP when scheduling its own terminals, especially terminals on the cell edge that are more likely to be interfered by the neighboring cell.

One kind of deployment that has recently received some interest is the *Centralized RAN* (C-RAN) [73], where the baseband processing of the eNodeBs is located in a central office, geographically separate from the actual cell sites. In such a scenario it is more straightforward to coordinate the scheduling between multiple geographically separate cells, either by introducing inter-eNodeB coordination within the central office or by simply deploying a massive eNodeB able to handle all the

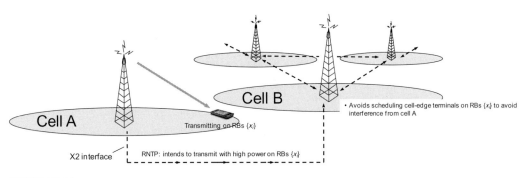

FIGURE 13.11

Illustration of downlink ICIC based on RNTP X2 signaling.

cells connected to the central office. As an eNodeB is anyway just a logical node, in terms of physical realization these two approaches are essentially the same thing.

13.4 HETEROGENEOUS NETWORK DEPLOYMENTS

The continuous increase in traffic within mobile-broadband systems and an equally continuous increase in terms of the data rates requested by end-users will impact how cellular networks are deployed in the future. In general, providing very high system capacity (traffic per m^2) and very high per-user data rates will require a densification of the radio-access network – that is, the deployment of additional network nodes. By increasing the number of cells, the traffic per m^2 can be increased without requiring a corresponding increase in the traffic that needs to be supported per network node. Also, by increasing the number of network nodes, the base-station-to-terminal distances will, in general, be shorter, implying a link-budget improvement and a corresponding improvement in achievable data rates.

A general densification of the macro-cell layer – that is, reducing the coverage area of each cell and increasing the total number of macro-cell sites[8] – as illustrated in the upper part of Figure 13.12, is a path that has already been taken by many operators. As an example, in many major cities the distance between macro-cell sites is often less than a few hundred meters in many cases.

An alternative or complement to a uniform densification of the macro-cell layer is to deploy additional lower-power nodes under the coverage area of a macro cell, as illustrated in the lower part of Figure 13.12. In such a *heterogeneous* or *multi-layered* network deployment, the underlaid *pico-cell* layer does not need to provide full-area coverage. Rather, pico sites can be deployed to increase capacity and achievable data rates where needed. Outside of the pico-layer coverage, terminals would access the network by means of the overlaid macro cell.

Another example of heterogeneous network deployment is the complementary use of so-called *home-eNodeBs*, also often referred to as *femto* base stations. A home-eNodeB corresponds to a small low-power base station deployed by the end-user, typically within the home, and connecting to the operator network using the end-user's wireline broadband connection.

---

[8]A macro cell is herein defined as a high-power cell with its antennas typically located above rooftop level.

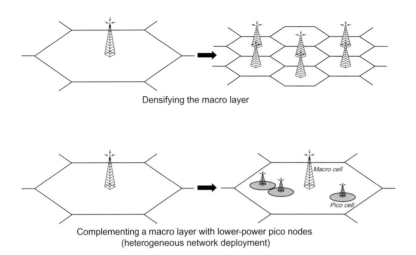

Densifying the macro layer

Complementing a macro layer with lower-power pico nodes
(heterogeneous network deployment)

FIGURE 13.12

Network densification to enhance system capacity and support higher data rates.

A home-eNodeB is often associated with a so-called *Closed Subscriber Group* (CSG), with only users that are members of the CSG being allowed to access the home-eNodeB. Thus, users not being members of the CSG have to access the radio-access network via the overlaid macro-cell layer even when in close proximity to a home-eNodeB. As discussed further below, this causes additional interference problems with home-eNodeB deployments, beyond those of ordinary heterogeneous network deployments.

13.4.1 Interference Handling in a Heterogeneous Deployment

In itself, the use of heterogeneous network deployments in mobile-communication systems – that is, complementing a macro layer with lower-power pico nodes to increase traffic capacity and/or achievable data rates in specific areas – is nothing new and has been used for a relatively long time in, for example, GSM networks. Different sets of carrier frequencies have then typically been used in the different cell layers, thereby avoiding strong interference between the layers.

However, for a wideband radio-access technology such as LTE, using different carrier frequencies for different cell layers may lead to an undesirable spectrum fragmentation. As an example, for an operator having access to 20 MHz of spectrum, a static frequency separation between two cell layers would imply that the total available spectrum had to be divided, with less than 20 MHz of spectrum being available in each layer. This could obviously reduce the maximum achievable data rates within each cell layer. Also, assigning a substantial part of the overall available spectrum to a cell layer with relatively low traffic may lead to inefficient spectrum utilization. Thus, with a wideband high-data-rate system such as LTE, it should preferably be possible to deploy a multi-layer network structure with the same spectrum being available in the different cell layers.

However, the simultaneous use of the same spectrum in different cell layers will obviously imply interference between the layers. Due to the difference in transmit power between the nodes of a

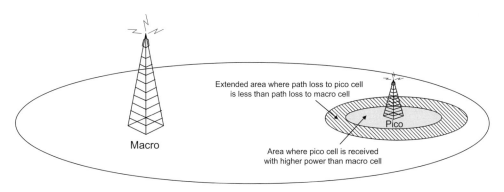

FIGURE 13.13

Illustration of high-interference area in a heterogeneous network deployment with range extension.

heterogeneous network deployment, such *inter-layer interference* may be more severe compared to inter-cell interference between cells of the same layer.

The characteristics of the inter-layer interference in a heterogeneous network deployment will also depend on the exact cell-selection strategy being used. Conventional cell selection is typically based on terminal measurements of the received power of some downlink signal, more specifically the cell-specific reference signals in the case of LTE. However, in a heterogeneous network deployment with cells with substantially different transmit power, including different power of the reference signals, selecting the cell that is received with the highest power implies that the terminal may often select the higher-power macro cell even if the path loss to a pico cell is significantly smaller. This will obviously not be optimal from an uplink coverage and capacity point of view.

It should also be noted that, even in terms of downlink system efficiency, it may not be optimal to select the cell with the highest received power in a heterogeneous network deployment. Although the high-power macro cell is received with higher power, this is at least partly due to the higher macro-cell transmit power. In that case, transmission from the macro cell is associated with a higher "cost" in terms of interference to other cells. Expressed alternatively, a transmission from the macro cell will prohibit the use of the same physical resource in *any* of the underlaid pico cells.

Alternatively, at the other extreme, cell selection could be based on estimates of the (uplink) path loss. In practice this can be achieved by applying a cell-specific offset to the received-power measurements used in conventional cell selection, an offset that would compensate for the difference in cell transmit power.[9] Such a cell-selection strategy would extend the area in which the pico-cell is selected, as illustrated in Figure 13.13. It is therefore also sometimes referred to as *range extension*.

Selecting the cell to which the path loss is the smallest – that is, applying range extension – would maximize the uplink received power/SINR, thus maximizing the achievable uplink data rates. Alternatively, for a given target received power, the terminal transmit power, and thus the interference to other cells, would be reduced, leading to higher overall uplink system efficiency. Also, it could allow for the same downlink physical resource to also be used in other pico cells, thereby also improving downlink system efficiency.

---

[9] Such offsets or *cell-selection biasing* is already supported by LTE.

However, due to the difference in transmit power between the cells of the different cell layers, there is an area (illustrated by the dashed region in Figure 13.13) where the pico cell is selected while, at the same time, the downlink transmission from the macro cell is received with substantially higher power than the actual desired downlink transmission from the pico cell. Within this area, there is thus potential for severe downlink inter-cell interference from the macro cell to pico-cell terminals, interference that may require special means to handle.

In the following we will assume that the transmissions from the macro cell and its underlaid pico cells are relatively time aligned, implying, for example, that the macro-cell PDCCH (PDSCH) transmissions will interfere with PDCCH (PDSCH) transmissions in the pico cell and vice versa.[10]

The interference from PDSCH transmissions in the macro cell to lower-power PDSCH transmissions within a pico cell can be relatively straightforwardly handled by scheduling coordination between the cells according to the same principles as inter-cell interference coordination between cells within a layer, as described in Section 13.3. As an example, the overlaid macro cell could simply avoid high-power PDSCH transmission in resource blocks in which a terminal in the high-interference region of a pico cell is to receive downlink data transmission. Such coordination can be more or less dynamic depending on to what extent and on what time scale the overlaid macro cell and the underlaid pico cells can be coordinated. It should also be noted that, for a pico cell on the border between two macro cells, it may be necessary to coordinate scheduling simultaneously with both macro cells.

Less obvious is how to handle interference due to the macro-cell transmissions that cannot be dynamically scheduled, such as the L1/L2 control signaling (PDCCH, PCFICH, and PHICH). Within a cell layer, for example between two macro cells, interference between such transmissions is not a critical issue as LTE, including its control channels, has been designed to allow for one-cell frequency reuse and a corresponding SIR down to and even below -5 dB. However, in a heterogeneous network deployment with extensive range expansion, the downlink interference from the macro cell can be more than 15 dB above the signal received from the pico cell, corresponding to an SIR below -15 dB. In such low SIR it will not be possible to correctly decode even the very robust control channels. Instead, means are needed to avoid macro-to-pico interference for the L1/L2 control channels.

One possible approach to handle the extended interference in a heterogeneous network deployment with range-extended pico cells is to use carrier aggregation in combination with cross-carrier scheduling, as described in Section 10.4.6. The basic principle of such an approach is illustrated in Figure 13.14 for the case of two layers (macro and pico) and two downlink carriers (f_M and f_P).

In terms of data (PDSCH) transmission, both carriers are available in both cell layers and interference between the layers is handled by "conventional" inter-cell interference coordination, as discussed above. As already mentioned, such interference coordination can be more or less dynamic depending on the time scale on which the macro cell and its underlaid pico cells can be coordinated. Also, the possibility for carrier aggregation allows for both carriers – that is, the total available spectrum – to be assigned for transmission to a single terminal. Thus, at least for carrier-aggregation-capable terminals, there is no spectrum fragmentation in terms of data (PDSCH) transmission.

On the other hand, in terms of L1/L2 control signaling there is at least partly a more semi-static frequency separation between the layers. More specifically, the macro cell should avoid high-power

---

[10]This also assumes the same size of the control (PDCCH) region for macro and pico cells.

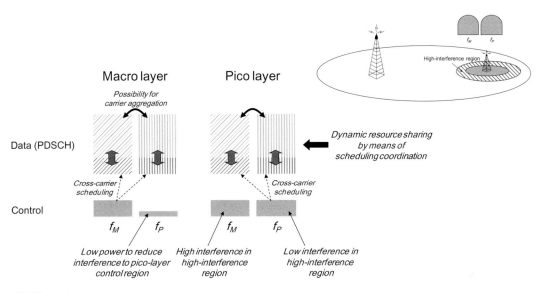

FIGURE 13.14

Carrier-aggregation approach to interference avoidance in a heterogeneous network deployment.

transmission within the control region on carrier f_P. In this way, interference to the control region of an underlaid pico cell is reduced on this carrier and the pico cells can use the carrier for control signaling to terminals in the high-interference region. Due to the possibility for cross-carrier scheduling, even if the macro cell only transmits control signaling, including downlink scheduling assignments, on carrier f_M, DL-SCH transmission on PDSCH can still be scheduled on both carriers as well as an aggregation of these. The same is true for the pico cell; even if the pico cell can only use carrier f_P for transmission of scheduling assignments to terminals in the high-interference region, DL-SCH transmissions can still be scheduled on both carriers.

It should be noted that, for terminals not in the high interference region, the pico cell could also use carrier f_M for L1/L2 control signaling. Similarly, the macro cell could use also carrier f_P for control signaling, assuming a reduced transmit power is used. Thus, the macro cell could use carrier f_P for lower-power control signaling, for example for terminals close to the macro-cell site.

The main drawback with the above-described carrier-aggregation-based approach to interference handling in a heterogeneous network deployment is that it requires terminal support for carrier aggregation in order to ensure full flexibility in spectrum usage. For terminals not capable of carrier aggregation, for example all pre-release-10 terminals, the frequency separation of control signaling between the layers implies a corresponding frequency separation in terms of data transmission as, for such terminals, scheduling assignments on carrier f_M can only schedule PDSCH transmissions on the same carrier. Thus, for such terminals there would be spectrum fragmentation.

If the carrier-aggregation approach described above cannot be used, one can instead apply more conventional interference coordination between the different layers, similar to interference coordination within one cell layer as described in Section 13.3, but extended to also cover interference between control-channel transmissions of the different cell layers.

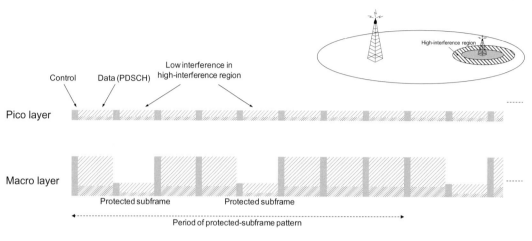

FIGURE 13.15

Single-carrier approach to interference avoidance in a heterogeneous network deployment.

With such an approach, the same carrier is used for transmission in both the macro layer and the underlaid pico cells. However, the power of macro-cell transmissions is restricted in some sub-frames. In contrast to conventional ICIC as described above, this restriction would not only apply to data (PDSCH) transmission, but also to the control region, as illustrated in Figure 13.15. In these *protected subframes* pico-cell terminals will thus experience less macro-cell interference for both data and control, and the pico cell can also use these subframes for transmission to terminals in the high-interference region. In the example illustrated in Figure 13.15, two of eight subframes are configured as protected subframes.

To support this second approach to interference handling in a heterogeneous network, signaling of *protected-subframe patterns* – that is, information about the set of protected subframes – is supported between eNodeBs of different cell layers. Note that the set of protected subframes could be more or less dynamic, once again depending on the time scale on which the macro cell and the underlaid pico cells can be coordinated.

It should be noted that the macro cell must not necessarily completely avoid control-signaling transmission in the protected subframes. In particular, it could be beneficial to retain the possibility for a limited amount of control signaling related to uplink transmissions, for example a limited amount of uplink scheduling grants and/or PHICH transmission, in order not to cause too much impact on the uplink scheduling. As long as the macro-cell control-signaling transmissions are limited and only occupy a small fraction of the overall control region, the interference to terminals in the high-interference region of the pico cell could be kept at an acceptable level. However, the signaling of protected-subframe patterns is also defined so that impact on the uplink scheduling is minimized even if no uplink scheduling grants and PHICH can be transmitted in protected subframes. This is achieved by having the protected-subframe patterns matched to the eight-subframe timing of the uplink hybrid-ARQ protocol. It should be noted that this implies that the pattern is not aligned to the 10 ms frame but only to a 40 ms four-frame structure for FDD. For TDD the periodicity also depends on the uplink-downlink configuration.

Clearly, the interference experienced by pico-cell terminals may vary significantly between protected and non-protected subframes. CSI measurements carried out jointly on both the protected and non-protected subframes will thus not accurately reflect the interference of either type of subframes. Thus, as part of the enhanced support for heterogeneous network deployments, it is possible to configure a terminal with different *CSI-measurement subsets*, confining the terminal CSI measurements to subsets of the full set of subframes with terminals reporting CSI for each subset separately. The corresponding CSI reports should then preferably reflect the interference level in protected and non-protected subframes respectively. However, as mentioned above, the set of protected subframes may vary dynamically and it may not be feasible to update the CSI measurement sets accordingly. Thus, in practice, the CSI-measurement set corresponding to protected subframes may consist of only a subset of the protected subframes, a set that should be relatively static. Similarly, the CSI-measurement set corresponding to non-protected subframes may consist of only a subset of the non-protected subframes.

It should also be noted that, in some cases, a pico cell could be located at the border between, and suffer interference from, two macro cells. If the macro cells have different configured and only partly overlapping sets of protected subframes, the pico-cell scheduling as well as the configuration of the CSI-measurement sets need to take the structure of the protected sets of both macro cells into account.

13.4.2 Interference Coordination in the Case of Home-eNodeB

In the case of home-eNodeB with CSG there are additional interference issues due to the fact that a terminal can be very close to the home-eNodeB and still have to communicate with the overlaid macro cell. Such a terminal may then be severely interfered on the downlink by any home-eNodeB transmission and may also cause severe uplink interference to the home-eNodeB. In principle, this can be solved by the same means as above – that is, by relying on interference coordination between the scheduling in the home-eNodeB layer and an overlaid macro, possibly extended by the carrier-aggregation approach.

A key difference in this case, though, is that the interference avoidance must be two-way – that is, one must not only avoid interference from the macro cell to home-eNodeB terminals in the high-interference outer region of the home-eNodeB coverage area, but also home-eNodeB interference to terminals close to the home-eNodeB but not being part of the home-eNodeB CSG.

A further complicating factor in a home-eNodeB scenario is that there are obvious limitations in the coordination between the home-eNodeB and an overlaid macro layer due to limited backhaul capabilities of a home-eNodeB. In essence, there is no X2 interface to an Home-eNodeB and, in practice, the configuration of protected /non-protected subframes must be more or less static, something that could obviously affect the overall system efficiency.

Access Procedures

The previous chapters have described the LTE uplink and downlink transmission schemes. However, prior to transmission of data, the terminal needs to connect to the network. This chapter describes the procedures necessary for a terminal to be able to access an LTE-based network.

14.1 ACQUISITION AND CELL SEARCH

Before an LTE terminal can communicate with an LTE network it has to do the following:

- Find and acquire synchronization to a cell within the network.
- Receive and decode the information, also referred to as the *cell system information*, needed to communicate with and operate properly within the cell.

The first of these steps, often simply referred to as *cell search*, is discussed in this section. The next section then discusses, in more detail, the means by which the network provides the cell system information.

Once the system information has been correctly decoded, the terminal can access the cell by means of the random-access procedure as described in Section 14.3.

14.1.1 Overview of LTE Cell Search

A terminal does not only need to carry out cell search at power-up – that is, when initially accessing the system. Rather, to support mobility, terminals need to continuously search for, synchronize to, and estimate the reception quality of neighboring cells. The reception quality of the neighboring cells, in relation to the reception quality of the current cell, is then evaluated to conclude if a *handover* (for terminals in RRC_CONNECTED) or *cell reselection* (for terminals in RRC_IDLE) should be carried out.

LTE cell search consists of the following basic parts:

- Acquisition of frequency and symbol synchronization to a cell.
- Acquisition of frame timing of the cell – that is, determine the start of the downlink frame.
- Determination of the physical-layer cell identity of the cell.

As already mentioned (e.g. in Chapter 10), there are 504 different physical-layer cell identities defined for LTE, where each cell identity corresponds to one specific downlink reference-signal sequence. The set of physical-layer cell identities is further divided into 168 *cell-identity groups*, with three cell identities within each group.

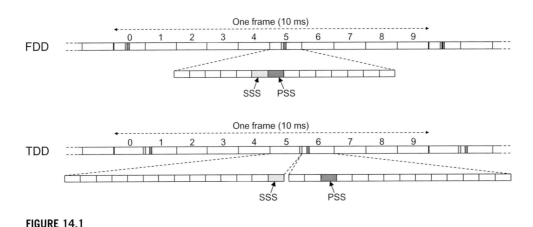

FIGURE 14.1

Time-domain position of PSS and SSS for FDD and TTD.

To assist the cell search, two special signals are transmitted on each downlink component carrier, the *Primary Synchronization Signal* (PSS) and the *Secondary Synchronization Signal* (SSS). Although having the same detailed structure, the time-domain positions of the synchronization signals within the frame differ somewhat depending on whether the cell is operating in FDD or TDD:

- In the case of FDD (upper part of Figure 14.1), the PSS is transmitted within the last symbol of the first slot of subframes 0 and 5, while the SSS is transmitted within the second last symbol of the same slot – that is, just prior to the PSS.
- In the case of TDD (lower part of Figure 14.1), the PSS is transmitted within the third symbol of subframes 1 and 6 – that is, within the DwPTS – while the SSS is transmitted in the last symbol of subframes 0 and 5 – that is, three symbols ahead of the PSS.

There are several reasons for the difference in the positions of the synchronization signals for FDD and TDD. For example, the difference allows for the detection of the duplex scheme used on a carrier if this is not known in advance.

Within one cell, the two PSSs within a frame are identical. Furthermore, the PSS of a cell can take three different values depending on the physical-layer cell identity of the cell. More specifically, the three cell identities within a cell-identity group always correspond to different PSS. Thus, once the terminal has detected and identified the PSS of the cell, it has found the following:

- Five-millisecond timing of the cell and thus also the position of the SSS, which has a fixed offset relative to the PSS.
- The cell identity within the cell-identity group. However, the terminal has not yet determined the cell-identity group itself – that is, the number of possible cell identities has been reduced from 504 to 168.

Thus, from the SSS, the position of which is known once the PSS has been detected, the terminal should find the following:

- frame timing (two different alternatives given the found position of the PSS); and
- the cell-identity group (168 alternatives).

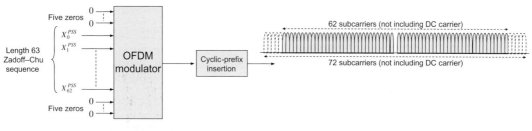

FIGURE 14.2

Definition and structure of PSS.

Furthermore, it should be possible for a terminal to do this by the reception of one single SSS. The reason is that, for example, in the case when the terminal is searching for cells on other carriers, the search window may not be sufficiently large to cover more than one SSS.

To enable this, each SSS can take 168 different values corresponding to the 168 different cell-identity groups. Furthermore, the set of values valid for the two SSSs within a frame (SSS_1 in sub-frame 0 and SSS_2 in subframe 5) are different, implying that, from the detection of a single SSS, the terminal can determine whether SSS_1 or SSS_2 has been detected and thus determine frame timing.

Once the terminal has acquired frame timing and the physical-layer cell identity, it has identified the cell-specific reference signal. The behavior is slightly different depending on whether it is an initial cell search or cell search for the purpose of neighboring cell measurements:

- In the case of initial cell search – that is, the terminal state is in RRC_IDLE mode – the reference signal will be used for channel estimation and subsequent decoding of the BCH transport channel to obtain the most basic set of system information.
- In the case of mobility measurements – that is, the terminal is in RRC_CONNECTED mode – the terminal will measure the received power of the reference signal. If the measurement fulfills a configurable condition, it will trigger sending of a *reference signal received power* (RSRP) measurement report to the network. Based on the measurement report, the network will conclude whether a handover should take place. The RSRP reports can also be used for component carrier management,[1] for example whether an additional component carrier should be configured or if the primary component carrier should be reconfigured.

14.1.2 PSS Structure

On a more detailed level, the three PSSs are three length-63 Zadoff–Chu (ZC) sequences extended with five zeros at the edges and mapped to the center 73 subcarriers (center six resource blocks) as illustrated in Figure 14.2. It should be noted though that the center subcarrier is actually not transmitted as it coincides with the DC subcarrier. Thus, only 62 elements of the length-63 ZC sequences are actually transmitted (element X_{32}^{PSS} is not transmitted).

---

[1] As discussed in Chapter 9, the specifications use the terms primary and secondary cells instead of primary and secondary component carriers.

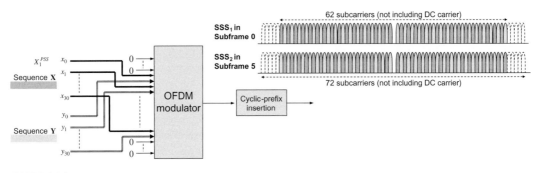

FIGURE 14.3

Definition and structure of SSS.

The PSS thus occupies 72 resource elements (not including the DC carrier) in subframes 0 and 5 (FDD) and subframes 1 and 6 (TDD). These resource elements are then not available for transmission of DL-SCH.

14.1.3 SSS Structure

Similar to PSS, the SSS occupies the center 72 resource elements (not including the DC carrier) in subframes 0 and 5 (for both FDD and TDD). As described above, the SSS should be designed so that:

- The two SSS (SSS_1 in subframe 0 and SSS_2 in subframe 5) take their values from sets of 168 possible values corresponding to the 168 different cell-identity groups.
- The set of values applicable for SSS_2 is different from the set of values applicable for SSS_1 to allow for frame-timing detection from the reception of a single SSS.

The structure of the two SSS is illustrated in Figure 14.3. SSS_1 is based on the frequency interleaving of two length-31 m-sequences X and Y, each of which can take 31 different values (actually 31 different shifts of the same m-sequence). Within a cell, SSS_2 is based on exactly the same two sequences as SSS_1. However, the two sequences have been swapped in the frequency domain, as outlined in Figure 14.3. The set of valid combinations of X and Y for SSS_1 has then been selected so that a swapping of the two sequences in the frequency domain is not a valid combination for SSS_1. Thus, the above requirements are fulfilled:

- The set of valid combinations of X and Y for SSS_1 (as well as for SSS_2) are 168, allowing for detection of the physical-layer cell identity.
- As the sequences X and Y are swapped between SSS_1 and SSS_2, frame timing can be found.

14.2 SYSTEM INFORMATION

By means of the basic cell-search procedure described in Section 14.1, a terminal synchronizes to a cell, acquires the physical-layer identity of the cell, and detects the cell frame timing. Once this has been achieved, the terminal has to acquire the cell *system information*. This is information that is repeatedly broadcast by the network and which needs to be acquired by terminals in order for them to be able to access and, in general, operate properly within the network and within a specific cell. The system information includes, among other things, information about the downlink and uplink cell bandwidths, the uplink/downlink configuration in the case of TDD, detailed parameters related to random-access transmission and uplink power control, etc.

In LTE, system information is delivered by two different mechanisms relying on two different transport channels:

- A limited amount of system information, corresponding to the so-called *Master-Information Block* (MIB), is transmitted using the BCH.
- The main part of the system information, corresponding to different so-called *System-Information Blocks* (SIBs), is transmitted using the downlink shared channel (DL-SCH).

It should be noted that system information in both the MIB and the SIBs corresponds to the BCCH logical channel. Thus, as also illustrated in Figure 8.7 in Chapter 8, BCCH can be mapped to both BCH and DL-SCH depending on the exact BCCH information.

14.2.1 **MIB and BCH Transmission**

As mentioned above, the MIB transmitted using BCH consists of a very limited amount of system information, mainly such information that is absolutely needed for a terminal to be able to read the remaining system information provided using DL-SCH. More specifically, the MIB includes the following information:

- *Information about the downlink cell bandwidth.* Four bits are available within the MIB to indicate the downlink bandwidth. Thus, up to 16 different bandwidths, measured in number of resource blocks, can be defined for each frequency band.
- *Information about the PHICH configuration of the cell.* As mentioned in Section 10.4.2, the terminal must know the PHICH configuration to be able to receive the L1/L2 control signaling on PDCCH which, in turn, is needed to receive DL-SCH. Thus, information about the PHICH configuration (three bits) is included in the MIB – that is, transmitted using BCH, which can be received and decoded without first receiving any PDCCH.
- *The System Frame Number (SFN)* or, more exactly, all bits except the two least significant bits of the SFN are included in the MIB. As described below, the terminal can indirectly acquire the two least significant bits of the SFN from the BCH decoding.

BCH physical-layer processing, such as channel coding and resource mapping, differs quite substantially from the corresponding processing and mapping for DL-SCH outlined in Chapter 10.

As can be seen in Figure 14.4, one BCH transport block, corresponding to the MIB, is transmitted every 40 ms. The BCH Transmissions Time Interval (TTI) thus equals 40 ms.

The BCH relies on a 16-bit CRC, in contrast to a 24-bit CRC used for all other downlink transport channels. The reason for the shorter BCH CRC is to reduce the relative CRC overhead, having the very small BCH transport-block size in mind.

BCH channel coding is based on the same rate-1/3 tail-biting convolutional code as is used for the PDCCH control channel. The reason for using convolutional coding for BCH, rather than the Turbo

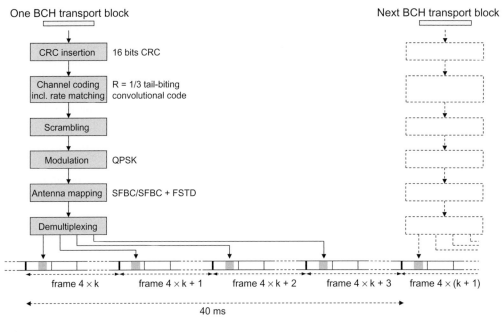

FIGURE 14.4

Channel coding and subframe mapping for the BCH transport channel.

code used for all other transport channels, is the small size of the BCH transport block. With such small blocks, tail-biting convolutional coding actually outperforms Turbo coding. The channel coding is followed by rate matching, in practice repetition of the coded bits, and bit-level scrambling. QPSK modulation is then applied to the coded and scrambled BCH transport block.

BCH multi-antenna transmission is limited to transmit diversity – that is, SFBC in the case of two antenna ports and combined SFBC/FSTD in the case of four antenna ports. Actually, as mentioned in Chapter 10, if two antenna ports are available within the cell, SFBC *must* be used for BCH. Similarly, if four antenna ports are available, combined SFBC/FSTD *must* be used. Thus, by blindly detecting what transmit-diversity scheme is used for BCH, a terminal can indirectly determine the number of cell-specific antenna ports within the cell and also the transmit-diversity scheme used for the L1/L2 control signaling.

As can also be seen from Figure 14.4, the coded BCH transport block is mapped to the first subframe of each frame in four consecutive frames. However, as can be seen in Figure 14.5 and in contrast to other downlink transport channels, the BCH is not mapped on a resource-block basis. Instead, the BCH is transmitted within the first four OFDM symbols of the second slot of subframe 0 and only over the 72 center subcarriers.[2] Thus, in the case of FDD, BCH follows immediately after the PSS and SSS in subframe 0. The corresponding resource elements are then not available for DL-SCH transmission.

The reason for limiting the BCH transmission to the 72 center subcarriers, regardless of the cell bandwidth, is that a terminal may not know the downlink cell bandwidth when receiving BCH. Thus,

---

[2] Not including the DC carrier.

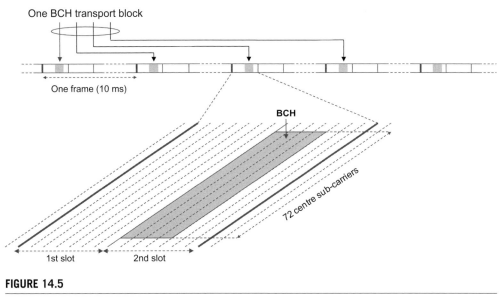

FIGURE 14.5

Detailed resource mapping for the BCH transport channel.

when first receiving BCH of a cell, the terminal can assume a cell bandwidth equal to the minimum possible downlink bandwidth – that is, six resource blocks corresponding to 72 subcarriers. From the decoded MIB, the terminal is then informed about the actual downlink cell bandwidth and can adjust the receiver bandwidth accordingly.

Clearly, the total number of resource elements to which the coded BCH is mapped is very large compared to the size of the BCH transport block, implying extensive repetition coding or, equivalently, massive processing gain for the BCH transmission. Such large processing gain is needed as it should be possible to receive and correctly decode the BCH also by terminals in neighboring cells, implying potentially very low receiver *Signal-to-Interference-and-Noise Ratio* (SINR) when decoding the BCH. At the same time, many terminals will receive BCH in much better channel conditions. Such terminals then do not need to receive the full set of four subframes over which a BCH transport block is transmitted to acquire sufficient energy for correct decoding of the transport block. Instead, already by receiving only a few or perhaps only a single subframe, the BCH transport block may be decodable.

From the initial cell search, the terminal has found only the cell frame timing. Thus, when receiving BCH, the terminal does not know to what set of four subframes a certain BCH transport block is mapped. Instead, a terminal must try to decode the BCH at four possible timing positions. Depending on which decoding is successful, indicated by a correct CRC check, the terminal can implicitly determine 40 ms timing or, equivalently, the two least significant bits of the SFN.[3] This is the reason why these bits do not need to be explicitly included in the MIB.

---

[3] BCH scrambling is defined with 40 ms periodicity, hence even if the terminal successfully decodes the BCH after observing only a single transmission instant, it can determine the 40 ms timing.

14.2.2 System-Information Blocks

As already mentioned, the MIB on the BCH only includes a very limited part of the system information. The main part of the system information is instead included in different *System-Information Blocks* (SIBs) that are transmitted using the DL-SCH. The presence of system information on DL-SCH in a subframe is indicated by the transmission of a corresponding PDCCH marked with a special *System-Information RNTI* (SI-RNTI). Similar to the PDCCH providing the scheduling assignment for "normal" DL-SCH transmission, this PDCCH also indicates the transport format and physical resource (set of resource blocks) used for the system-information transmission.

LTE defines a number of different SIBs characterized by the type of information that is included within them:

- SIB1 includes information mainly related to whether a terminal is allowed to camp on the cell. This includes information about the operator/operators of the cell, if there are restrictions with regards to what users may access the cell, etc. SIB1 also includes information about the allocation of subframes to uplink/downlink and configuration of the special subframe in the case of TDD. Finally, SIB1 includes information about the time-domain scheduling of the remaining SIBs (SIB2 and beyond).
- SIB2 includes information that terminals need in order to be able to access the cell. This includes information about the uplink cell bandwidth, random-access parameters, and parameters related to uplink power control.
- SIB3 mainly includes information related to cell reselection.
- SIB4–SIB8 include neighboring-cell-related information, including information related to neighboring cells on the same carrier, neighboring cells on different carriers, and neighboring non-LTE cells, such as WCDMA/HSPA, GSM, and CDMA2000 cells.
- SIB9 contains the name of the home-eNodeB.
- SIB10–SIB12 contain public warning messages, for example earthquake information.
- SIB13 contains information necessary for MBMS reception (see also Chapter 15).

Not all the SIBs need to be present. For example, SIB9 is not relevant for an operator-deployed node and SIB13 is not necessary if MBMS is not provided in the cell.

Similar to the MIB, the SIBs are broadcasted repeatedly. How often a certain SIB needs to be transmitted depends on how quickly terminals need to acquire the corresponding system information when entering the cell. In general, a lower-order SIB is more time critical and is thus transmitted more often compared to a higher-order SIB. SIB1 is transmitted every 80 ms, whereas the transmission period for the higher-order SIBs is flexible and can be different for different networks.

The SIBs represent the basic system information to be transmitted. The different SIBs are then mapped to different *System-Information messages* (SIs), which correspond to the actual transport blocks to be transmitted on DL-SCH. SIB1 is always mapped, by itself, on to the first system-information message SI-1,[4] whereas the remaining SIBs may be group-wise multiplexed on to the same SI subject to the following constraints:

- The SIBs mapped to the same SI must obviously have the same transmission period. Thus, as an example, two SIBs with a transmission period of 320 ms can be mapped to the same SI, whereas an SIB with a transmission period of 160 ms must be mapped to a different SI.

---

[4] Strictly speaking, as SIB1 is not multiplexed with any other SIBs, it is not even said to be mapped to an SI. Rather, SIB1 in itself directly corresponds to the transport block.

FIGURE 14.6

Example of mapping of SIBs to SIs.

- The total number of information bits that is mapped to a single SI must not exceed what is possible to transmit within a transport block.

It should be noted that the transmission period for a given SIB might be different in different networks. For example, different operators may have different requirements concerning the period when different types of neighboring-cell information needs to be transmitted. Furthermore, the amount of information that can fit into a single transport block very much depends on the exact deployment situation, such as cell bandwidth, cell size, and so on.

Thus, in general, the SIB-to-SI mapping for SIBs beyond SIB1 is flexible and may be different for different networks or even within a network. An example of SIB-to-SI mapping is illustrated in Figure 14.6. In this case, SIB2 is mapped to SI-2 with a transmission period of 160 ms. SIB3 and SIB4 are multiplexed into SI-3 with a transmission period of 320 ms, whereas SIB5, which also requires a transmission period of 320 ms, is mapped to a separate SI (SI-4). Finally, SIB6, SIB7, and SIB8 are multiplexed into SI-5 with a transmission period of 640 ms. Information about the detailed SIB-to-SI mapping, as well as the transmission period of the different SIs, is provided in SIB1.

Regarding the more detailed transmission of the different system-information messages there is a difference between the transmission of SI-1, corresponding to SIB1, and the transmission of the remaining SIs.

The transmission of SI-1 has only a limited flexibility. More specifically, SI-1 is always transmitted within subframe 5. However, the bandwidth or, in general, the set of resource blocks over which SI-1 is transmitted, as well as other aspects of the transport format, may vary and is signaled on the associated PDCCH.

For the remaining SIs, the scheduling on DL-SCH is more flexible in the sense that each SI can, in principle, be transmitted in any subframe within *time windows* with well-defined starting points and durations. The starting point and duration of the time window of each SI are provided in SIB-1. It should be noted that an SI does not need to be transmitted on consecutive subframes within the time window, as is illustrated in Figure 14.7. Within the time window, the presence of system information in a subframe is indicated by the SI-RNTI on PDCCH, which also provides the frequency-domain scheduling as well as other parameters related to the system-information transmission.

Different SIs have different non-overlapping time windows. Thus, a terminal knows what SI is being received without the need for any specific identifier for each SI.

In the case of a relatively small SI and a relatively large system bandwidth, a single subframe may be sufficient for the transmission of the SI. In other cases, multiple subframes may be needed for the transmission of a single SI. In the latter case, instead of segmenting each SI into sufficiently small

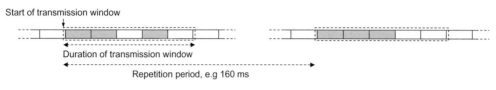

Start of transmission window

Duration of transmission window

Repetition period, e.g 160 ms

FIGURE 14.7

Transmission window for the transmission of an SI.

blocks that are separately channel coded and transmitted in separate subframes, the complete SI is channel coded and mapped to multiple, not necessarily consecutive, subframes.

Similar to the case of the BCH, terminals that are experiencing good channel conditions may then be able to decode the complete SI after receiving only a subset of the subframes to which the coded SI is mapped, while terminals in bad positions need to receive more subframes for proper decoding of the SI. This approach has two benefits:

- Similar to BCH decoding, terminals in good positions need to receive fewer subframes, implying the possibility for reduced terminal power consumption.
- The use of larger code blocks in combination with Turbo coding leads to improved channel-coding gain.

Strictly speaking the single transport block containing the SI is not transmitted over multiple sub-frames. Rather, the subsequent SI transmissions are seen as autonomous hybrid-ARQ retransmissions of the first SI transmission – that is, retransmissions taking place without any explicit feedback sign-aling provided on the uplink.

For terminals capable of carrier aggregation, the system information for the primary component carrier is obtained as described above. For secondary component carriers, the terminal does not need to read the system-information blocks but assumes that the information obtained for the primary component carrier also holds for the secondary component carriers. System information specific for the secondary component carrier is provided through dedicated RRC signaling as part of the procedure to configure an additional secondary component carrier. Using dedicated signaling instead of reading the system information on the secondary component carrier enables faster activation of secondary component carriers as the terminal otherwise would have to wait until the relevant system information had been transmitted.

14.3 **RANDOM ACCESS**

A fundamental requirement for any cellular system is the possibility for the terminal to request a connection setup, commonly referred to as *random access*. In LTE, random access is used for several purposes, including:

- for initial access when establishing a radio link (moving from RRC_IDLE to RRC_CONNECTED; see Chapter 8 for a discussion on different terminal states);
- to re-establish a radio link after radio-link failure;

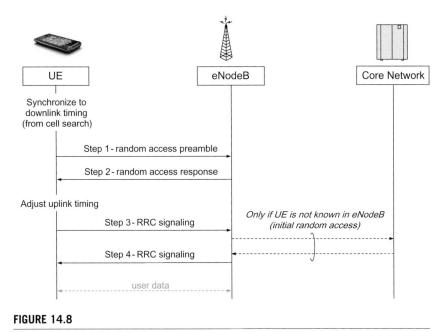

FIGURE 14.8

Overview of the random-access procedure.

- for handover when uplink synchronization needs to be established to the new cell;
- to establish uplink synchronization if uplink or downlink data arrives when the terminal is in RRC_CONNECTED and the uplink is not synchronized;
- for the purpose of positioning using positioning methods based on uplink measurements;
- as a scheduling request if no dedicated scheduling-request resources have been configured on PUCCH (see Chapter 13 for a discussion on uplink scheduling procedures).

Acquisition of uplink timing is a main objective for all the cases above; when establishing an initial radio link (that is, when moving from RRC_IDLE to RRC_CONNECTED), the random-access procedure also serves the purpose of assigning a unique identity, the C-RNTI, to the terminal.

A terminal may perform random access on its primary component carrier only.[5] Either a contention-based or a contention-free scheme can be used.

Contention-based random access uses a four-step procedure, illustrated in Figure 14.8, with the following steps:

1. The transmission of a random-access preamble, allowing the eNodeB to estimate the transmission timing of the terminal. Uplink synchronization is necessary as the terminal otherwise cannot transmit any uplink data.
2. The network transmits a timing advance command to adjust the terminal transmit timing, based on the timing estimate obtained in the first step. In addition to establishing uplink synchronization,

---

[5]The primary component carrier is UE specific, as already discussed; hence, from an eNodeB perspective, random access may occur on multiple component carriers.

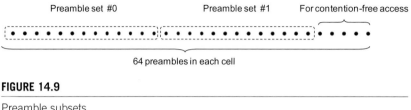

FIGURE 14.9

Preamble subsets.

the second step also assigns uplink resources to the terminal to be used in the third step in the random-access procedure.

3. The transmission of the mobile-terminal identity to the network using the UL-SCH similar to normal scheduled data. The exact content of this signaling depends on the state of the terminal, in particular whether it is previously known to the network or not.

4. The final step consists of transmission of a contention-resolution message from the network to the terminal on the DL-SCH. This step also resolves any contention due to multiple terminals trying to access the system using the same random-access resource.

Only the first step uses physical-layer processing specifically designed for random access. The subsequent three steps utilize the same physical-layer processing as used for normal uplink and downlink data transmission. In the following, each of these steps is described in more detail.

Contention-free random access can only be used for re-establishing uplink synchronization upon downlink data arrival, handover, and positioning. Only the first two steps of the procedure above are used as there is no need for contention resolution in a contention-free scheme.

14.3.1 Step 1: Random-Access Preamble Transmission

The first step in the random-access procedure is the transmission of a random-access preamble. The main purpose of the preamble transmission is to indicate to the base station the presence of a random-access attempt and to allow the base station to estimate the delay between the eNodeB and the terminal. The delay estimate will be used in the second step to adjust the uplink timing.

The time–frequency resource on which the random-access preamble is transmitted is known as the *Physical Random-Access Channel* (PRACH). The network broadcasts information to all terminals in which time–frequency resource random-access preamble transmission is allowed (that is, the PRACH resources, in SIB-2). As part of the first step of the random-access procedure, the terminal selects one preamble to transmit on the PRACH.

In each cell, there are 64 preamble sequences available. Two subsets of the 64 sequences are defined as illustrated in Figure 14.9, where the set of sequences in each subset is signaled as part of the system information. When performing a (contention-based) random-access attempt, the terminal selects at random one sequence in one of the subsets. As long as no other terminal is performing a random-access attempt using the same sequence at the same time instant, no collisions will occur and the attempt will, with a high likelihood, be detected by the eNodeB.

The subset to select the preamble sequence from is given by the amount of data the terminal would like to (and from a power perspective can) transmit on the UL-SCH in the third random-access

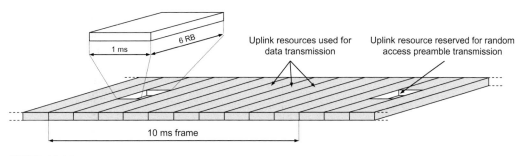

Uplink resources used for data transmission

Uplink resource reserved for random access preamble transmission

6 RB

1 ms

10 ms frame

FIGURE 14.10

Principal illustration of random-access preamble transmission.

step. Hence, from the preamble the terminal used, the eNodeB, will get some guidance on the amount of uplink resources to be granted to the terminal.

If the terminal has been requested to perform a contention-free random access, for example for handover to a new cell, the preamble to use is explicitly indicated from the eNodeB. To avoid collisions, the eNodeB should preferably select the contention-free preamble from sequences outside the two subsets used for contention-based random access.

14.3.1.1 *PRACH Time–Frequency Resources*

In the frequency domain, the PRACH resource, illustrated in Figure 14.10, has a bandwidth corresponding to six resource blocks (1.08 MHz). This nicely matches the smallest uplink cell bandwidth of six resource blocks in which LTE can operate. Hence, the same random-access preamble structure can be used, regardless of the transmission bandwidth in the cell.

In the time domain, the length of the preamble region depends on configured preamble, as will be discussed further below. The basic random-access resource is 1 ms in duration, but it is also possible to configure longer preambles. Also, note that the eNodeB uplink scheduler in principle can reserve an arbitrary long-random-access region by simply avoiding scheduling terminals in multiple subsequent subframes.

Typically, the eNodeB avoids scheduling any uplink transmissions in the time–frequency resources used for random access, resulting in the random-access preamble being *orthogonal* to user data. This avoids interference between UL-SCH transmissions and random-access attempts from different terminals. However, from a specification perspective, nothing prevents the uplink scheduler from scheduling transmissions in the random-access region. Hybrid-ARQ retransmissions are examples of this; synchronous non-adaptive hybrid-ARQ retransmissions may overlap with the random-access region and it is up to the implementation to handle this, either by moving the retransmissions in the frequency domain as discussed in Chapter 12 or by handling the interference at the eNodeB receiver.

For FDD, there is at most one random-access region per subframe – that is, multiple random-access attempts are not multiplexed in the frequency domain. From a delay perspective, it is better to spread out the random-access opportunities in the time domain to minimize the average waiting time before a random-access attempt can be initialized.

For TDD, multiple random-access regions can be configured in a single subframe. The reason is the smaller number of uplink subframes per radio frame in TDD. To maintain the same

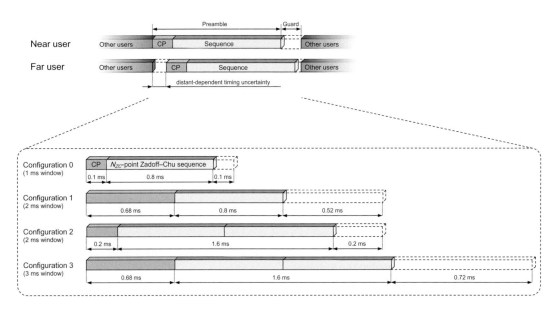

FIGURE 14.11

Different preamble formats.

random-access capacity as in FDD, frequency-domain multiplexing is sometimes necessary. The number of random-access regions is configurable and can vary from one per 20 ms to one per 1 ms for FDD; for TDD up to six attempts per 10 ms radio frame can be configured.

14.3.1.2 *Preamble Structure and Sequence Selection*

The preamble consists of two parts:

- Preamble sequence.
- Cyclic prefix.

Furthermore, the preamble transmission uses a guard period to handle the timing uncertainty. Prior to starting the random-access procedure, the terminal has obtained downlink synchronization from the cell-search procedure. However, as uplink synchronization has not yet been established prior to random access, there is an uncertainty in the uplink timing[6] as the location of the terminal in the cell is not known. The uplink timing uncertainty is proportional to the cell size and amounts to 6.7 µs/km. To account for the timing uncertainty and to avoid interference with subsequent subframes not used for random access, a guard time is used as part of the preamble transmission – that is, the length of the actual preamble is shorter than 1 ms.

Including a cyclic prefix as part of the preamble is beneficial as it allows for frequency-domain processing at the base station (discussed further below), which can be advantageous from a complexity perspective. Preferably, the length of the cyclic prefix is approximately equal to the length

---

[6]The start of an uplink frame at the terminal is defined relative to the start of a downlink frame received at the terminal.

of the guard period. With a preamble sequence length of approximately 0.8 ms, there is 0.1 ms cyclic prefix and 0.1 ms guard time. This allows for cell sizes up to 15 km and is the typical random-access configuration, configuration 0 in Figure 14.11. To handle larger cells, where the timing uncertainty is larger, preamble configurations 1–3 can be used. Some of these configurations also support a longer preamble sequence to increase the preamble energy at the detector, which can be beneficial in larger cells. The preamble configuration used in a cell is signaled as part of the system information. Finally, note that guard times larger than those in Figure 14.11 can easily be created by not scheduling any uplink transmissions in the subframe following the random-access resource.

The preamble formats in Figure 14.11 are applicable to both FDD and TDD. However, for TDD, there is an additional fourth preamble configuration for random access. In this configuration, the random-access preamble is transmitted in the UpPTS field of the special subframe instead of in a normal subframe. Since this field is at most two OFDM symbols long, the preamble and the possible guard time are substantially shorter than the preamble formats described above. Hence, format 4 is applicable to very small cells only. The location of the UpPTS, next to the downlink-to-uplink switch for TDD, also implies that the interference from distant base stations may interfere with this short random-access format, which limits its usage to small cells and certain deployment scenarios.

14.3.1.3 *PRACH Power Setting*
The basis for setting the transmission power of the random-access preamble is a downlink path-loss estimate obtained from measuring the cell-specific reference signals on the primary downlink component carrier. From this path-loss estimate, the initial PRACH transmission power is obtained by adding a configurable offset.

The LTE random-access mechanism allows power ramping where the actual PRACH transmission power is increased for each unsuccessful random-access attempt. For the first attempt, the PRACH transmission power is set to the initial PRACH power. In most cases, this is sufficient for the random-access attempts to be successful. However, if the random-access attempt fails (random-access failures are detected at the second of four random-access steps, as described in the following sections), the PRACH transmission power for the next attempt is increased by a configurable step size to increase the likelihood of the next attempt being successful.

Since the random-access preamble is orthogonal to the user data, the need for power ramping to control intra-cell interference is smaller than in other systems with non-orthogonal random access and in many cases the transmission power is set such that the first random-access attempt with a high likelihood is successful. This is beneficial from a delay perspective.

14.3.1.4 *Preamble Sequence Generation*
The preamble sequences are generated from cyclic shifts of root Zadoff–Chu sequences [74]. Zadoff–Chu sequences are also used for creating the uplink reference signals as described in Chapter 11, where the structure of those sequences is described. From each root Zadoff–Chu sequence, $X_{ZC}^{(u)}(k)$, $[N_{ZC}/N_{CS}]$ cyclically shifted[7] sequences are obtained by cyclic shifts of N_{CS} each, where N_{ZC} is the length of the root Zadoff–Chu sequence. The generation of the random-access preamble is illustrated

---

[7]The cyclic shift is in the time domain. Similar to the uplink reference signals and control signaling, this can equivalently be described as a phase rotation.

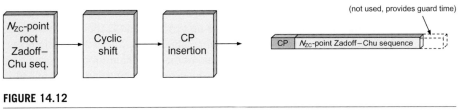

FIGURE 14.12

Random-access preamble generation.

in Figure 14.12. Although the figure illustrates generation in the time domain, frequency-domain generation can equally well be used in an implementation.

Cyclically shifted Zadoff–Chu sequences possess several attractive properties. The amplitude of the sequences is constant, which ensures efficient power amplifier utilization and maintains the low PAR properties of the single-carrier uplink. The sequences also have ideal cyclic auto-correlation, which is important for obtaining an accurate timing estimation at the eNodeB. Finally, the cross-correlation between different preambles based on cyclic shifts of the same Zadoff–Chu root sequence is zero at the receiver as long as the cyclic shift N_{CS} used when generating the preambles is larger than the maximum round-trip propagation time in the cell plus the maximum delay spread of the channel. Therefore, due to the ideal cross-correlation property, there is no intra-cell interference from multiple random-access attempts using preambles derived from the same Zadoff–Chu root sequence.

To handle different cell sizes, the cyclic shift N_{CS} is signaled as part of the system information. Thus, in smaller cells, a small cyclic shift can be configured, resulting in a larger number of cyclically shifted sequences being generated from each root sequence. For cell sizes below 1.5 km, all 64 preambles can be generated from a single root sequence. In larger cells, a larger cyclic shift needs to be configured and to generate the 64 preamble sequences, multiple root Zadoff–Chu sequences must be used in the cell. Although the larger number of root sequences is not a problem in itself, the zero cross-correlation property only holds between shifts of the *same* root sequence and from an interference perspective it is therefore beneficial to use as few root sequences as possible.

Reception of the random-access preamble is discussed further below, but in principle it is based on correlation of the received signal with the root Zadoff–Chu sequences. One disadvantage of Zadoff–Chu sequences is the difficulties in separating a frequency offset from the distance-dependent delay. A frequency offset results in an additional correlation peak in the time domain, a correlation peak that corresponds to a spurious terminal-to-base-station distance. In addition, the true correlation peak is attenuated. At low-frequency offsets, this effect is small and the detection performance is hardly affected. However, at high Doppler frequencies, the spurious correlation peak can be larger than the true peak. This results in erroneous detection; the correct preamble may not be detected or the delay estimate may be incorrect.

To avoid the ambiguities from spurious correlation peaks, the set of preamble sequences generated from each root sequence can be restricted. Restrictions imply that only some of the sequences that can be generated from a root sequence are used to define random-access preambles. Whether restriction should be applied or not to the preamble generation is signaled as part of the system information. The location of the spurious correlation peak relative to the "true" peak depends on the root sequence and hence different restrictions have to be applied to different root sequences. The restrictions to apply are broadcast as part of the system information in the cell.

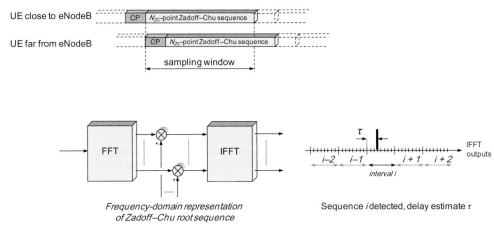

FIGURE 14.13

Random-access preamble detection in the frequency domain.

14.3.1.5 *Preamble Detection*

The base-station processing is implementation specific, but due to the cyclic prefix included in the preamble, low-complexity frequency-domain processing is possible. An example hereof is shown in Figure 14.13. Samples taken in a time-domain window are collected and converted into the frequency-domain representation using an FFT. The window length is 0.8 ms, which is equal to the length of the Zadoff-Chu sequence without a cyclic prefix. This allows handling timing uncertainties up to 0.1 ms and matches the guard time defined for the basic preamble configuration.

The output of the FFT, representing the received signal in the frequency domain, is multiplied by the complex-conjugate frequency-domain representation of the root Zadoff-Chu sequence and the result is fed through an IFFT. By observing the IFFT outputs, it is possible to detect which of the shifts of the root Zadoff-Chu sequence has been transmitted and its delay. Basically, a peak of the IFFT output in interval i corresponds to the ith cyclically shifted sequence and the delay is given by the position of the peak within the interval. This frequency-domain implementation is computationally efficient and allows simultaneous detection of multiple random-access attempts using different cyclically shifted sequences generated from the same root Zadoff-Chu sequence; in the case of multiple attempts there will simply be a peak in each of the corresponding intervals.

14.3.2 **Step 2: Random-Access Response**

In response to the detected random-access attempt, the eNodeB will, as the second step of the random-access procedure, transmit a message on the DL-SCH, containing:

- The index of the random-access preamble sequences the network detected and for which the response is valid.
- The timing correction calculated by the random-access preamble receiver.

- A scheduling grant, indicating resources the terminal will use for the transmission of the message in the third step.
- A temporary identity, the TC-RNTI, used for further communication between the terminal and the network.

If the network detects multiple random-access attempts (from different terminals), the individual response messages of multiple terminals can be combined in a single transmission. Therefore, the response message is scheduled on the DL-SCH and indicated on a PDCCH using an identity reserved for random-access response, the RA-RNTI. All terminals that have transmitted a preamble monitor the L1/L2 control channels for random-access response within a configurable time window. The timing of the response message is not fixed in the specification in order to be able to respond to sufficiently many simultaneous accesses. It also provides some flexibility in the base-station implementation. If the terminal does not detect a random-access response within the time window, the attempt will be declared as failed and the procedure will repeat from the first step again, possibly with an increased preamble transmission power.

As long as the terminals that performed random access in the same resource used different preambles, no collision will occur and from the downlink signaling it is clear to which terminal(s) the information is related. However, there is a certain probability of contention – that is, multiple terminals using the same random-access preamble at the same time. In this case, multiple terminals will react upon the same downlink response message and a collision occurs. Resolving these collisions is part of the subsequent steps, as discussed below. Contention is also one of the reasons why hybrid ARQ is not used for transmission of the random-access response. A terminal receiving a random-access response intended for another terminal will have incorrect uplink timing. If hybrid ARQ were used, the timing of the hybrid-ARQ acknowledgement for such a terminal would be incorrect and may disturb uplink control signaling from other users.

Upon reception of the random-access response in the second step, the terminal will adjust its uplink transmission timing and continue to the third step. If contention-free random access using a dedicated preamble is used, then this is the last step of the random-access procedure as there is no need to handle contention in this case. Furthermore, the terminal already has a unique identity allocated in the form of a C-RNTI.

14.3.3 **Step 3: Terminal Identification**

After the second step, the uplink of the terminal is time synchronized. However, before user data can be transmitted to/from the terminal, a unique identity within the cell, the C-RNTI, must be assigned to the terminal. Depending on the terminal state, there may also be a need for additional message exchange for setting up the connection.

In the third step, the terminal transmits the necessary messages to the eNodeB using the UL-SCH resources assigned in the random-access response in the second step. Transmitting the uplink message in the same manner as scheduled uplink data instead of attaching it to the preamble in the first step is beneficial for several reasons. First, the amount of information transmitted in the absence of uplink synchronization should be minimized, as the need for a large guard time makes such transmissions relatively costly. Secondly, the use of the "normal" uplink transmission scheme for message transmission allows the grant size and modulation scheme to be adjusted to, for example, different radio conditions. Finally, it allows for hybrid ARQ with soft combining for the uplink message. The latter is an important aspect, especially in coverage-limited scenarios, as it allows for the use of one or several retransmissions to collect sufficient energy for the uplink signaling to ensure a sufficiently high probability of successful transmission. Note that RLC retransmissions are not used for the uplink RRC signaling in step 3.

An important part of the uplink message is the inclusion of a terminal identity, as this identity is used as part of the contention-resolution mechanism in the fourth step. If the terminal is in the RRC_CONNECTED state – that is, connected to a known cell and therefore has a C-RNTI assigned – this C-RNTI is used as the terminal identity in the uplink message.[8] Otherwise, a core-network terminal identifier is used and the eNodeB needs to involve the core network prior to responding to the uplink message in step 3.

UE-specific scrambling is used for transmission on UL-SCH, as described in Chapter 11. However, as the terminal may not yet have been allocated its final identity, the scrambling cannot be based on the C-RNTI. Instead, a temporary identity is used (TC-RNTI).

14.3.4 Step 4: Contention Resolution

The last step in the random-access procedure consists of a downlink message for contention resolution. Note that, from the second step, multiple terminals performing simultaneous random-access attempts using the same preamble sequence in the first step listen to the same response message in the second step and therefore have the same temporary identifier. Hence, in the fourth step, each terminal receiving the downlink message will compare the identity in the message with the identity transmitted in the third step. Only a terminal which observes a match between the identity received in the fourth step and the identity transmitted as part of the third step will declare the random-access procedure successful. If the terminal has not yet been assigned a C-RNTI, the TC-RNTI from the second step is promoted to the C-RNTI; otherwise the terminal keeps its already assigned C-RNTI.

The contention-resolution message is transmitted on the DL-SCH, using the temporary identity from the second step for addressing the terminal on the L1/L2 control channel. Since uplink synchronization has already been established, hybrid ARQ is applied to the downlink signaling in this step. Terminals with a match between the identity they transmitted in the third step and the message received in the fourth step will also transmit a hybrid-ARQ acknowledgement in the uplink.

Terminals that do not find a match between the identity received in the fourth step and the respective identity transmitted as part of the third step are considered to have failed the random-access procedure and need to restart the procedure from the first step. Obviously, no hybrid-ARQ feedback is transmitted from these terminals. Furthermore, a terminal that has not received the downlink message in step 4 within a certain time from the transmission of the uplink message in step 3 will declare the random-access procedure as failed and need to restart from the first step.

14.4 PAGING

Paging is used for network-initiated connection setup when the terminal is in RRC_IDLE. In LTE, the same mechanism as for "normal" downlink data transmission on the DL-SCH is used and the mobile terminal monitors the L1/L2 control signaling for downlink scheduling assignments related to paging.

---

[8]The terminal identity is included as a MAC control element on the UL-SCH.

Table 14.1 Paging Cycles and Paging Subframes

		Number of Paging Subframes per Paging Cycle							
		1/32	1/16	1/8	1/4	1/2	1	2	4
Paging subframes in a	FDD	9	9	9	9	9	9	4, 9	0, 4, 5, 9
paging frame	TDD	0	0	0	0	0	0	0, 5	0, 1, 5, 6

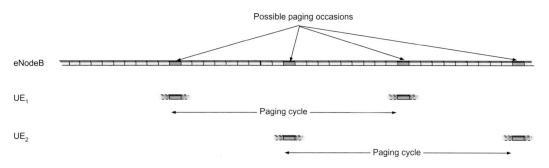

FIGURE 14.14

Illustration of paging cycles.

Since the location of the terminal typically is not known on a cell level, the paging message is typically transmitted across multiple cells in the so-called *tracking area* (the tracking area is controlled by the MME; see [9] for a discussion on tracking areas).

An efficient paging procedure should allow the terminal to sleep with no receiver processing most of the time and to briefly wake up at predefined time intervals to monitor paging information from the network. Therefore, a paging cycle is defined, allowing the terminal to sleep most of the time and only briefly wake up to monitor the L1/L2 control signaling. If the terminal detects a group identity used for paging (the P-RNTI) when it wakes up, it will process the corresponding downlink paging message transmitted on the PCH. The paging message includes the identity of the terminal(s) being paged, and a terminal not finding its identity will discard the received information and sleep according to the DRX cycle. Obviously, as the uplink timing is unknown during the DRX cycles, no hybrid-ARQ acknowledgements can be transmitted and consequently hybrid ARQ with soft combining is not used for paging messages.

The network configures in which subframes a terminal should wake up and listen for paging. Typically, the configuration is cell specific, although there is a possibility to complement the setting by UE-specific configuration. In which frame a given terminal should wake up and search for the P-RNTI on a PDCCH is determined by an equation taking as input the identity of the terminal as well as a cell-specific and (optionally) a UE-specific paging cycle. The identity used is the so-called IMSI, an identity coupled to the subscription, as an idle mode terminal does not have a C-RNTI allocated, and the paging cycle for a terminal can range from once per 256 up to once per 32 frames. The subframe within a frame to monitor for paging is also derived from the IMSI. Since different terminals have different IMSI, they will compute different paging instances. Hence, from a network perspective, paging may be transmitted more often than once per 32 frames, although not all terminals can be paged at all paging occasions as they are distributed across the possible paging instances as shown in Figure 14.14.

Paging messages can only be transmitted in some subframes, ranging from one subframe per 32 frames up to a very high paging capacity with paging in four subframes in every frame. The configurations are shown in Table 14.1. Note that, from a network perspective, the cost of a short paging cycle is minimal as resources not used for paging can be used for normal data transmission and are not wasted. However, from a terminal perspective, a short paging cycle increases the power consumption as the terminal needs to wake up frequently to monitor the paging instants.

In addition to initiating connection to terminals being in RRC_IDLE, paging can also be used to inform terminals in RRC_IDLE as well as RRC_CONNECTED about changes of system information. A terminal being paged for this reason knows that the system information will change and therefore needs to acquire the update system information as described in Section 14.2.

Multimedia Broadcast/Multicast Services

In the past, cellular systems have mostly focused on transmission of data intended for a single user and not on multicast/broadcast services. Broadcast networks, exemplified by the radio and TV broadcasting networks, have on the other hand focused on covering very large areas with the same content and have offered no or limited possibilities for transmission of data intended for a single user. *Multimedia Broadcast Multicast Services* (MBMS) support multicast/broadcast services in a cellular system, thereby combining the provision of multicast/broadcast and unicast services within a single network.

With MBMS, the same content is transmitted to multiple users located in a specific area, known as the *MBMS service area* and typically comprising multiple cells. In each cell participating in the transmission, a point-to-multipoint radio resource is configured and all users subscribing to the MBMS service simultaneously receive the same transmitted signal. No tracking of users' movement in the radio-access network is performed and users can receive the content without notifying the network.

When providing multicast/broadcast services for mobile devices there are several aspects to take into account, of which two deserve special attention and will be elaborated upon further below: good coverage and low terminal power consumption.

The coverage, or more accurately the data rate possible to provide, is basically determined by the link quality of the worst-case user, as no user-specific adaptation of transmission parameters can be used in a multicast/broadcast system providing the same information to multiple users. As discussed in Chapter 3, OFDM transmission provides specific benefits for provision of multi-cell multicast/broadcast services. If the transmissions from the different cells are time synchronized, the resulting signal will, from a terminal point of view, appear as a transmission from a single point over a time-dispersive channel. As mentioned in Chapter 7, for LTE this kind of transmission is referred to as an *MBMS Single-Frequency Network* (MBSFN). MBSFN transmission provides several benefits:

- Increased received signal strength, especially at the border between cells involved in the MBSFN transmission, as the terminal can utilize the signal energy received from multiple cells.
- Reduced interference level, once again especially at the border between cells involved in the MBSFN transmission, as the signals received from neighboring cells will not appear as interference but as useful signals.
- Additional diversity against fading on the radio channel as the information is received from several, geographically separated locations, typically making the overall aggregated channel appear highly time-dispersive or, equivalently, highly frequency selective.

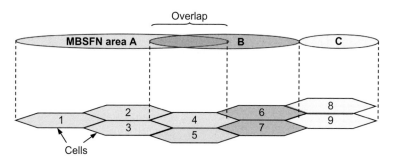

FIGURE 15.1

Example of MBSFN areas.

Altogether, this allows for significant improvements in the multicast/broadcast reception quality, especially at the border between cells involved in the MBSFN transmission, and, as a consequence, significant improvements in the achievable multicast/broadcast data rates.

Providing for power-efficient reception in the terminal in essence implies that the structure of the overall transmission should be such that data for a service-of-interest is provided in short high-data-rate bursts rather than longer low-data-rate bursts. This allows the terminal to occasionally wake up to receive data with long periods of DRX in between. In LTE, this is catered for by time-multiplexing unicast and broadcast transmissions, as well as by the scheduling of different MBMS services, as discussed further below.

15.1 ARCHITECTURE

An *MBSFN area* is a specific area where one or several cells transmit the same content. For example, in Figure 15.1, cells 8 and 9 both belong to MBSFN area C. Not only can an MBSFN area consist of multiple cells, a single cell can also be part of multiple, up to eight, MBSFN areas, as shown in Figure 15.1, where cells 4 and 5 are part of both MBSFN areas A and B. Note that, from an MBSFN reception point of view, the individual cells are invisible, although the terminal needs to be aware of the different cells for other purposes, such as reading system information and notification indicators, as discussed below. The MBSFN areas are static and do not vary over time.

The usage of MBSFN transmission obviously requires not only time synchronization among the cells participating in an MBSFN area, but also usage of the same set of radio resources in each of the cells for a particular service. This coordination is the responsibility of the *Multi-cell/multicast Coordination Entity* (MCE), which is a logical node in the radio-access network handling allocation of radio resources and transmission parameters (time–frequency resources and transport format) across the cells in the MBSFN area. As shown in Figure 15.2, the MCE[1] can control multiple eNodeBs, each handling one or more cells.

---

[1]There is an alternative architecture supported where MCE functionality is included in every eNodeB. However, as there is no communication between MCEs in different eNodeBs, the MBSFN area would in this case be limited to the set of cells controlled by a single eNodeB.

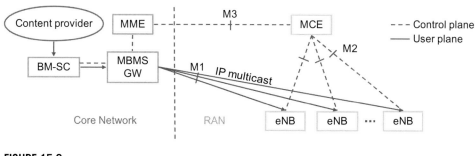

FIGURE 15.2

LTE MBMS architecture.

The *Broadcast Multicast Service Center* (BM-SC), located in the core network, is responsible for authorization and authentication of content providers, charging, and the overall configuration of the data flow through the core network. The MBMS gateway (MBMS-GW) is a logical node handling multicast of IP packets from the BM-SC to all eNodeBs involved in transmission in the MBSFN area. It also handles session control signaling via the MME.

From the BM-SC, the MBMS data is forwarded using IP multicast, a method of sending an IP packet to multiple receiving network nodes in a single transmission, via the MBMS gateway to the cells from which the MBMS transmission is to be carried out. Hence, MBMS is not only efficient from a radio-interface perspective, but it also saves resources in the transport network by not having to send the same packet to multiple nodes individually unless necessary. This can lead to significant savings in the transport network.

15.2 OVERALL CHANNEL STRUCTURE AND PHYSICAL-LAYER PROCESSING

The basis for MBSFN transmission is the *Multicast Channel* (MCH), a transport channel type supporting MBSFN transmission. Two types of logical channels can be multiplexed and mapped to the MCH:

- *Multicast Traffic Channel* (MTCH)
- *Multicast Control Channel* (MCCH).

The MTCH is the logical channel type used to carry MBMS data corresponding to a certain MBMS service. If the number of services to be provided in an MBSFN area is large, multiple MTCHs can be configured. Obviously, as no acknowledgements are transmitted by the terminals, no RLC retransmissions can be used and consequently the RLC unacknowledged mode is used.

The MCCH is the logical channel type used to carry control information necessary for reception of a certain MBMS service, including the subframe allocation and modulation-and-coding scheme for each MCH. There is one MCCH per MBSFN area. Similarly to the MTCH, the RLC uses unacknowledged mode.

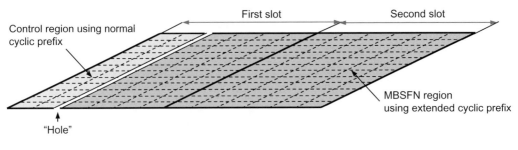

FIGURE 15.3

Resource-block structure for MBSFN subframes, assuming normal cyclic prefix for the control region.

One or several MTCHs and, if applicable,[2] one MCCH are multiplexed at the MAC layer to form an MCH transport channel. As already described in Chapter 8, the MAC header contains information about the logical-channel multiplexing, in this specific case the MTCH/MCCH multiplexing, such that the terminal can demultiplex the information upon reception. The MCH is transmitted using MBSFN in one MBSFN area.

The transport-channel processing for MCH is, in most respects, the same as that for DL-SCH as described in Section 10.1, with some exceptions:

- In the case of MBSFN transmission, the same data is to be transmitted with the same transport format using the same physical resource from multiple cells typically belonging to different eNodeBs. Thus, the MCH transport format and resource allocation cannot be dynamically adjusted by the eNodeB. As described above, the transport format is instead determined by the MCE and signaled to the terminals as part of the information sent on the MCCH.
- As the MCH transmission is simultaneously targeting multiple terminals and therefore no feedback is used, hybrid ARQ is not applicable in the case of MCH transmission.
- As already mentioned, multi-antenna transmission (transmit diversity and spatial multiplexing) does not apply to MCH transmission.

Furthermore, as also mentioned in Chapter 10, the PMCH scrambling should be *MBSFN-area specific* – that is, identical for all cells involved in the MBSFN transmission.

The MCH is mapped to the PMCH physical channel and transmitted in MBSFN subframes, illustrated in Figure 15.3. As discussed in Chapter 9, an MBSFN subframe consists of two parts: a *control region*, used for transmission of regular unicast L1/L2 control signaling; and an *MBSFN region*, used for transmission of the MCH.[3] Unicast control signaling may be needed in an MBSFN subframe, for example to schedule uplink transmissions in a later subframe, but is also used for MBMS-related signaling, as discussed further below.

---

[2] One MCCH per MBSFN area is needed, but it does not have to occur in every MCH TTI, nor on all MCHs in the MBSFN area.

[3] As discussed in Chapter 9, MBSFN subframes can be used for multiple purposes and not all of them have to be used for MCH transmission.

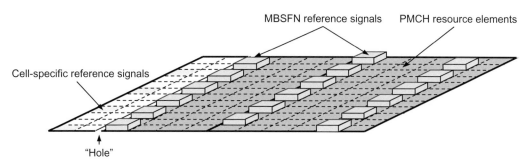

FIGURE 15.4

Reference-signal structure for PMCH reception.

As already discussed in Chapter 3, in the case of MBSFN-based multicast/broadcast transmission, the cyclic prefix should not only cover the main part of the actual channel time dispersion but also the timing difference between the transmissions received from the cells involved in the MBSFN transmission. Therefore, MCH transmissions, which can take place in the MBSFN region only, use an extended cyclic prefix. If a normal cyclic prefix is used for normal subframes, and therefore also in the control region of MBSFN subframes, there will be a small "hole" between the two parts of an MBSFN subframe, as illustrated in Figure 15.3. The reason is to keep the start timing of the MBSFN region fixed, irrespective of the cyclic prefix used for the control region.

As already mentioned, the MCH is transmitted by means of MBSFN from the set of cells that are part of the corresponding MBSFN area. Thus, as seen from the terminal point of view, the radio channel that the MCH has propagated over is the aggregation of the channels of each cell within the MBSFN area. For channel estimation for coherent demodulation of the MCH, the terminal can thus not rely on the normal cell-specific reference signals transmitted from each cell. Rather, in order to enable coherent demodulation for MCH, special MBSFN reference symbols are inserted within the MBSFN part of the MBSFN subframe, as illustrated in Figure 15.4. These reference symbols are transmitted by means of MBSFN over the set of cells that constitute the MBSFN area – that is, they are transmitted at the same time–frequency position and with the same reference-symbol values from each cell. Channel estimation using these reference symbols will thus correctly reflect the overall aggregated channel corresponding to the MCH transmissions of all cells that are part of the MBSFN area.

MBSFN transmission in combination with specific MBSFN reference signals can be seen as transmission using a specific antenna port, referred to as *antenna port 4*.

A terminal can assume that all MBSFN transmissions within a given subframe correspond to the same MBSFN area. Hence, a terminal can interpolate over all MBSFN reference symbols within a given MBSFN subframe when estimating the aggregated MBSFN channel. In contrast, MCH transmissions in different subframes may, as already discussed, correspond to different MBSFN areas. Consequently, a terminal cannot necessarily interpolate the channel estimates across multiple subframes.

As can be seen in Figure 15.4, the frequency-domain density of MBSFN reference symbols is higher than the corresponding density of cell-specific reference symbols. This is needed as the aggregated channel of all cells involved in the MBSFN transmission will be equivalent to a highly time-dispersive or, equivalently, highly frequency-selective channel. Consequently, a higher frequency-domain reference-symbol density is needed.

There is only a single MBSFN reference signal in MBSFN subframes. Thus, multi-antenna transmission such as transmit diversity and spatial multiplexing is not supported for MCH transmission. The main argument for not supporting any standardized transmit diversity scheme for MCH transmission is that the high frequency selectivity of the aggregated MBSFN channel in itself provides substantial (frequency) diversity.

15.3 SCHEDULING OF MBMS SERVICES

Good coverage throughout the MBSFN area is, as already explained, one important aspect of providing broadcast services. Another important aspect, as mentioned in the introduction, is to provide for energy-efficient reception. In essence, for a given service, this translates into transmission of short high-rate bursts in between which the terminal can enter a DRX state to reduce power consumption. LTE therefore makes extensive use of time-multiplexing of MBMS services and the associated signaling, as well as provides a mechanism to inform the terminal *when* in time a certain MBMS service is transmitted. Fundamental to the description of this mechanism are the *Common Subframe Allocation* (CSA) period and the *MCH Scheduling Period* (MSP).

All MCHs that are part of the same MBSFN area occupy a pattern of MBSFN subframes known as the *Common Subframe Allocation* (CSA). The CSA is periodic, as illustrated in Figure 15.5.

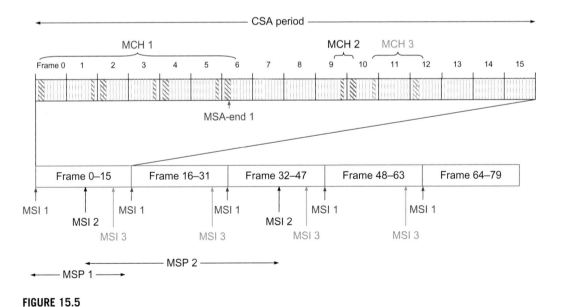

FIGURE 15.5

Example of scheduling of MBMS services.

Obviously, the subframes used for transmission of the MCH must be configured as MBSFN subframes, but the opposite does not hold – MBSFN subframes can be configured for other purposes as well, for example to support the backhaul link in the case of relaying, as described in Chapter 16. Furthermore, the allocation of MBSFN subframes for MCH transmission should obviously be identical across the MBSFN area as there otherwise will not be any MBSFN gain. This is the responsibility of the MCE.

Transmission of a specific MCH follows the *MCH subframe allocation* (MSA). The MSA is periodic and at the beginning of each *MCH Scheduling Period* (MSP), a MAC control element is used to transmit the *MCH Scheduling Information* (MSI). The MSI indicates which subframes are used for a certain MTCH in the upcoming scheduling period. Not all possible subframes need to be used; if a smaller number than allocated to an MCH is required by the MTCH(s), the MSI indicates the last MCH subframe to be used for this particular MTCH (*MSA end* in Figure 15.5), while the remaining subframes are not used for MBMS transmission. The different MCHs are transmitted in consecutive order within a CSA period – that is, all subframes used by MCH n in a CSA are transmitted before the subframes used for MCH $n + 1$ in the same CSA period.

The fact that the transport format is signaled as part of the MCCH implies that the MCH transport format may differ between MCHs but must remain constant across subframes used for the same MCH. The only exception is subframes used for the MCCH and MSI, where the MCCH-specific transport format, signaled as part of the system information, is used instead.

In the example in Figure 15.5, the scheduling period for the first MCH is 16 frames, corresponding to one CSA period, and the scheduling information for this MCH is therefore transmitted once every 16 frames. The scheduling period for the second MCH, on the other hand, is 32 frames, corresponding to two CSA periods, and the scheduling information is transmitted once every 32 frames. The MCH scheduling periods can range from 80 ms to 10.24 s.

To summarize, for each MBSFN area, the MCCH provides information about the CSA pattern, the CSA period, and, for each MCH in the MBSFN area, the transport format and the scheduling period. This information is necessary for the terminal to properly receive the different MCHs. However, the MCCH is a logical channel and itself mapped to the MCH, which would result in a chicken-and-egg problem – the information necessary for receiving the MCH is transmitted on the MCH. Hence, in TTIs when the MCCH (or MSI) is multiplexed into the MCH, the MCCH-specific transport format is used for the MCH. The MCCH-specific transport format is provided as part of the system information (SIB13; see Chapter 14 for a discussion about system information). The system information also provides information about the scheduling and modifications periods of the MCCH (but not about CSA period, CSA pattern, and MSP, as those quantities are obtained from the MCCH itself). Reception of a specific MBMS service can thus be described by the following steps:

- Receive SIB13 to obtain knowledge on how to receive the MCCH for this particular MBSFN area.
- Receive the MCCH to obtain knowledge about the CSA period, CSA pattern, and MSP for the service of interest.
- Receive the MSI at the beginning of each MSP. This provides the terminal with information on which subframes the service of interest can be found in.

After the second step above, the terminal has acquired the CSA period, CSA pattern, and MSP. These parameters typically remain fixed for a relatively long time. The terminal therefore only needs to receive the MSI and the subframes in which the MTCH carrying the service of interest are located,

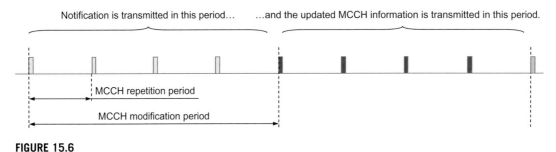

FIGURE 15.6

MCCH transmission schedule.

as described in the third bullet above. This greatly helps to reduce the power consumption in the terminal as it can sleep in most of the subframes.

Occasionally there may be a need to update the information provided on the MCCH, for example when starting a new service. Requiring the terminal to repeatedly receive the MCCH comes at a cost in terms of terminal power consumption. Therefore, a fixed schedule for MCCH transmission is used in combination with a change-notification mechanism, as described below.

The MCCH information is transmitted repeatedly with a fixed repetition period and changes to the MCCH information can only occur at specific time instants. When (parts of) the MCCH information is changed, which can only be done at the beginning of a new modification period, as shown in Figure 15.6, the network notifies the terminals about the upcoming MCCH information change in the preceding MCCH modification period. The notification mechanism uses the PDCCH for this purpose. An eight-bit bitmap, where each bit represents a certain MBSFN area, is transmitted on the PDCCH in an MBSFN subframe using DCI format 1C and a reserved identifier, the M-RNTI. The notification bitmap is only transmitted when there are any changes in the services provided (in release 10, notification is also used to indicate a counting request in an MBSFN area) and follows the modification period described above.

The purpose of the concept of notification indicators and modification periods is to maximize the amount of time the terminal may sleep to save battery power. In the absence of any changes to the MCCH information, a terminal currently not receiving MBMS may enter DRX and only wake up when the notification indicator is transmitted. As a PDCCH in an MBSFN subframe spans at most two OFDM symbols, the duration during which the terminal needs to wake up to check for notifications is very short, translating to a high degree of power saving. Repeatedly transmitting the MCCH is useful to support mobility; a terminal entering a new area or a terminal missing the first transmission does not have to wait until the start of a new modification period to receive the MCCH information.

Relaying

The possibility of a terminal communicating with the network, and the data rate that can be used, depends on several factors, the path loss between the terminal and the base station being one. The link performance of LTE is already quite close to the Shannon limit and from a pure link-budget perspective, the highest data rates supported by LTE require a relatively high signal-to-noise ratio. Unless the link budget can be improved, for example with different types of beam-forming solutions, a denser infrastructure is required to reduce the terminal-to-base-station distance and thereby improve the link budget.

A denser infrastructure is mainly a deployment aspect, but in later releases of LTE, various tools enhancing the support for low-power base stations are included. One of these tools is *relaying*, which can be used to reduce the distance between the terminal and the infrastructure, resulting in an improved link budget and an increased possibility for high data rates. In principle this reduction in terminal-to-infrastructure distance could be achieved by deploying traditional base stations with a wired connection to the rest of the network. However, relays with a shorter deployment time can often be an attractive alternative, as there is no need to deploy a specific backhaul.

A wide range of relay types can be envisioned, some of which could already be deployed in release 8.

Amplify-and-forward relays, commonly referred to as *repeaters*, simply amplify and forward the received analog signals and are, on some markets, relatively common as a tool for handling coverage holes. Traditionally, once installed, repeaters continuously forward the received signal regardless of whether there is a terminal in their coverage area or not, although more advanced repeaters can be considered as well. Repeaters are transparent to both the terminal and the base station and can therefore be introduced in existing networks. The fact that the basic principle of a repeater is to amplify whatever it receives, including noise and interference as well as the useful signal, implies that repeaters are mainly useful in high-SNR environments. Expressed differently, the SNR at the output of the repeater can never be higher than at the input.

Decode-and-forward relays decode and re-encode the received signal prior to forwarding it to the served users. The decode-and-re-encode process results in this class of relays not amplifying noise and interference, as is the case with repeaters. They are therefore also useful in low-SNR environments. Furthermore, independent rate adaptation and scheduling for the base station–relay and relay–terminal links is possible. However, the decode-and-re-encode operation implies a larger delay than for an amplify-and-forward repeater, longer than the LTE subframe duration of 1 ms. As for repeaters, many different options exist depending on supported features (support of more than two hops, support for mesh structures, etc) and, depending on the details of those features, a decode-and-forward relay may or may not be transparent to the terminal.

4G LTE/LTE-Advanced for Mobile Broadband.
© 2011 Erik Dahlman, Stefan Parkvall & Johan Sköld. Published by Elsevier Ltd. All rights reserved.

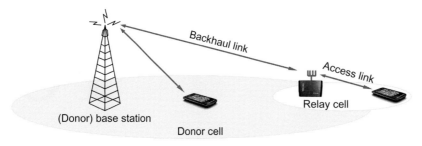

FIGURE 16.1

Access and backhaul links.

16.1 RELAYS IN LTE

LTE release 10 introduces support for a decode-and-forward relaying scheme (repeaters require no additional standardization support other than RF requirements and are available already in release 8). A basic requirement in the development of LTE relaying solutions was that the relay should be transparent to the terminal – that is, the terminal should not be aware of whether it is connected to a relay or to a conventional base station. This ensures that release-8/9 terminals can also be served by relays, despite relays being introduced in release 10. Therefore, so-called *self-backhauling* was taken as the basis for the LTE relaying solution. In essence, from a logical perspective, a relay is an eNodeB wirelessly connected to the rest of the radio-access network by using the LTE radio interface. It is important to note that, even though the relay from a terminal perspective is identical to an eNodeB, the physical implementation may differ significantly, from a traditional base station, for example in terms of output power.

In conjunction with relaying, the terms *backhaul link* and *access link* are often used to refer to the base station–relay connection and the relay–terminal connection respectively. The cell to which the relay is connected using the backhaul link is known as the *donor cell* and the donor cell may, in addition to one or several relays, also serve terminals not connected via a relay. This is illustrated in Figure 16.1.

Since the relay communicates both with the donor cell and terminals served by the relay, interference between the access and backhaul links must be avoided. Otherwise, since the power difference between access-link transmissions and backhaul-link reception at the relay can easily be more than 100 dB, the possibility of receiving the backhaul link may be completely ruined. Similarly, transmissions on the backhaul link may cause significant interference to the reception of the access link. These two cases are illustrated in Figure 16.2. Therefore, isolation between the access and backhaul links is required, isolation that can be obtained in one or several of the frequency, time, and/or spatial domains.

Depending on the spectrum used for access and backhaul links, relaying can be classified into *outband* and *inband* types.

Outband relaying implies that the backhaul operates in a spectrum separate from that of the access link, using the same radio interface as the access link. Provided that the frequency separation between the backhaul and access links is sufficiently large, interference between the backhaul and access links can be avoided and the necessary isolation is obtained in the frequency domain. Consequently, no enhancements to the release-8 radio interface are needed to operate an outband relay. There are no

Access-to-backhaul interference   Backhaul-to-access interference

FIGURE 16.2

Interference between access and backhaul links.

restrictions on the activity on the access and backhaul links and the relay can in principle operate with full duplex.

Inband relaying implies that the backhaul and access links operate in the same spectrum. Depending on the deployment and operation of the relay, this may, as the access and backhaul link share the same spectrum, require additional mechanisms to avoid interference between the access and backhaul links. Unless this interference can be handled by proper antenna arrangements, for example with the relay deployed in a tunnel with the backhaul antenna placed outside the tunnel, a mechanism to separate activity on the access and backhaul links in the time domain is required. Such a mechanism was introduced as part of release 10 and will be described in more detail in the following. Since the backhaul and access links are separated in the time domain, there is a dependency on the transmission activity and the two links cannot operate simultaneously.

16.2 OVERALL ARCHITECTURE

From an architectural perspective, a relay can, on a high level, be thought of as having a "base-station side" and a "terminal side". Towards terminals, it behaves as a conventional eNodeB using the access link, and a terminal is not aware of whether it is communicating with a relay or a "traditional" base station. Relays are therefore transparent for the terminals and terminals from the first LTE release, release 8, can also benefit from relays. This is important from an operator's perspective, as it allows a gradual introduction of relays without affecting the existing terminal fleet.

Towards the donor cell, a relay initially operates as a terminal, using the LTE radio interface to connect to the donor cell. Once connection is established and the relay is configured, the relay uses a subset of the "terminal side" functionality for communication on the backhaul link. In this phase, the relay-specific enhancements described in this chapter may be used for the backhaul.

In release 10, the focus is on two-hop relaying and scenarios with a relay connected to the network via another relay are not considered. Furthermore, relays are stationary – that is, handover of a relay from one donor cell to another donor cell is not supported. The case for using mobile relays is not yet clear and therefore it was decided in release 10 not to undertake the relatively large task of adapting existing core-network procedures to handle cells that are moving over time, something that could have been a consequence of a mobile relay.

The overall LTE relaying architecture is illustrated in Figure 16.3. One key aspect of the architecture is that the donor eNodeB acts as a proxy between the core network and the relay. From a relay perspective, is appears as if it is connected directly to the core network as the donor eNodeB appears

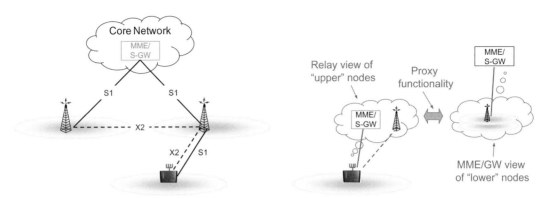

FIGURE 16.3

LTE relaying architecture.

as an MME for the S1 interface and an eNodeB for X2 towards the relay. From a core-network perspective, on the other hand, the relay cells appear as if they belong to the donor eNodeB. It is the task of the proxy in the donor eNodeB to connect these two views. The use of a proxy is motivated by the desire to minimize the impact to the core network from the introduction of relays, as well as to allow for features such as tight coordination of radio-resource management between the donor eNodeB and the relay.

16.3 BACKHAUL DESIGN FOR INBAND RELAYING

In the case of inband relaying, the backhaul and access links operate in the same spectrum. As discussed in the previous section, a mechanism to separate activity on the access and backhaul links in the time domain is required unless sufficient isolation between the two links can be achieved in other ways, for example through appropriate antenna arrangements. Such a mechanism should ensure that the relay is not transmitting on the access link at the same time as it is receiving on the backhaul link (and vice versa).

An obvious way to handle this is to "blank" some subframes on the access link to provide the relay with the possibility to communicate with the donor eNodeB on the backhaul link. In the uplink, the scheduler in the relay can in principle schedule such that there is no access-link activity in certain subframes. These subframes can then be used for uplink transmissions on the backhaul link as the relay does not need to receive anything on the access link in these subframes. However, blanking subframes on the access downlink is not possible. Although a release-10 terminal in principle could have been designed to cope with blank subframes, terminals from earlier releases expect at least cell-specific reference signals to be present in all downlink subframes. Hence, to preserve the possibility of also serving release-8/9 terminals, which was an important requirement during standardization of release 10, the design of the backhaul link must be based on the assumption that the access link can operate with release-8 functionality only.

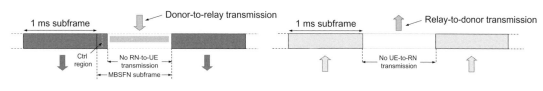

FIGURE 16.4

Multiplexing between access and backhaul links.

Fortunately, already from the first release LTE included the possibility of configuring MBSFN subframes (see Chapter 9). In an MBSFN subframe, terminals expect cell-specific reference signals and (possibly) L1/L2 control signaling to be transmitted only in the first one or two OFDM symbols, while the remaining part of the subframe can be empty. By configuring some of the access-link subframes as MBSFN subframes, the relay can stop transmitting in the latter part of these subframes and receive transmissions from the donor cell. As seen in Figure 16.4, the gap during which the relay can receive transmissions from the donor cell is shorter than the full subframe duration. In particular, as the first OFDM symbols in the subframe are unavailable for reception of donor-cell transmissions, L1/L2 control signaling from the donor to the relay cannot be transmitted using the regular PDCCH. Instead, a relay-specific control channel, the R-PDCCH, is introduced in release 10.

Not only are transmission gaps in the access downlink required in order to receive transmissions from the donor cell, but also reception gaps in the access link are needed in order to transmit on the backhaul from the relay to the donor cell. As already mentioned, such gaps can be created through proper scheduling of uplink transmissions.

The detailed specifications of the physical-layer enhancements introduced in release 10 to support the backhaul can be found in [75].

16.3.1 Access-Link Hybrid-ARQ Operation

The access-link gaps discussed above, MBSFN subframes in the downlink and scheduling gaps in the uplink, used in order to be able to receive and transmit respectively on the backhaul link, affect the hybrid-ARQ operation. Note that hybrid ARQ is used on both the access and backhaul links. Since compatibility with release 8 was a fundamental requirement in the development of the LTE relaying solution, there are no changes to access-link hybrid-ARQ operation.

For uplink transmissions on PUSCH, hybrid-ARQ acknowledgements are transmitted on PHICH. Since the PHICH can be transmitted by the relay even in MBSFN subframes, the operation is identical to that in earlier releases of the LTE standard. However, although the hybrid-ARQ acknowledgement can be received, the subframe where the retransmission should take place (8 ms after the initial transmission for FDD, configuration dependent for TDD) may be used by the backhaul link and not be available for the access link. In that case the corresponding uplink hybrid-ARQ process needs to be suspended by transmitting a positive acknowledgement on the PHICH, irrespective of the outcome of the decoding. By using PDCCH, a retransmission can instead be requested in a later subframe available for the same hybrid-ARQ process, as described in Chapter 12. The hybrid-ARQ round-trip time will be larger in those cases (for example, 16 ms instead of 8 ms for FDD).

Downlink transmissions on PDSCH trigger hybrid-ARQ acknowledgements to be sent on PUCCH and, for proper operation, the relay should be able to receive those acknowledgements. The possibility to receive PUCCH on the access link depends on the backhaul operation, more specifically on the allocation of subframes for backhaul communication.

In FDD, backhaul subframes are configured such that an uplink subframe occurs 4 ms after a downlink subframe. This is chosen to match the access-link hybrid-ARQ timing relations, where an uplink subframe follows 4 ms after a downlink subframe As the relay cannot transmit on the access link simultaneously with the backhaul link, there is no access-link transmission in subframe n and, consequently, no hybrid-ARQ transmission in subframe $n + 4$. Hence, the inability to receive access-link hybrid-ARQ acknowledgements in some subframes is of no concern as the corresponding downlink subframes cannot be used for access-link transmission anyway. Downlink retransmissions are not an issue as they are asynchronous and can be scheduled in any suitable downlink subframe on the access link.

In TDD, the relay node may not be able to receive hybrid-ARQ feedback on PUCCH in uplink subframes used for transmission on the backhaul link. One possibility is to restrict the downlink scheduler such that no terminals transmit PUCCH in uplink subframes the relay cannot receive. However, such a restriction may be too limiting. Alternatively, the relay can schedule without restrictions in the downlink and ignore the hybrid-ARQ acknowledgement. Retransmissions can then either be handled blindly – that is, the relay has to make an educated "guess" on whether a retransmission is required based on, for example, CSI feedback – or RLC retransmissions are used to handle missing packets. Another possibility is to configure repetition of the hybrid-ARQ acknowledgements such that at least some of the repeated acknowledgements are receivable by the relay.

16.3.2 Backhaul-Link Hybrid-ARQ Operation

For the backhaul link, the underlying principle in the design is to maintain the same timing relations as in release 8 for scheduling grants and hybrid-ARQ acknowledgements. As the donor cell may schedule both relays and terminals, such a principle simplifies the scheduling implementation, as scheduling and retransmission decisions for terminals and relays are taken at the same point in time. It also simplifies the overall structure, as release-8 solutions can be reused for the relay backhaul design.

For FDD, the subframes configured for downlink backhaul transmission therefore follow a period of 8 ms in order to match the hybrid-ARQ round-trip time to the extent possible. This also ensures that the PUCCH can be received in the access link, as discussed in the previous section. However, as the possible configurations of MBSFN subframes have an inherent 10 ms structure while the hybrid-ARQ timing follows an 8 ms periodicity, there is an inherent mismatch between the two. Hence, as illustrated in Figure 16.5, some backhaul subframes may be spaced 16 ms apart, as subframes 0, 4, 5, and 9 cannot be configured as MBSFN subframes (see Chapter 9). Uplink backhaul subframes follow 4 ms after a downlink backhaul subframe, following the principle discussed in the previous paragraph.

For TDD, there is an inherent 10 ms component in the hybrid-ARQ timing relations, which matches the 10 ms MBSFN structure and makes it possible to keep a regular spacing of the backhaul transmission attempts. Subframes 0, 1, 5, and 6 cannot be configured as MBSFN subframes. Hence, TDD configuration 0, where subframes 0 and 5 are the only downlink subframes, cannot be used in a relay cell since this configuration does not support any MBSFN subframes. For configuration 5, there

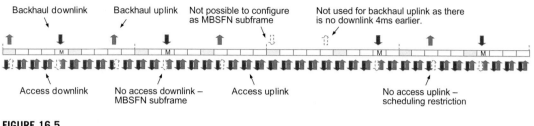

FIGURE 16.5

Example of backhaul configuration for FDD.

is only a single uplink subframe and in order to support both the backhaul and access links at least two uplink subframes are needed. Therefore, of the seven TDD configurations supported in LTE, only configurations 1, 2, 3, 4, and 6 are supported in relay cells. For each TDD configuration, one or several backhaul configurations are supported, as shown in Table 16.1.

The underlying timing principles for hybrid-ARQ acknowledgements and uplink scheduling grants are, as mentioned above, to keep the same principles as for the access link. However, obviously backhaul transmissions may occur in backhaul subframes only. Therefore, for TDD, the acknowledgement of a transport block on the backhaul link in subframe n is transmitted in subframe $n + k$, where $k \geq 4$ and is selected such that $n + k$ is an uplink *backhaul* subframe if the acknowledgement is to be transmitted from the relay and a downlink *backhaul* subframe if the acknowledgement is transmitted from the eNodeB.

The numbering of uplink hybrid-ARQ processes on the backhaul link is similar to the TDD numbering on the access link, where uplink hybrid-ARQ process numbers are assigned sequentially to the available backhaul occasions as shown in Figure 16.6, taking into account the same processing times as for the access link (see Chapter 12). This is in contrast to the FDD access link, where the uplink hybrid-ARQ process number can be directly derived from the subframe number. The reason for adopting a somewhat different strategy is to minimize the maximum hybrid-ARQ round-trip time. Due to the fact that a pure 8 ms periodicity does not always match the MBSFN allocation, the actual uplink round-trip time, unlike the FDD access link, is not constant but, similar to the TDD access link, is dependent on the subframe number.

16.3.3 **Backhaul Downlink Control Signaling**

The gap during which the relay can receive transmissions from the donor cell is, as seen in Figure 16.4, shorter than the full subframe duration. In particular, as the first OFDM symbols in the subframe are unavailable for reception of transmissions from the donor cell, L1/L2 control signaling from the donor to the relay cannot be transmitted using the regular PDCCH.[1] Instead, a relay-specific control channel, the R-PDCCH, is introduced in release 10.

---

[1] In principle, the PDCCH could be received if the subframe structures of the access and backhaul links are offset by two to three OFDM symbols, but with the drawback that relay and donor cells would not be time aligned, which is beneficial, for example, in heterogeneous deployments.

Table 16.1 Supported Backhaul Configurations for TDD

Backhaul Subframe Configuration	Uplink–Downlink Configuration in Relay Cell	Backhaul DL:UL Ratio	Subframe Number									
			0	1	2	3	4	5	6	7	8	9
0	1	1:1					D			U		
1						U						D
2		2:1					D			U	D	
3						U	D					D
4		2:2				U	D				U	D
5	2	1:1	U							D		
6						D			U			
7		2:1	U				D			D		
8						D			U			D
9			U			D	D			D		
10		3:1				D				U	D	D
11	3	2:1				U				D		D
12		3:1				U				D	D	D
13		1:1				U						D
14		2:1				U				D		D
15	4					U					D	D
16		3:1				U				D	D	D
17		4:1				U	D			D	D	D
18	6	1:1					U					D

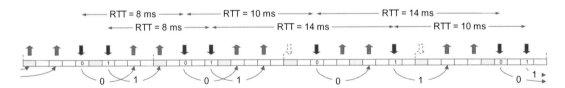

FIGURE 16.6

Example of hybrid-ARQ process numbering for FDD.

The R-PDCCH carries downlink scheduling assignments and uplink scheduling grants, using the same DCI formats as for the PDCCH. However, there is no support for power control commands using DCI formats 3/3A. The main function of DCI formats 3/3A is to support semi-persistent scheduling, a feature mainly targeting overhead reduction for low-rate services and not supported for the backhaul link.

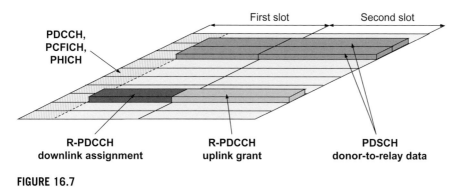

PDCCH,
PCFICH,
PHICH

First slot   Second slot

R-PDCCH
downlink assignment

R-PDCCH
uplink grant

PDSCH
donor-to-relay data

FIGURE 16.7

Example of R-PDCCH transmission.

In the time domain, the R-PDCCH is, as already mentioned, received in the "MBSFN region" of the subframe, while in the frequency domain, transmission of the R-PDCCH occurs in a set of semi-statically allocated resource blocks. From a latency perspective it is beneficial to locate transmissions of downlink scheduling assignments as early as possible in the subframe. As discussed in Chapter 10, this was the main motivation for dividing normal subframes into a control region and a data region. In principle, a similar approach could be taken for the R-PDCCH, namely dividing the set of resource blocks used for R-PDCCH transmission into a control part and a data part. However, as it is not possible to exploit fractions of a subframe for transmission of PDSCH to terminals connected directly to the donor cell, transmission of a single R-PDCCH could block usage of a relatively large number of resource blocks. From an overhead and scheduling flexibility perspective, a structure where the frequency span of the R-PDCCH is minimized (while still providing sufficient diversity) and resources are allocated mainly in the time dimension is preferable. In the release-10 design of the R-PDCCH, these seemingly contradicting requirements have been addressed through a structure where downlink assignments are located in the first slot and uplink grants, which are less time critical, in the second slot of a subframe (see Figure 16.7). This structure allows the time-critical downlink assignments to be decoded early. To handle the case when there is no uplink grant to transmit to the relay, the R-PDCCH resources in the second slot may be used for PDSCH transmission *to the same relay*.

Coding, scrambling, and modulation for the R-PDCCH follows the same principles as for the PDCCH (see Chapter 10), with the same set of aggregation levels supported (one, two, four, and eight CCEs). However, the mapping of the R-PDCCH to time–frequency resources is, for obvious reasons, different. Two different mapping methods, illustrated in Figure 16.8, are supported:

- Without cross-PDCCH interleaving
- With cross-PDCCH interleaving.

Without cross-interleaving, one R-PDCCH is mapped to one set of virtual resource blocks, where the number of resource blocks (one, two, four, or eight) depends on the aggregation level. No other R-PDCCHs are transmitted using the same set of resource blocks. If the resource blocks are located sufficiently apart in the frequency domain, frequency diversity can be obtained, at least for the higher aggregation levels. Non-interleaved mapping is, for example, useful for beam-forming of the backhaul transmissions or when applying frequency-selective scheduling to the R-PDCCH. Either

Non-interleaved R-PDCCHs   Cross-interleaved R-PDCCHs

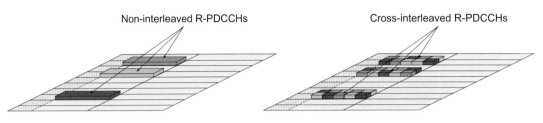

FIGURE 16.8

R-PDCCH mapping types: no cross-interleaving (left) and cross-interleaving (right).

cell-specific reference signals (CRS) or demodulation reference signals (DM-RS) can be used for demodulation.

Cross-interleaved mapping is similar to the strategy used for the PDCCH and reuses most of the PDCCH processing structures except for the mapping to resource elements. A set of R-PDCCHs is multiplexed together, interleaved, and mapped to a set of resource blocks allocated for R-PDCCH transmission. Obviously, as transmissions to multiple relays may share the same set of resource blocks, cell-specific reference signals are the only possibility for demodulation. The motivation for this mapping method is to obtain frequency diversity also for the lowest aggregation level. However, it also comes at the cost of blocking additional resource blocks from PDSCH transmission as, even at low aggregation levels, several resource blocks in the frequency domain are used for the R-PDCCH.

For both mapping cases, cross-interleaved as well as non-cross-interleaved, a set of candidate R-PDCCHs is monitored by the relay node. The set of resource blocks upon which the relay monitors for R-PDCCH transmission is configurable by the donor cell by signaling a set of virtual resource blocks using resource allocation type 0, 1, or 2 (see Chapter 10 for a discussion on resource allocation types). The sets may or may not overlap across multiple relay nodes. In the subframes used for backhaul reception, the relay attempts to receive and decode each of the R-PDCCHs candidates as illustrated in Figure 16.9 and, if valid downlink control information is found, applies this information to downlink reception or uplink transmission. This approach is in essence similar to the blind decoding procedure used in the terminals, although there are some differences. First, there are no common search spaces for the relays as there is no need to receive broadcast information. Any information necessary for relay operation is transmitted using dedicated signaling. Secondly, the search spaces for the non-interleaved mapping are not time varying as in terminals, but remain static in time.

The number of blind decoding attempts is the same as for a terminal – that is, six, six, two, and two attempts for aggregation levels one, two, four, and eight respectively. However, note that an R-PDCCH can be transmitted in either the first or second slot. Hence, the total number of decoding attempts performed by a relay is 64.[2]

No PHICH channel is defined for the backhaul. The main reason for the PHICH in release 8 was efficient support of non-adaptive retransmissions for delay-sensitive low-rate applications such as voice-over IP. The backhaul from a relay, on the other hand, typically uses a higher data rate as multiple terminals are served by the relay. Hence, as control signaling overhead is less of an issue, the

---

[2]Two slots and two DCI formats per transmission mode results in $2 \cdot 2 \cdot (6 + 6 + 2 + 2) = 64$.

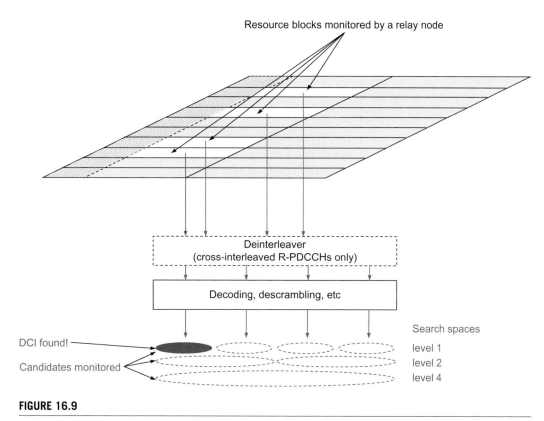

FIGURE 16.9

Illustration of the principle of R-PDCCH monitoring.

PHICH was omitted from the backhaul in order to simplify the overall design. Retransmissions are still supported through the use of the R-PDCCH.

16.3.4 **Reference Signals for the Backhaul Link**

Backhaul reception at the relay can use cell-specific reference signals (CRS) or demodulation reference signals (DM-RS), described in Chapter 10. Different reference-signal types can be used for R-PDCCH and PDSCH, but if the R-PDCCH is received using demodulation reference signals, then demodulation reference signals should be used for PDSCH as well. This is a reasonable restriction as DM-RS for R-PDCCH is motivated by beam-forming. If beam-forming is used for the R-PDCCH there is no incentive not to use beam-forming also for the PDSCH. The opposite scenario, CRS for R-PDCCH and DM-RS for the PDSCH, does make sense though. One example is interleaved mapping of the control signaling, where multiple R-PDCCHs are multiplexed and individual beam-forming cannot be used, together with beam-forming of the PDSCH. The different combinations of reference signals supported for the backhaul link are summarized in Table 16.2.

Table 16.2 Combinations of Reference Signals and R-PDCCH Mapping Schemes

Reference Signal Type Used For Demodulation Of		R-PDCCH Mapping Scheme
R-PDCCH	**PDSCH**	
CRS	CRS	Cross-interleaved or non-cross-interleaved
CRS	DM-RS	Cross-interleaved or non-cross-interleaved
DM-RS	DM-RS	Non-cross-interleaved

Note also that in the case of (global) time alignment between the donor and relay cell, the last OFDM symbol cannot be received by the relay as it is needed for reception-transmission switching. Hence, the demodulation reference signals on the last OFDM symbols in the subframe cannot be received. For transmission ranks up to 4 this is not a problem, as the necessary reference signals are also available earlier in the subframe. However, for spatial multiplexing with five or more layers, the first set of reference signals in the subframe is used for the lower layers while the second set of reference signals, located at the end of the subframe and that cannot be received, is used for the higher layers. This implies that reference signals for rank 5 and higher cannot be received by the relay, and backhaul transmissions are therefore restricted to at most four-layer spatial multiplexing, irrespective of the timing relation used.

16.3.5 Backhaul–Access Link Timing

To ensure that the relay is able to receive transmissions from the donor cell, some form of timing relation between the downlink transmissions in the donor and relay cells must be defined, including any guard time needed to allow the relay to switch between access-link transmission to backhaul-link reception and vice versa.

A natural choice for the timing of the access link is to synchronize it to the frame timing of the backhaul link as observed by the relay. From this backhaul downlink timing reference, the timing of the access-link transmission is derived as shown at the bottom of Figure 16.10. The backhaul uplink timing is subject to the normal timing advance controlled by the donor cell, ensuring that the backhaul uplink transmissions are time aligned with other uplink transmissions received by the donor base station.

In the backhaul downlink, the first OFDM symbol in the data region is left unused to provide the guard time for relay switching, and a small time offset is used to distribute the guard between Tx–Rx and Rx–Tx switching at the relay. This case is shown at the bottom of Figure 16.11. Locating the guard symbol at the beginning of the data region instead of at the end is beneficial as the guard symbol is needed at the relay side only and can therefore still be used for transmission of PDCCHs to terminals in the donor cell. In principle, the guard time comes "for free" from a donor cell perspective and the freedom in shifting the relay node frame timing relative to the donor cell timing is used to move the "free" guard period to where it is needed.

The backhaul uplink is subject to the normal timing advance controlled by the donor cell, ensuring that the backhaul uplink transmissions are time aligned with other uplink transmissions received by the donor base station. Similarly to the guard time needed to switch from access-link transmission to backhaul-link reception, which influenced the downlink timing relation between the access and

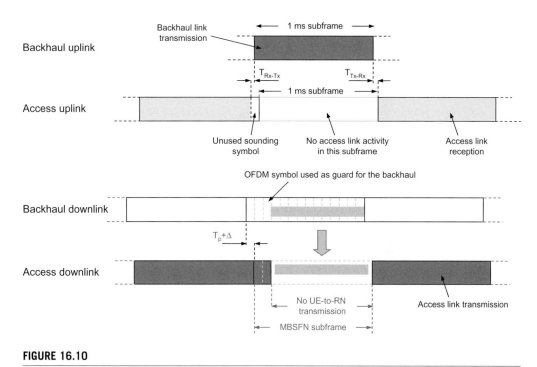

FIGURE 16.10

Backhaul timing relations when the relay cell timing is derived from the backhaul timing.

backhaul links, there may also be the need for a guard time in the uplink direction to switch from access-link reception to backhaul-link transmission. However, unlike the downlink case, how to handle this is not standardized but left for implementation, noting that functionality already present in release 8 is sufficient for providing the necessary guard time.

In principle, if the relay could switch from access-link reception to backhaul-link transmission within the cyclic prefix, no provisions for additional switching time would be necessary. However, the switching time is implementation dependent and typically larger than the cyclic prefix. For larger switching times, one possibility is to use the shortened transmission format on the access link, originally intended for sounding, as shown in the top part of Figure 16.10. By configuring all the terminals in the relay cell to reserve the last OFDM symbol of the preceding subframe for sounding-reference signals but not to transmit any sounding-reference signals, a guard period of one OFDM symbol is created. This guard time can then be divided into Rx–Tx and Tx–Rx switching times through a time offset between the frame timing of the backhaul and access links.

For some deployments, it is desirable to align the access-link transmission timing of the relay with the transmission timing of the donor cell – that is, to use a global timing reference for all the cells. One example hereof is TDD. In such deployments, the necessary guard times are obtained in a slightly different manner compared to the case of using reception timing of the backhaul downlink. In this case it is obviously not possible to obtain the necessary guard time by shifting the subframe timing at the relay. Hence, the guard time for switching from backhaul-link reception to access-link

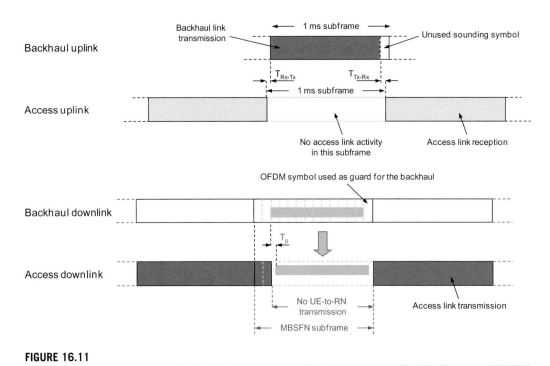

FIGURE 16.11

Backhaul timing relations in access-link transmission in the relay and donor cells are time synchronized.

transmission will also be visible at the donor cell, as the last OFDM symbol in the resource blocks used for the backhaul transmission cannot be used for other transmissions in the relay cell. If the time for Tx–Rx switching is longer than the donor-cell-to-relay-node propagation delay, then the first OFDM symbol has to be left unused as well. This case is shown in the bottom part of Figure 16.11.

In the backhaul uplink, the guard time necessary, similar to the previous timing case, is obtained through configuration of (unused) sounding instances. However, unlike the previous case, sounding is configured in the *backhaul* link, as shown at the top of Figure 16.11. Note that this implies that sounding cannot be used for the backhaul link as the OFDM symbol intended as a sounding-reference symbol is used as guard time.

In the case of TDD operation, guard time for the access–backhaul switch can, in addition to the methods discussed above, be obtained from the guard period required for TDD operation itself. This is shown in Figure 16.12 and is a matter of using the appropriate settings of timing advance and timing offsets.

Backhaul downlink transmissions consist of data transmitted on the PDSCH and L1/L2 control signaling transmitted on the R-PDCCH, as already discussed. Both these types of transmission obviously must follow one of the timing scenarios discussed above. In order to allow for different implementations and deployments, the LTE specifications provide not only the possibility to configure which of the two access–backhaul downlink timing relations to use, but also flexibility in terms of the time span of the channels transmitted on the backhaul link.

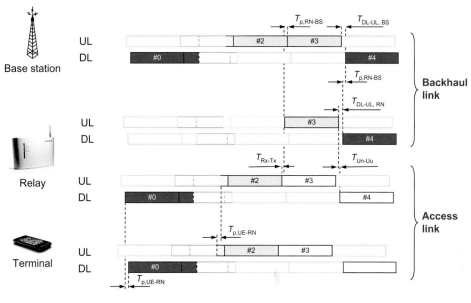

FIGURE 16.12

Example of uplink timing relation for TDD.

PDSCH transmissions intended for a relay can be semi-statically configured to start on the second, third, or fourth OFDM symbol to cater for different control region sizes in the donor cell and relay cells. The PDSCH transmission ends at the last or second last OFDM symbol, depending on which of the two timing cases above is used.

R-PDCCH transmissions intended for a relay always start at the fourth OFDM symbol. A fixed starting position was chosen to simplify the overall structure. Since the amount of resource blocks occupied by an R-PDCCH is relatively small compared to the PDSCH, the overhead reduction possible with a configurable starting positions is small and does not justify the additional specification and testing complexity.

FIGURE 16.7

Spectrum and RF Characteristics

17

Spectrum flexibility is, as mentioned in Chapter 7, a key feature of LTE radio access and is set out in the LTE design targets [10]. It consists of several components, including deployment in different-sized spectrum allocations and deployment in diverse frequency ranges, both in paired and unpaired frequency bands.

There are a number of frequency bands identified for mobile use and specifically for IMT today. Most of these bands were already defined for operation with WCDMA/HSPA, and LTE is the next technology to be deployed in those bands. Both paired and unpaired bands are included in the LTE specifications. The additional challenge with LTE operation in some bands is the possibility of using channel bandwidths up to 20 MHz with a single carrier and even beyond that with aggregated carriers.

The use of OFDM in LTE gives flexibility both in terms of the size of the spectrum allocation needed and in the instantaneous transmission bandwidth used. The OFDM physical layer also enables frequency-domain scheduling, as briefly discussed in Chapter 7. Beyond the physical layer implications described in Chapters 10 and 11, these properties also impact the RF implementation in terms of filters, amplifiers, and all other RF components that are used to transmit and receive the signal. This means that the RF requirements for the receiver and transmitter will have to be expressed with flexibility in mind.

17.1 SPECTRUM FOR LTE

LTE can be deployed both in existing IMT bands and in future bands that may be identified. The possibility of operating radio-access technology in different frequency bands is, in itself, nothing new. For example, quad-band GSM terminals are common, capable of operating in the 850, 900, 1800, and 1900 MHz bands. From a radio-access functionality perspective, this has no or limited impact and the LTE physical-layer specifications [61–64] do not assume any specific frequency band. What may differ, in terms of specification, between different bands are mainly the more specific RF requirements, such as the allowed maximum transmit power, requirements/limits on out-of-band (OOB) emission, and so on. One reason for this is that external constraints, imposed by regulatory bodies, may differ between different frequency bands.

17.1.1 Spectrum Defined for IMT Systems by the ITU-R

The global designations of spectrum for different services and applications are done within the ITU-R. The *World Administrative Radio Congress* WARC-92 identified the bands 1885–2025 and

2110–2200 MHz as intended for implementation of IMT-2000. Of these 230 MHz of 3G spectrum, 2 × 30 MHz were intended for the satellite component of IMT-2000 and the rest for the terrestrial component. Parts of the bands were used during the 1990s for deployment of 2G cellular systems, especially in the Americas. The first deployment of 3G in 2001–2002 by Japan and Europe were done in this band allocation, and for that reason it is often referred to as the IMT-2000 "core band".

Additional spectrum for IMT-2000 was identified at the World Radio-communication Conference WRC-2000, where it was considered that an additional need for 160 MHz of spectrum for IMT-2000 was forecasted by the ITU-R. The identification includes the bands used for 2G mobile systems at 806–960 and 1710–1885 MHz, and "new" 3G spectrum in the bands at 2500–2690 MHz. The identification of bands assigned for 2G was also recognition of the evolution of existing 2G mobile systems into 3G. Additional spectrum was identified at WRC'07 for IMT, encompassing both IMT-2000 and IMT-Advanced. The bands added were 450–470, 698–806, 2300–2400, and 3400–3600 MHz, but the applicability of the bands varies on a regional and national basis.

The somewhat diverging arrangement between regions of the frequency bands assigned to 3G means that there is not one single band that can be used for 3G roaming worldwide. Large efforts have, however, been put into defining a minimum set of bands that can be used to provide roaming. In this way, multi-band devices can provide efficient worldwide roaming for 3G.

17.1.2 Frequency Bands for LTE

The frequency bands where LTE will operate are in both paired and unpaired spectrum, requiring flexibility in the duplex arrangement. For this reason, LTE supports both FDD and TDD, as discussed in the previous chapters.

Release 8 of the 3GPP specifications for LTE includes 19 frequency bands for FDD and nine for TDD. The paired bands for FDD operation are numbered from 1 to 21 [76], as shown in Table 17.1, while the unpaired bands for TDD operation are numbered from 33 to 41, as shown in Table 17.2. Note that the frequency bands for UTRA FDD use the same numbers as the paired LTE bands, but are labeled with Roman numerals. All bands for LTE are summarized in Figures 17.1 and 17.2, which also show the corresponding frequency allocation defined by the ITU.

Some of the frequency bands are partly or fully overlapping. In most cases this is explained by regional differences in how the bands defined by the ITU are implemented. At the same time, a high degree of commonality between the bands is desired to enable global roaming. The set of bands have first been specified as bands for UTRA, with each band originating in global, regional, and local spectrum developments. The complete set of UTRA bands was then transferred to the LTE specifications in release 8 and additional ones have been added in later releases.

Bands 1, 33, and 34 are the same paired and unpaired bands that were defined first for UTRA in release 99 of the 3GPPP specifications, also called the 2 GHz "core band". *Band 2* was added later for operation in the US PCS1900 band and *Band 3* for 3G operation in the GSM1800 band. The unpaired *Bands 35, 36, and 37* are also defined for the PCS1900 frequency ranges, but are not deployed anywhere today. *Band 39* is an extension of the unpaired Band 33 from 20 to 40 MHz for use in China.

Band 4 was introduced as a new band for the Americas following the addition of the 3G bands at WRC-2000. Its downlink overlaps completely with the downlink of Band 1, which facilitates roaming and eases the design of dual Band 1 + 4 terminals. *Band 10* is an extension of Band 4 from 2 × 45 to 2 × 60 MHz.

Table 17.1 Paired Frequency Bands Defined by 3GPP for LTE

Band	Uplink Range (MHz)	Downlink Range (MHz)	Main Region(s)
1	1920–1980	2110–2170	Europe, Asia
2	1850–1910	1930–1990	Americas (Asia)
3	1710–1785	1805–1880	Europe, Asia (Americas)
4	1710–1755	2110–2155	Americas
5	824–849	869–894	Americas
6	830–840	875–885	Japan (only for UTRA)
7	2500–2570	2620–2690	Europe, Asia
8	880–915	925–960	Europe, Asia
9	1749.9–1784.9	1844.9–1879.9	Japan
10	1710–1770	2110–2170	Americas
11	1427.9–1447.9	1475.9–1495.9	Japan
12	698–716	728–746	USA
13	777–787	746–756	USA
14	788–798	758–768	USA
17	704–716	734–746	USA
18	815–830	860–875	Japan
19	830–845	875–890	Japan
20	832–862	791–821	Europe
21	1447.9–1462.9	1495.9–1510.9	Japan

Table 17.2 Unpaired Frequency Bands Defined by 3GPP for LTE

Band	Frequency Range (MHz)	Main Region(s)
33	1900–1920	Europe, Asia (not Japan)
34	2010–2025	Europe, Asia
35	1850–1910	(Americas)
36	1930–1990	(Americas)
37	1910–1930	–
38	2570–2620	Europe
39	1880–1920	China
40	2300–2400	Europe, Asia
41	2496–2690	USA

Band 9 overlaps with Band 3, but is intended only for Japan. The specifications are drafted in such a way that implementation of roaming dual Band 3 + 9 terminals is possible. The 1500 MHz frequency band is also identified in 3GPP for Japan as *Bands 11* and *21*. It is allocated globally to mobile service on a co-primary basis and was previously used for 2 G in Japan.

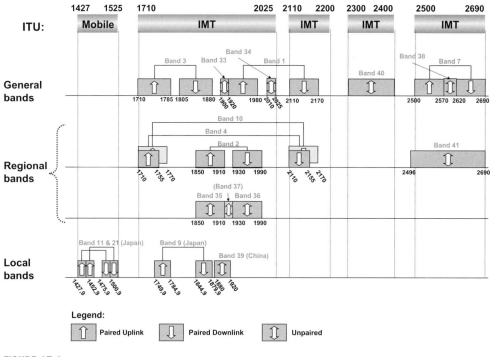

FIGURE 17.1

Operating bands specified for LTE in 3GPP above 1 GHz and the corresponding ITU allocation.

With WRC-2000, the band 2500–2690 MHz was identified for IMT-2000 and it is identified as *Band 7* in 3GPP for FDD and *Band 38* for TDD operation in the "center gap" of the FDD allocation. The band has a slightly different arrangement in North America, where a US-specific *Band 41* is defined. *Band 40* is an unpaired band specified for the new frequency range 2300–2400 MHz identified for IMT and has a widespread allocation globally.

WRC-2000 also identified the frequency range 806–960 MHz for IMT-2000, complemented by the frequency range 698–806 MHz in WRC'07. As shown in Figure 17.2, several bands are defined for FDD operation in this range. Band 8 uses the same band plan as GSM900. *Bands 5, 18, and 19* overlap, but are intended for different regions. Band 5 is based on the US cellular band, while Bands 18 and 19 are restricted to Japan in the specifications. 2 G systems in Japan had a very specific band plan and Bands 18 and 19 are a way of partly aligning the Japanese spectrum plan in the 810–960 MHz range to that in other parts of the world. Note that Band 6 was originally defined in this frequency range for Japan, but it is not used for LTE.

Bands 12, 13, 14, and 17 make up the first set of bands defined for what is called the *digital dividend* – that is, for spectrum previously used for broadcasting. This spectrum is partly migrated to be used by other wireless technologies, since TV broadcasting is migrating from analog to more spectrum-efficient digital technologies. Another regional band for the digital dividend is *Band 20* that is defined in Europe.

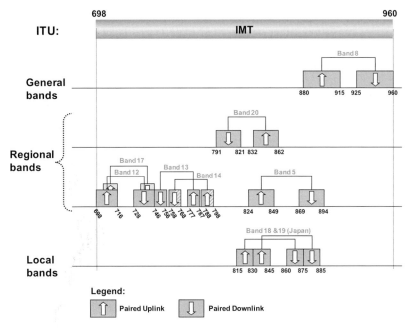

FIGURE 17.2

Operating bands specified for LTE in 3GPP below 1 GHz and the corresponding
ITU allocation.

17.1.3 New Frequency Bands

Additional frequency bands are continuously specified for UTRA and LTE. WRC'07 identified additional frequency bands for IMT, which encompasses both IMT-2000 and IMT-Advanced. Several bands were defined by WRC'07 that will be available partly or fully for deployment on a global basis:

- *450–470 MHz* was identified for IMT globally. It is already allocated to mobile service globally, but it is only 20 MHz wide.
- *698–806 MHz* was allocated to mobile service and identified to IMT to some extent in all regions. Together with the band at 806–960 MHz identified at WRC-2000, it forms a wide frequency range from 698 to 960 MHz that is partly identified to IMT in all regions, with some variations.
- *2300–2400 MHz* was identified for IMT on a worldwide basis in all three regions.
- *3400–3600 MHz* was allocated to the mobile service on a primary basis in Europe and Asia and partly in some countries in the Americas. There is also satellite use in the bands today.

Additional bands for IMT are not on the agenda for WRC'12, but are likely to be treated at WRC'16. For the frequency ranges below 1 GHz identified at WRC-07, 3GPP has already specified several operating bands, as shown in Figure 17.2. The bands with the widest use are Bands 5 and 8, while most of the other bands have limited regional use. With the identification of bands down to 698 MHz for IMT use and the switchover from analog to digital TV broadcasting, Bands 12, 13, 14, and 17 are defined in the USA and Band 20 in Europe for the digital dividend. There is a recent development in the Asia-Pacific area, where a harmonized band plan is agreed [77] for the digital dividend at 698–806 MHz, consisting either of a 2 × 45 MHz paired band or a ~100 MHz unpaired band.

Work in 3GPP is ongoing for the frequency band 3.4–3.8 GHz [78]. In Europe, a majority of countries already license the band at 3.4–3.6 GHz for both Fixed Wireless Access and mobile use. Licensing of 3.6–3.8 GHz for Wireless Access is more limited. There is a European spectrum decision for 3.4–3.8 GHz with "flexible usage modes" for deployment of fixed, nomadic, and mobile networks. Frequency arrangements considered in the decision include FDD use with 100 MHz block offset between paired blocks and/or TDD use. In Japan, not only 3.4–3.6 GHz but also 3.6–4.2 GHz will be available to terrestrial mobile services such as IMT to use after 2010. The band 3.4–3.6 GHz has also been licensed for wireless access in Latin America. 3GPP is specifying both LTE TDD and LTE FDD modes for the band.

Several Mobile Satellite Service (MSS) operators in the USA are planning to deploy an Ancillary Terrestrial Component (ATC) using LTE. For this purpose two new frequency bands are defined, one 2 × 20 MHz band for the S-band MSS operators at 2 GHz and one 2 × 34 MHz band for the L-band MSS operators at 1.5 GHz.

The US PCS1900 band will be extended with an additional 2 × 5 MHz block called the *G block*. In order to accommodate the new block, work is ongoing in 3GPP to create an extended PCS1900 band that will be 2 × 65 MHz, compared to the original Band 2, which is 2 × 60 MHz.

17.2 FLEXIBLE SPECTRUM USE

Most of the frequency bands identified above for deployment of LTE are existing IMT-2000 bands and some bands also have legacy systems deployed, including WCDMA/HSPA and GSM. Bands are also in some regions defined in a "technology neutral" manner, which means that coexistence between different technologies is a necessity.

The fundamental LTE requirement to operate in different frequency bands [79] does not, in itself, impose any specific requirements on the radio-interface design. There are, however, implications for the RF requirements and how those are defined, in order to support the following:

- *Coexistence between operators in the same geographical area in the band*. These other operators may deploy LTE or other IMT-2000 technologies, such as UMTS/HSPA or GSM/EDGE. There may also be non-IMT-2000 technologies. Such coexistence requirements are to a large extent developed within 3GPP, but there may also be regional requirements defined by regulatory bodies in some frequency bands.
- *Co-location of base station equipment between operators*. There are in many cases limitations to where base-station equipment can be deployed. Often, sites must be shared between operators or an operator will deploy multiple technologies in one site. This puts additional requirements on both base-station receivers and transmitters.
- *Coexistence with services in adjacent frequency bands and across country borders*. The use of the RF spectrum is regulated through complex international agreements, involving many interests. There will therefore be requirements for coordination between operators in different countries and for coexistence with services in adjacent frequency bands. Most of these are defined in different

regulatory bodies. Sometimes the regulators request that 3GPP includes such coexistence limits in the 3GPP specifications.

- *Coexistence between operators of TDD systems* in the same band is provided by inter-operator synchronization, in order to avoid interference between downlink and uplink transmissions of different operators. This means that all operators need to have the same downlink/uplink configurations and frame synchronization, not in itself an RF requirement, but it is implicitly assumed in the 3GPP specifications. RF requirements for unsynchronized systems become very strict.
- *Release-independent frequency-band principles.* Frequency bands are defined regionally and new bands are added continuously. This means that every new release of 3GPP specifications will have new bands added. Through the "release independence" principle, it is possible to design terminals based on an early release of 3GPP specifications that support a frequency band added in a later release.

17.3 FLEXIBLE CHANNEL BANDWIDTH OPERATION

The frequency allocations in Figures 17.1 and 17.2 are up to 2 × 75 MHz, but the spectrum available for a single operator may be from 2 × 20 MHz down to 2 × 5 MHz for FDD and down to 1 × 5 MHz for TDD. Furthermore, the migration to LTE in frequency bands currently used for other radio-access technologies must often take place gradually to ensure that a sufficient amount of spectrum remains to support the existing users. Thus, the amount of spectrum that can initially be migrated to LTE can be relatively small, but may then gradually increase, as shown in Figure 17.3. The variation of

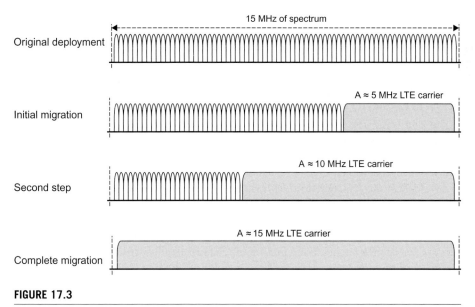

FIGURE 17.3

Example of how LTE can be migrated step-by-step into a spectrum allocation with an original GSM deployment.

Table 17.3 Channel Bandwidths Specified in LTE	
Channel Bandwidth, $BW_{channel}$ (MHz)	**Number of Resource Blocks (N_{RB})**
1.4	6
3	15
5	25
10	50
15	75
20	100

possible spectrum scenarios implies a requirement for spectrum flexibility for LTE in terms of the transmission bandwidths supported.

The spectrum flexibility requirement points out the need for LTE to be scalable in the frequency domain. This flexibility requirement is stated in [10] as a list of LTE spectrum allocations from 1.25 to 20 MHz. Note that the final channel bandwidths selected differ slightly from this initial assumption.

As shown in Chapter 9, the frequency-domain structure of LTE is based on resource blocks consisting of 12 subcarriers with a total bandwidth of $12 \times 15\,kHz = 180\,kHz$. The basic radio-access specification including the physical-layer and protocol specifications enables *transmission bandwidth configurations* from six up to 110 resource blocks on one LTE RF carrier. This allows for channel bandwidths ranging from 1.4 MHz up to beyond 20 MHz in steps of 180 kHz and is fundamental to providing the required spectrum flexibility.

In order to limit implementation complexity, only a limited set of bandwidths are defined in the RF specifications. Based on the frequency bands available for LTE deployment today and in the future, as described above, and considering the known migration and deployment scenarios in those bands, a limited set of six channel bandwidths is specified. The RF requirements for the base station and UE are defined only for those six channel bandwidths. The channel bandwidths range from 1.4 to 20 MHz, as shown in Table 17.3. The lower bandwidths, 1.4 and 3 MHz, are chosen specifically to ease migration to LTE in spectrum where CDMA2000 is operated, and also to facilitate migration of GSM and TD-SCDMA to LTE. The specified bandwidths target relevant scenarios in different frequency bands. For this reason, the set of bandwidths available for a specific band is not necessarily the same as in other bands. At a later stage, if new frequency bands are made available that have other spectrum scenarios requiring additional channel bandwidths, the corresponding RF parameters and requirements can be added in the RF specifications, without actually having to update the physical-layer specifications. The process of adding new channel bandwidths in this way is similar to adding new frequency bands.

Figure 17.4 illustrates in principle the relationship between the channel bandwidth and the number of resource blocks N_{RB} for one RF carrier. Note that for all channel bandwidths except 1.4 MHz, the resource blocks in the transmission bandwidth configuration fill up 90% of the channel bandwidth. The spectrum emissions shown in Figure 17.4 are for a pure OFDM signal, while the actual transmitted emissions will also depend on the transmitter RF chain and other components. The emissions outside the channel bandwidth are called *unwanted emissions* and the requirements for those are discussed further below.

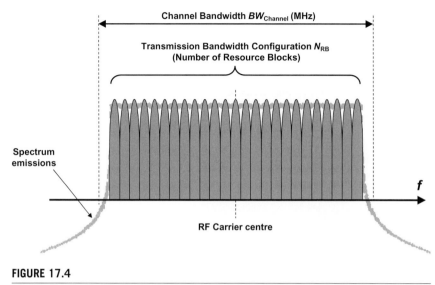

FIGURE 17.4

The channel bandwidth for one RF carrier and the corresponding transmission bandwidth configuration.

17.4 CARRIER AGGREGATION FOR LTE

The possibility in release 10 to aggregate two or more component carriers in order to support wider transmission bandwidths has several implications for the RF characteristics. The impact for the base station and UE RF characteristics are also quite different. Release 10 has some restrictions on carrier aggregation in the RF specification, compared to what has been specified for physical layer and signaling.

There is, from an RF point of view, a substantial difference between the two types of Carrier Aggregation (CA) defined for LTE (see also Section 7.3.1 for more details):

- *Intra-Band Carrier Aggregation* implies that two or more carriers within the same operating band are aggregated (see also the first two examples in Figure 7.4). RF requirements are restricted in release 10 to contiguous intra-band aggregation and a maximum of two carriers. Since aggregated carriers from an RF perspective have similar RF properties as a corresponding wider carrier being transmitted and received, there are many implications for the RF requirements. This is especially true for the UE. For the base station, it corresponds in practice to a multicarrier configuration (non-aggregated) already supported in earlier releases, which also means that the impact is less than for the UE.
- *Inter-Band Carrier Aggregation* implies that carriers in different operating bands are aggregated (see also the last example in Figure 7.4). Many RF properties within a band can, to a large extent, remain the same as for a single carrier case. There is, however, impact for the UE, due to the possibility for intermodulation and cross-modulation within the UE device when multiple transmitter and receiver chains are operated simultaneously. For the base station it has very little impact, since in practice it corresponds to a base station supporting multiple bands, which is a configuration not really treated in RF specifications.

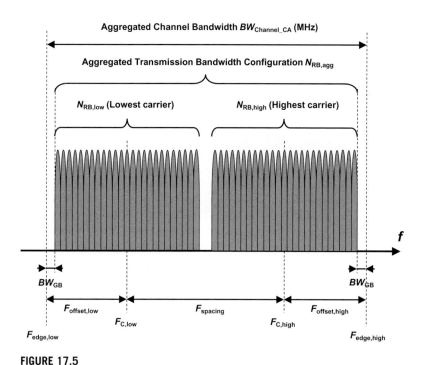

FIGURE 17.5

Definitions for Intra-Band Carrier Aggregation RF parameters, for an example with two aggregated carriers.

Intra-band carrier aggregation is limited to two component carriers and to one paired band (Band 1) and one unpaired (Band 40) band in release 10. Inter-band carrier aggregation is limited to the generic case of aggregating carriers between Bands 1 and 5. The next band pair for which a carrier aggregation capability is specified is a "European" scenario for Bands 3 and 7, which is planned for later inclusion in release 10. The band or set of bands over which carriers are aggregated is defined as a UE capability called *E-UTRA CA Band*. For the base station the band or set of bands defines what is called a *Carrier Aggregation Configuration* for the base station.

For intra-band carrier aggregation, the definitions of $BW_{channel}$ and N_{RB} shown in Figure 17.4 still apply for each component carrier, while new definitions are needed for the Aggregated Channel Bandwidth ($BW_{Channel\_CA}$) and the Aggregated Transmission bandwidth Configuration ($N_{RB,agg}$) shown in Figure 17.5. In connection with this, a new capability is defined for the UE called *Carrier Aggregation Bandwidth Class*. There are six classes, where each class corresponds to a range for $N_{RB,agg}$ and a maximum number of component carriers, as shown in Table 17.4. The classes corresponding to aggregation of more than two component carriers or consisting of more than 200 RBs are under study for later releases.

A fundamental parameter for intra-band CA is the channel spacing. A tighter channel spacing than the nominal spacing for any two single carriers could potentially lead to an increase in spectral efficiency, since there would be a smaller unused "gap" between carriers. On the other hand, there is also a requirement for the possibility to support legacy single-carrier terminals of earlier releases.

Table 17.4 UE Carrier Aggregation Bandwidth Classes

Channel Aggregation Bandwidth Classes	Aggregated Transmission BW Configuration	Number of Component Carriers
A	≤100	1
B	≤100	2
C	101–200	2
D, E, F	Under study (201–500)	Under study

An additional complication is that the component carriers should be on the same 15 kHz subcarrier raster in order to allow reception of multiple adjacent component carriers using a single FFT instead of an FFT per subcarrier.[1] As discussed in Section 9.4, this property, together with the fact that the frequency numbering scheme is on a 100 kHz raster, results in the spacing between two component carriers having to be a multiple of 300 kHz, which is the least common denominator of 15 and 100 kHz.

For the specification, RF requirements are based on a nominal channel spacing that is derived from the channel bandwidth of the two adjacent carriers $BW_{Channel(1)}$ and $BW_{Channel(2)}$ as follows:[2]

$$F_{Spacing,Nominal} = \left\lfloor \frac{BW_{Channel(1)} + BW_{Channel(2)} - 0.1\left|BW_{Channel(1)} - BW_{Channel(2)}\right|}{2 \cdot 0.3} \right\rfloor 0.3. \quad (17.1)$$

In order to allow for a tighter packing of component carriers, the value of $F_{Spacing}$ can be adjusted to any multiple of 300 kHz that is smaller than the nominal spacing, as long as the carriers do not overlap.

RF requirements for LTE are normally defined relative to the channel bandwidth edges. For intra-band CA, this is generalized so that requirements are defined relative to the edges of the Aggregated Channel Bandwidth, identified in Figure 17.5 as $F_{edge,low}$ and $F_{edge,high}$. In this way many RF requirements can be reused, but with new reference points in the frequency domain. The aggregated channel bandwidth for both UE and base station is defined as:

$$BW_{Channel\_CA} = F_{edge,high} - F_{edge,low} \quad (17.2)$$

The location of the edges is defined relative to the carriers at the edges through a new parameter F_{offset} (see Figure 17.5) using the following relation to the carrier center positions F_C of the lowest and highest carriers:

$$F_{edge,low} = F_{C,low} - F_{offset,low} \quad (17.3)$$

$$F_{edge,high} = F_{C,high} + F_{offset,high} \quad (17.4)$$

---

[1] In case of independent frequency errors between component carriers, multiple FFTs and frequency-tracking functionality may be needed anyway.
[2] $\lfloor \ldots \rfloor$ denotes the "floor" operator, which rounds the number down.

The value of F_{offset} for the edge carriers and the corresponding location of the edges are, however, not defined in the same way for UE and base station.

For the base station, there are legacy scenarios where the base station receives and transmits adjacent independent carriers, supporting legacy terminals of earlier releases using single carriers. This scenario will also have to be supported for a configuration of aggregated carriers. In addition, for backward compatibility reasons, a fundamental parameter such as channel bandwidth and the corresponding reference points (the channel edge) for all RF requirements will have to remain the same. The implication is that the channel edges shown in Figure 17.4 for each component carrier will also remain as reference points when the carriers are aggregated. This results in the following base station definition of F_{offset}, for carrier aggregation, which is "inherited" from the single carrier scenario:

$$F_{\text{offset}} = \frac{BW_{\text{channel}}}{2} \quad \text{(for base station).} \tag{17.5}$$

Unlike the base station, the UE is not restricted by legacy operation, but rather from the nonlinear properties of the PA and the resulting unwanted emissions mask. At both edges of the aggregated channel bandwidth, a guard band BW_{GB} will be needed, in order for the emissions to reach a level where the out-of-band emissions limits in terms of an emission mask are applied. Whether a single wide carrier or multiple aggregated carriers of the same or different sizes are transmitted, the guard band needed will have to be the same at both edges, since the emission mask roll-off is the same. A problem with the backwards-compatible base station definition is that the resulting guard BW_{GB} is proportional to the channel BW and would therefore be *different* if carriers of different channel BW are aggregated.

For this reason, a different definition is used for the UE, based on a "symmetrical" guard band. For the edge carriers (low and high), F_{offset} is half of the transmission bandwidth configuration, plus a symmetrical guard band BW_{GB}:

$$F_{\text{offset}} = \frac{0.18\,\text{MHz} \cdot N_{\text{RB}}}{2} + BW_{\text{GB}} \quad \text{(for UE),} \tag{17.6}$$

where 0.18 MHz is the bandwidth of one resource block and BW_{GB} is proportional to the channel BW of the largest component carrier. For the CA bandwidth classes defined in release 10 and where the edge carriers have the same channel bandwidth, F_{offset} will be the same for terminals and base stations and $BW_{\text{Channel\_CA}}$ will be the same.

It may look like an anomaly that the definitions may potentially lead to slightly different aggregated channel BW for the UE and the base station, but this is in fact not a problem. UE and base-station requirements are defined separately and do not have to cover the same frequency ranges. The aggregated channel BW for both UE and base station do, however, have to be within an operator's license block in the operating band.

Once the frequency reference point is set, the actual RF requirements are to a large extent the same as for a single carrier configuration. Which requirements are affected is explained for each requirement in the discussion later in this chapter.

17.5 **MULTI-STANDARD RADIO BASE STATIONS**

Traditionally the RF specifications have been developed separately for the different 3GPP radio-access technologies GSM/EDGE, UTRA, and E-UTRA (LTE). The rapid evolution of mobile radio and the need to deploy new technologies alongside the legacy deployments has, however, lead to implementation of different Radio-Access Technologies (RAT) at the same sites, often sharing antennas and other parts of the installation. A natural further step is then to also share the base-station equipment between multiple RATs. This requires multi-RAT base stations.

The evolution to multi-RAT base stations is also fostered by the evolution of technology. While multiple RATs have traditionally shared parts of the site installation, such as antennas, feeders, backhaul or power, the advance of both digital baseband and RF technologies enables a much tighter integration. A base station consisting of two separate implementations of both baseband and RF, together with a passive combiner/splitter before the antenna, could in theory be considered a multi-RAT base station. 3GPP has, however, made a narrower, but more forward-looking definition.

In a *Multi-Standard Radio* (MSR) base station, both the receiver and the transmitter are capable of simultaneously processing multiple carriers of different RATs in common *active* RF components. The reason for this stricter definition is that the true potential of multi-RAT base stations, and the challenge in terms of implementation complexity, comes from having a common RF. This principle is illustrated in Figure 17.6 with an example base station capable of both GSM/EDGE and LTE. Much of the GSM/EDGE and LTE baseband functionality may be separate in the base station, but is possibly implemented in the same hardware. The RF must, however, be implemented in the same active components as shown in the figure.

The main advantages of an MSR base station implementation are twofold:

- Migration between RATs in a deployment, for example from GSM/EDGE to LTE, is possible using the same base station hardware. In the example in Figure 17.6, a migration is performed in

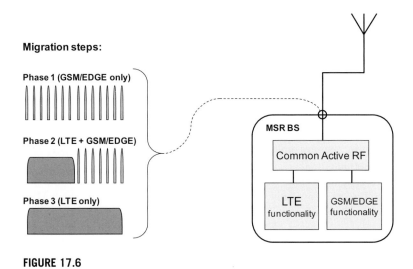

FIGURE 17.6

Example of migration from GSM to LTE using an MSR base station for all migration phases.

three phases using the same MSR base station. In the first phase, the base station is deployed in a network for GSM/EDGE-only operation. In the second phase, the operator migrates part of the spectrum to LTE. The same MSR base station will now operate one LTE carrier, but still supports the legacy GSM/EDGE users in half of the band available. In the third phase, when the GSM/EDGE users have migrated from the band, the operator can configure the MSR base station to LTE-only operation with double the channel bandwidth.

- A single base station designed as an MSR base station can be deployed in various environments for single-RAT operation for each RAT supported, as well as for multi-RAT operation where that is required by the deployment scenario. This is also in line with the recent technology trends seen in the market, with fewer and more generic base-station designs. Having fewer varieties of base station is an advantage both for the base-station vendor and for the operator, since a single solution can be developed and implemented for a variety of scenarios.

The single-RAT 3GPP radio-access standards, with requirements defined independently per RAT, do not support such migration scenarios with an implementation where common base-station RF hardware is shared between multiple access technologies, and hence a separate set of requirements for multi-standard radio equipment is needed.

An implication of a common RF for multiple RATs is that carriers are no longer received and transmitted independently of each other. For this reason, a common RF specification must be used to specify the MSR base station. 3GPP in release 9 has developed MSR specifications for the core RF requirements [80] and for test requirements [81]. Those specifications support GSM/EDGE,[3] UTRA and E-UTRA, and all combinations thereof. To support all possible RAT combinations, the MSR specifications have many generic requirements applicable regardless of RAT combination, together with specific single-access-technology-specific requirements to secure the integrity of the systems in single-RAT operation.

The MSR concept has a substantial impact for many requirements, while others remain completely unchanged. A fundamental concept introduced for MSR base stations is *RF bandwidth*, which is defined as the total bandwidth over the set of carriers transmitted and received. Many receiver and transmitter requirements for GSM/EDGE and UTRA are specified relative to the carrier center and for LTE in relation to the channel edges. For an MSR base station, they are instead specified relative to the *RF bandwidth edges*, in a way similar to carrier aggregation in release 10. In the same way as for carrier aggregation, a parameter F_{offset} is also introduced to define the location of the RF bandwidth edges relative to the edge carriers. For GSM/EDGE carriers, F_{offset} is set to 200 kHz, while it is in general half the channel bandwidth for UTRA and E-UTRA. By introducing the RF bandwidth concept and introducing generic limits, the requirements for MSR shift from being carrier centric towards being frequency block centric, thereby embracing technology neutrality by being independent of the access technology or operational mode.

While E-UTRA and UTRA carriers have quite similar RF properties in terms of bandwidth and power spectral density, GSM/EDGE carriers are quite different. The operating bands for which MSR base stations are defined are therefore divided into three *Band Categories* (BC):

- BC1 – All paired bands where UTRA FDD and E-UTRA FDD can be deployed.

---

[3]The MSR specifications are not applicable to single-RAT operation of GSM/EDGE.

- BC2 – All paired bands where in addition to UTRA FDD and E-UTRA FDD, GSM/EDGE can also be deployed.
- BC3 – All unpaired bands where UTRA TDD and E-UTRA TDD can be deployed.

Since the carriers of different RATs are not transmitted and received independently, it is necessary to perform parts of the testing with carriers of multiple RATs being activated. This is done through a set of multi-RAT *Test Configurations* defined in [81], specifically tailored to stress transmitter and receiver properties. These test configurations are of particular importance for the unwanted emission requirements for the transmitter and for testing of the receiver susceptibility to interfering signals (blocking, etc.). An advantage of the multi-RAT test configurations is that the RF performance of multiple RATs can be tested simultaneously, thereby avoiding repetition of test cases for each RAT. This is of particular importance for the very time-consuming tests of requirements outside the operating band over the complete frequency range up to 12.75 GHz.

The requirement with the largest impact from MSR is the spectrum mask, or the *operating band unwanted emissions* requirement, as it is called. The spectrum mask requirement for MSR base stations is applicable for multi-RAT operation where the carriers at the RF bandwidth edges are either GSM/EDGE, UTRA, or E-UTRA carriers of different channel bandwidths. The mask is generic and applicable to all cases and covers the complete operating band of the base station. There is an exception for the 150 kHz closest to the RF bandwidth edge, where the mask is aligned with the GSM/EDGE modulation spectrum for the case when a GSM/EDGE carrier or a 1.4/3 MHz E-UTRA carrier is transmitted adjacent to the edge.

An important aspect of MSR is the declaration by the base station vendor of the supported RF bandwidth, power levels, multicarrier capability, etc. All testing is based on the capability of the base station through a declaration of the supported *Capability Set* (CS), which defines all supported single RATs and multi-RAT combinations. There are currently six capability sets CS1–CS6 defined in the MSR test specification [81], allowing full flexibility for implementing and deploying base stations compliant to the MSR specification. For a large part of the base station RF requirements, multi-RAT testing is not necessary and the actual test limits are unchanged for the MSR base station. In these cases, both the requirements and the test cases are simply incorporated through direct references to the corresponding single-RAT specifications.

Carrier aggregation as described above in Section 17.4 is also applicable to MSR base stations. Since the MSR specification already has most of the concepts and definitions in place for defining multi-carrier RF requirements, whether aggregated or not, the changes of the MSR requirements compared to non-aggregated carriers are very minor.

17.6 OVERVIEW OF RF REQUIREMENTS FOR LTE

The RF requirements define the receiver and transmitter RF characteristics of a base station or UE. The base station is the physical node that transmits and receives RF signals on one or more antenna connectors. Note that a base station is not the same thing as an eNodeB, which is the corresponding logical node in the LTE Radio-Access Network. The terminal is denoted UE in the description below, as it is in all RF specifications.

The set of RF requirements defined for LTE is fundamentally the same as those defined for UTRA or any other radio system. Some requirements are also based on regulatory requirements and are more concerned with the frequency band of operation and/or the place where the system is deployed, than with the type of system.

What is particular to LTE is the flexible bandwidth and the related multiple channel bandwidths of the system, which makes some requirements more complex to define. These properties have special implications for the transmitter requirements on unwanted emissions, where the definition of the limits in international regulation depends on the channel bandwidth. Such limits are harder to define for a system where the base station may operate with multiple channel bandwidths and where the UE may vary its channel bandwidth of operation. The properties of the flexible OFDM-based physical layer also have implications for specifying the transmitter modulation quality and how to define the receiver selectivity and blocking requirements.

The type of transmitter requirements defined for the UE is very similar to what is defined for the base station, and the definitions of the requirements are often similar. The output power levels are, however, considerably lower for a UE, while the restrictions on the UE implementation are much higher. There is tight pressure on cost and complexity for all telecommunications equipment, but this is much more pronounced for terminals, due to the scale of the total market, which is more than one *billion* devices per year. In cases where there are differences in how requirements are defined between UE and base station, they are treated separately in this chapter.

The detailed background of the RF requirements for LTE is described in [82,83], with further details of the additional requirements in release 10 (for LTE-Advanced) in [84,85]. The RF requirements for the base station are specified in [86] and for the UE in [76]. The RF requirements are divided into transmitter and receiver characteristics. There are also *performance characteristics* for base station and UE that define the receiver baseband performance for all physical channels under different propagation conditions. These are not strictly RF requirements, though the performance will also depend on the RF to some extent.

Each RF requirement has a corresponding test defined in the LTE test specifications for the base station [87] and the UE [74]. These specifications define the test setup, test procedure, test signals, test tolerances, etc. needed to show compliance with the RF and performance requirements.

17.6.1 **Transmitter Characteristics**

The transmitter characteristics define RF requirements for the wanted signal transmitted from the UE and base station, but also for the unavoidable unwanted emissions outside the transmitted carrier(s). The requirements are fundamentally specified in three parts:

- *Output power level* requirements set limits for the maximum allowed transmitted power, for the dynamic variation of the power level and in some cases for the transmitter OFF state.
- *Transmitted signal quality* requirements define the "purity" of the transmitted signal and also the relation between multiple transmitter branches.
- *Unwanted emissions* requirements set limits to all emissions outside the transmitted carrier(s) and are tightly coupled to regulatory requirements and coexistence with other systems.

A list of the UE and base-station transmitter characteristics arranged according to the three parts defined above is shown in Table 17.5. A more detailed description of the requirements can be found later in this chapter.

Table 17.5 Overview of LTE Transmitter Characteristics

	Base-Station Requirement	**UE Requirement**
Output power level	Maximum output power	Transmit power
	Output power dynamics	Output power dynamics
	On/Off power (TDD only)	Power control
Transmitted signal quality	Frequency error	Frequency error
	Error Vector Magnitude (EVM)	Transmit modulation quality
	Time alignment between transmitter branches	
Unwanted emissions	Operating band unwanted emissions	Spectrum emission mask
	Adjacent Channel Leakage Ratio (ACLR)	Adjacent Channel Leakage Ratio (ACLR)
	Spurious emissions	Spurious emissions
	Occupied bandwidth	Occupied bandwidth
	Transmitter intermodulation	Transmit intermodulation

17.6.2 Receiver Characteristics

The set of receiver requirements for LTE is quite similar to what is defined for other systems such as UTRA, but many of them are defined differently, due to the flexible bandwidth properties. The receiver characteristics are fundamentally specified in three parts:

- *Sensitivity and dynamic range* requirements for receiving the wanted signal.
- *Receiver susceptibility to interfering signals* defines receivers' susceptibility to different types of interfering signals at different frequency offsets.
- *Unwanted emissions* limits are also defined for the receiver.

A list of the UE and base-station receiver characteristics arranged according to the three parts defined above is shown in Table 17.6. A more detailed description of each requirement can be found later in this chapter.

17.6.3 Regional Requirements

There are a number of regional variations to the RF requirements and their application. The variations originate in different regional and local regulations of spectrum and its use. The most obvious regional variation is the different frequency bands and their use, as discussed above. Many of the regional RF requirements are also tied to specific frequency bands.

When there is a regional requirement on, for example, spurious emissions, this requirement should be reflected in the 3GPP specifications. For the base station it is entered as an optional requirement and is marked as "regional". For the UE, the same procedure is not possible, since a UE may roam between different regions and will therefore have to fulfill all regional requirements that are tied to an operating band in the regions where the band is used. For LTE, this becomes more complex than for UTRA, since there is an additional variation in the transmitter (and receiver) bandwidth used, making some regional requirements difficult to meet as a mandatory requirement. The concept of *network*

Table 17.6 Overview of LTE Receiver Characteristics

	Base-Station Requirement	UE Requirement
Sensitivity and dynamic range	Reference sensitivity	Reference sensitivity power level
	Dynamic range	Maximum input level
	In-channel selectivity	
Receiver susceptibility to interfering signals	Out-of-band blocking	Out-of-band blocking
		Spurious response
	In-band blocking	In-band blocking
	Narrowband blocking	Narrowband blocking
	Adjacent channel selectivity	Adjacent channel selectivity
	Receiver intermodulation	Intermodulation characteristics
Unwanted emissions from the receiver	Receiver spurious emissions	Receiver spurious emissions

signaling of RF requirements is therefore introduced for LTE, where a UE can be informed at call setup of whether some specific RF requirements apply when the UE is connected to a network.

17.6.4 Band-Specific UE Requirements Through Network Signaling

For the UE, the channel bandwidths supported are a function of the LTE operating band, and also have a relation to the transmitter and receiver RF requirements. The reason is that some RF requirements may be difficult to meet under conditions with a combination of maximum power and high number of transmitted and/or received resource blocks.

Some additional RF requirements apply for the UE when a specific Network Signaling Value (NS_x) is signaled to the UE as part of the cell handover or broadcast message. For implementation reasons, these requirements are associated with restrictions and variations to RF parameters such as UE output power, maximum channel bandwidth, and number of transmitted resource blocks. The variations of the requirements are defined together with the Network Signaling Value (NS_x) in the UE RF specification [76], where each value corresponds to a specific condition. The default value for all bands is NS_01. All NS_x values are connected to an allowed power reduction called *Additional Maximum Power Reduction* (A-MPR) and apply for transmission using a certain minimum number of resource blocks, depending also on the channel bandwidth. The following are examples of UE requirements that have a related Network Signaling Value for some bands:

- *NS_03, NS_04, or NS_06* is signaled when specific FCC requirements [88] on UE unwanted emissions apply for operation in a number of US bands.
- *NS_05* is signaled for protection of the PHS band in Japan when UE operates in the 2 GHz band (Band 1).

In some bands the NS_x signaling is also applied for testing of receiver sensitivity, since the active transmitted signal can affect the receiver performance.

17.6.5 **Base-Station Classes**

In the base-station specifications, there is one set of RF requirements that is generic, applicable to what is called "general purpose" base stations. This is the original set of requirements developed in 3GPP release 8. It has no restrictions on base-station output power and can be used for any deployment scenario. When the RF requirements were derived, however, the scenarios used were macro scenarios [89]. For this reason, in release 9 additional base-station classes were introduced that were intended for pico-cell and femto-cell scenarios. It is then also clarified from release 9 that the original set of RF parameters is for macro-cell scenarios. The terms macro, pico, and femto are not used in 3GPP to identify the base-station classes, instead the following terminology is used:

- *Wide Area base stations*. This type of base station is intended for macro-cell scenarios, defined with a minimum coupling loss between BS and UE of 70 dB.
- *Local Area base stations*. This type of base station is intended for pico-cell scenarios, defined with a minimum coupling loss between BS and UE of 45 dB. Typical deployments are indoor offices and indoor/outdoor hotspots, with the BS mounted on walls or ceilings.
- *Home base stations*. This type of base station is intended for femto-cell scenarios, which are not explicitly defined. Minimum coupling loss between BS and UE of 45 dB is also assumed here. Home BS can be used both for open access and in closed subscriber groups.

The Local Area and Home base station classes have modifications to a number of requirements compared to Wide Area base stations, mainly due to the assumption of a lower minimum coupling loss:

- Maximum base station power is limited to 24 dBm output power for Local Area base stations and to 20 dBm for Home base stations, counting the power over all antennas (up to four). There is no maximum base station power defined for Wide Area base stations.
- Home base stations have an additional requirement for protecting systems operating on adjacent channels. The reason is that a UE connected to a base station belonging to another operator on the adjacent channel may be in close proximity to the Home base station. To avoid an interference situation where the adjacent UE is blocked, the Home base station must make measurements on the adjacent channel to detect adjacent base-station operations. If an adjacent base-station transmission (UTRA or LTE) is detected under certain conditions, the maximum allowed Home base-station output power is reduced in proportion to how weak the adjacent base-station signal is, in order to avoid interference to the adjacent base station.
- The spectrum mask (operating band unwanted emissions) has lower limits for Local Area and Home base stations, in line with the lower maximum power levels.
- Unwanted emission limits for protecting Home base-station operation (from other Home base stations) are lower, since a stricter through-the-wall indoor interference scenario is assumed. Limits for co-location for Local Area are, however, less strict, corresponding to the relaxed reference sensitivity for the base station.
- Receiver reference sensitivity limits are higher (more relaxed) for Local Area and Home base stations. Receiver dynamic range and in-channel selectivity are also adjusted accordingly.
- All Local Area and Home base-station limits for receiver susceptibility to interfering signals are adjusted to take the higher receiver sensitivity limit and the lower assumed minimum coupling loss (base station-to-UE) into account.

17.7 OUTPUT POWER LEVEL REQUIREMENTS

17.7.1 Base-Station Output Power and Dynamic Range

There is no general maximum output power requirement for base stations. As mentioned in the discussion of base-station classes above, there is, however, a maximum power limit of 24 dBm output power for Local Area base stations and of 20 dBm for Home base stations, counting the power over all antennas. In addition to this there is a tolerance specified, defining how much the actual maximum power may deviate from the power level declared by the manufacturer.

The base station also has a specification of the total power control dynamic range for a resource element, defining the power range over which it should be possible to configure. There is also a dynamic range requirement for the Total base-station power.

For TDD operation, a power mask is defined for the base-station output power, defining the Off power level during the uplink subframes and the maximum time for the *transmitter transient period* between the transmitter On and Off states.

17.7.2 UE Output Power and Dynamic Range

The UE output power level is defined in three steps:

- *UE power class* defines a *nominal* maximum output power for QPSK modulation. It may be different in different operating bands, but the main UE power class is today set at 23 dBm for all bands.
- *Maximum Power Reduction (MPR)* defines an allowed reduction of maximum power level for certain combinations of modulation used and the number of resource blocks that are assigned.
- *Additional Maximum Power Reduction (A-MPR)* may be applied in some regions and is usually connected to specific transmitter requirements such as regional emission limits. For each such set of requirement, there is an associated network signaling value NS_x that identifies the allowed A-MPR and the associated conditions, as explained above.

The UE has a definition of the transmitter Off power level, applicable to conditions when the UE is not allowed to transmit. There is also a general On/Off time mask specified, plus specific time masks for PRACH, SRS, subframe boundary and PUCCH/PUSCH/SRS.

The UE transmit power control is specified through requirements for the *absolute power tolerance* for the initial power setting, the *relative power tolerance* between two subframes, and the *aggregated power tolerance* for a sequence of power-control commands.

17.8 TRANSMITTED SIGNAL QUALITY

The requirements for transmitted signal quality specify how much the transmitted base station or UE signal deviates from an "ideal" modulated signal in the signal and the frequency domains. Impairments on the transmitted signal are introduced by the transmitter RF parts, with the nonlinear properties of the power amplifier being a major contributor. The signal quality is measured for base station and UE through *EVM* and *frequency error*. An additional UE requirement is UE in-band emissions.

17.8.1 EVM and Frequency Error

While the theoretical definitions of the signal quality measures are quite straightforward, the actual assessment is a very elaborate procedure, described in great detail in the 3GPP specification. The reason is that it becomes a multidimensional optimization problem, where the best match for the timing, the frequency, and the signal constellation are found.

The *Error Vector Magnitude* (EVM) is a measure of the error in the modulated signal constellation, taken as the root mean square of the error vectors over the active subcarriers, considering all symbols of the modulation scheme. It is expressed as a percentage value in relation to the power of the ideal signal. The EVM fundamentally defines the maximum SINR that can be achieved at the receiver, if there are no additional impairments to the signal between transmitter and receiver.

Since a receiver can remove some impairments of the transmitted signal such as time dispersion, the EVM is assessed after cyclic prefix removal and equalization. In this way, the EVM evaluation includes a standardized model of the receiver. The frequency offset resulting from the EVM evaluation is averaged and used as a measure of the *frequency error* of the transmitted signal.

17.8.2 UE In-Band Emissions

In-band emissions are emissions within the channel bandwidth. The requirement limits how much a UE can transmit into non-allocated resource blocks within the channel bandwidth. Unlike the out-of-band emissions, the in-band emissions are measured after cyclic prefix removal and FFT, since this is how a UE transmitter affects a real base-station receiver.

17.8.3 Base-Station Time Alignment

Several LTE features require the base station to transmit from two ore more antennas, such as transmitter diversity and MIMO. For carrier aggregation, the carriers may also be transmitted from different antennas. In order for the UE to properly receive the signals from multiple antennas, the timing relation between any two transmitter branches is specified in terms of a maximum time alignment error between transmitter branches. The maximum allowed error depends on the feature or combination of features in the transmitter branches.

17.9 UNWANTED EMISSIONS REQUIREMENTS

Unwanted emissions from the transmitter are divided into *out-of-band (OOB) emission* and *spurious emissions* in ITU-R recommendations [90]. OOB emissions are defined as emissions on a frequency close to the RF carrier, which results from the modulation process. Spurious emissions are emissions outside the RF carrier that may be reduced without affecting the corresponding transmission of information. Examples of spurious emissions are harmonic emissions, intermodulation products, and frequency conversion products. The frequency range where OOB emissions are normally defined is called the *OOB domain*, whereas spurious emission limits are normally defined in the *spurious domain*.

ITU-R also defines the boundary between the OOB and spurious domains at a frequency separation from the carrier center of 2.5 times the necessary bandwidth, which corresponds to 2.5 times the channel bandwidth for LTE. This division of the requirements is easily applied for systems that have a fixed channel bandwidth. It does, however, become more difficult for LTE, which is a flexible bandwidth system, implying that the frequency range where requirements apply would then vary with the channel bandwidth. The approach taken for defining the boundary in 3GPP is slightly different for base-station and UE requirements.

With the recommended boundary between OOB emissions and spurious emissions set at 2.5 times the channel bandwidth, third- and fifth-order intermodulation products from the carrier will fall inside the OOB domain, which will have a bandwidth of twice the channel bandwidth. For the OOB domain, two overlapping requirements are defined for both base station and UE: *Spectrum Emission Mask* (SEM) and *Adjacent Channel Leakage Ratio* (ACLR). The details of these are further explained below.

17.9.1 Implementation Aspects

As shown in Chapter 3, the spectrum of an OFDM signal decays rather slowly outside of the transmission bandwidth configuration. Since the transmitted signal for LTE occupies 90% of the channel bandwidth, it is not possible to directly meet the unwanted emission limits directly outside the channel bandwidth with a "pure" OFDM signal. The techniques used for achieving the transmitter requirements are, however, not specified or mandated in LTE specifications. Time-domain windowing is one method commonly used in OFDM-based transmission systems to control spectrum emissions. Filtering is always used, both time-domain digital filtering of the baseband signal and analog filtering of the RF signal.

The nonlinear characteristics of the *Power Amplifier* (PA) used to amplify the RF signal must also be taken into account, since it is the source of intermodulation products outside the channel bandwidth. Power back-off to give a more linear operation of the PA can be used, but at the cost of a lower power efficiency. The power back-off should therefore be kept to a minimum. For this reason, additional linearization schemes can be employed. These are especially important for the base station, where there are fewer restrictions on implementation complexity and use of advanced linearization schemes is an essential part of controlling spectrum emissions. Examples of such techniques are feed-forward, feedback, predistortion and postdistortion.

17.9.2 Spectrum Emission Mask

The spectrum emission mask defines the permissible out-of-band spectrum emissions outside the necessary bandwidth. As explained above, how to take the flexible channel bandwidth into account when defining the frequency boundary between OOB emissions and spurious domains is done differently for the LTE base station and UE. Consequently, the spectrum emission masks are also based on different principles.

17.9.2.1 *Base-Station Operating Band Unwanted Emission Limits*

For the LTE base station, the problem of the implicit variation of the boundary between OOB and spurious domain with the varying channel bandwidth is handled by not defining an explicit boundary. The solution is a unified concept of *operating band unwanted emissions* for the LTE base station instead of the spectrum mask usually defined for OOB emissions. The operating band unwanted emissions requirement applies over the whole base station transmitter operating band, plus an

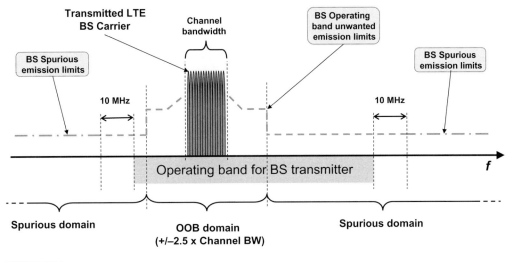

FIGURE 17.7

Frequency ranges for operating band unwanted emissions and spurious emissions applicable to an LTE base station.

additional 10 MHz on each side, as shown in Figure 17.7. All requirements outside of that range are set by the regulatory spurious emission limits, based on the ITU-R recommendations [90]. As seen in the figure, a large part of the operating band unwanted emissions are defined over a frequency range that for smaller channel bandwidths can be both in spurious and OOB domains. This means that the limits for the frequency ranges that may be in the spurious domain also have to align with the regulatory limits from the ITU-R. The shape of the mask is generic for all channel bandwidth from 5 to 20 MHz, with a mask that consequently has to align with the ITU-R limits starting 10 MHz from the channel edges. Special masks are defined for the smaller 1.4 and 3 MHz channel bandwidths. The operating band unwanted emissions are defined with a 100 kHz measurement bandwidth.

There are also special limits defined to meet a specific regulation set by the FCC [88] for the operating bands used in the USA and by the ECC for some European bands. These are specified as separate limits in addition to the operating band unwanted emission limits.

17.9.2.2 *UE Spectrum Emission Mask*

For implementation reasons, it is not possible to define a generic UE spectrum mask that does not vary with the channel bandwidth, so the frequency ranges for OOB limits and spurious emissions limits do not follow the same principle as for the base station. The SEM extends out to a separation Δf_{OOB} from the channel edges, as illustrated in Figure 17.8. For 5 MHz channel bandwidth, this point corresponds to 250% of the necessary bandwidth as recommended by the ITU-R, but for higher channel bandwidths it is set closer than 250%.

The SEM is defined as a general mask and a set of additional masks that can be applied to reflect different regional requirements. Each additional regional mask is associated with a specific network signaling value NS_x.

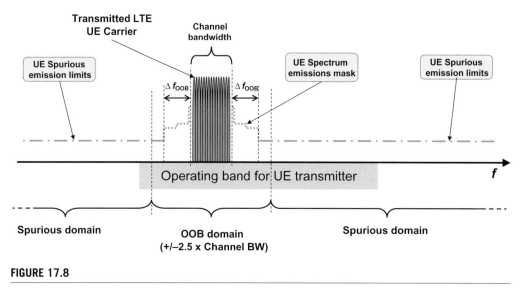

FIGURE 17.8

Frequency ranges for spectrum emission mask and spurious emissions applicable to an LTE UE.

17.9.3 Adjacent Channel Leakage Ratio

In addition to a spectrum emissions mask, the OOB emissions are defined by an *Adjacent Channel Leakage Ratio* (ACLR) requirement. The ACLR concept is very useful for analysis of coexistence between two systems that operate on adjacent frequencies. The ACLR defines the ratio of the power transmitted within the assigned channel bandwidth to the power of the unwanted emissions transmitted on an adjacent channel. There is a corresponding receiver requirement called *Adjacent Channel Selectivity* (ACS), which defines a receiver's ability to suppress a signal on an adjacent channel.

The definitions of ACLR and ACS are illustrated in Figure 17.9 for a wanted and an interfering signal received in adjacent channels. The interfering signal's leakage of unwanted emissions at the wanted signal receiver is given by the ACLR and the ability of the receiver of the wanted signal to suppress the interfering signal in the adjacent channel is defined by the ACS. The two parameters when combined define the total leakage between two transmissions on adjacent channels. That ratio is called the *Adjacent Channel Interference Ratio* (ACIR) and is defined as the ratio of the power transmitted on one channel to the total interference received by a receiver on the adjacent channel, due to both transmitter (ACLR) and receiver (ACS) imperfections.

This relation between the adjacent channel parameters is [91]:

$$ACIR = \frac{1}{\dfrac{1}{ACLR} + \dfrac{1}{ACS}}. \tag{17.7}$$

ACLR and ACS can be defined with different channel bandwidths for the two adjacent channels, which is the case for some requirements set for LTE due to the bandwidth flexibility. The equation

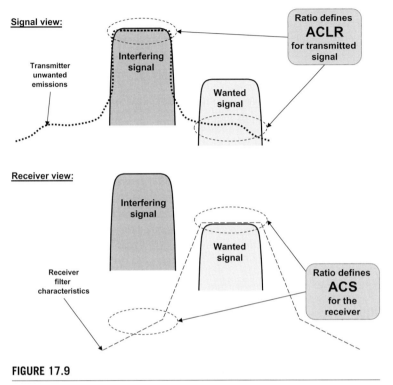

FIGURE 17.9

Illustration of ACLR and ACS, with example characteristics for an "aggressor" interferer and a receiver for a "victim" wanted signal.

above will also apply for different channel bandwidths, but only if the same two channel bandwidths are used for defining all three parameters ACIR, ACLR, and ACS used in the equation.

The ACLR limits for LTE UE and base station are derived based on extensive analysis [89] of LTE coexistence with LTE or other systems on adjacent carriers.

The LTE base-station requirements on ACLR and operating band unwanted emissions both cover the OOB domain, but the operating band unwanted emission limits are set slightly more relaxed compared to the ACLR, since they are defined in a much narrower measurement bandwidth of 100 kHz. This allows for some variations in the unwanted emissions due to intermodulation products from varying power allocation between resource blocks within the channel. For an LTE base station, there are ACLR requirements both for an adjacent channel with a UTRA receiver and with an LTE receiver of the same channel bandwidth.

ACLR limits for the UE are set both with assumed UTRA and LTE receivers on the adjacent channel. As for the base station, the limits are also set stricter than the corresponding SEM, thereby accounting for variations in the spectrum emissions resulting from variations in resource-block allocations.

17.9.4 Spurious Emissions

The limits for base station spurious emissions are taken from international recommendations [90], but are only defined in the region outside the frequency range of operating band unwanted emissions limits as illustrated in Figure 17.7 – that is, at frequencies that are separated from the base-station transmitter operating band by at least 10 MHz. There are also additional regional or optional limits for protection of other systems that LTE may coexist with or even be co-located with. Examples of other systems considered in those additional spurious emissions requirements are GSM, UTRA FDD/TDD, CDMA2000, and PHS.

UE spurious emission limits are defined for all frequency ranges outside the frequency range covered by the SEM. The limits are generally based on international regulations [90], but there are also additional requirements for coexistence with other bands when the mobile is roaming. The additional spurious emission limits can have an associated network signaling value.

In addition, there are base-station and UE emission limits defined for the receiver. Since receiver emissions are dominated by the transmitted signal, the receiver spurious emission limits are only applicable when the transmitter is Off, and also when the transmitter is On for an LTE FDD base station that has a separate receiver antenna connector.

17.9.5 Occupied Bandwidth

Occupied bandwidth is a regulatory requirement that is specified for equipment in some regions, such as Japan and the USA. It is originally defined by the ITU-R as a maximum bandwidth, outside of which emissions do not exceed a certain percentage of the total emissions. The occupied bandwidth is for LTE equal to the channel bandwidth, outside of which a maximum of 1% of the emissions are allowed (0.5% on each side).

In the case of carrier aggregation, the occupied bandwidth is equal to the aggregated channel bandwidth.

17.9.6 Transmitter Intermodulation

An additional implementation aspect of an RF transmitter is the possibility of intermodulation between the transmitted signal and another strong signal transmitted in the proximity of the base station or UE. For this reason there is a requirement for *transmitter intermodulation.*

For the base station, the requirement is based on a stationary scenario with a co-located other base-station transmitter, with its transmitted signal appearing at the antenna connector of the base station being specified, but attenuated by 30 dB. Since it is a stationary scenario, there are no additional unwanted emissions allowed, implying that all unwanted emission limits also have to be met with the interferer present.

For the UE, there is a similar requirement based on a scenario with another UE transmitted signal appearing at the antenna connector of the UE being specified, but attenuated by 40 dB. The requriement specifies the minimum attenuation of the resulting intermodulation product below the transmitted signal.

17.10 SENSITIVITY AND DYNAMIC RANGE

The primary purpose of the *reference sensitivity requirement* is to verify the receiver *Noise Figure*, which is a measure of how much the receiver's RF signal chain degrades the SNR of the received signal. For this reason, a low-SNR transmission scheme using QPSK is chosen as reference channel for the reference sensitivity test. The reference sensitivity is defined at a receiver input level where the throughput is 95% of the maximum throughput for the reference channel.

For the base station, reference sensitivity could potentially be defined for a single resource block up to a group covering all resource blocks. For reasons of complexity, a maximum granularity of 25 resource blocks has been chosen, which means that for channel bandwidths larger than 5 MHz, sensitivity is verified over multiple adjacent 5 MHz blocks, while it is only defined over the full channel for smaller channel bandwidths.

For the UE, reference sensitivity is defined for the full channel bandwidth signals and with all resource blocks allocated for the wanted signal. For the higher channel bandwidths (>5 MHz) in some operating bands, the nominal reference sensitivity needs to be met with a minimum number of allocated resource blocks. For larger allocation, a certain relaxation is allowed.

The intention of the *dynamic range requirement* is to ensure that the receiver can also operate at received signal levels considerably higher than the reference sensitivity. The scenario assumed for base-station dynamic range is the presence of increased interference and corresponding higher wanted signal levels, thereby testing the effects of different receiver impairments. In order to stress the receiver, a higher SNR transmission scheme using 16QAM is applied for the test. In order to further stress the receiver to higher signal levels, an interfering AWGN signal at a level 20 dB above the assumed noise floor is added to the received signal. The dynamic range requirement for the UE is specified as a *maximum signal level* at which the throughput requirement is met.

17.11 RECEIVER SUSCEPTIBILITY TO INTERFERING SIGNALS

There is a set of requirements for base station and UE, defining the receiver's ability to receive a wanted signal in the presence of a stronger interfering signal. The reason for the multiple requirements is that, depending on the frequency offset of the interferer from the wanted signal, the interference scenario may look very different and different types of receiver impairments will affect the performance. The intention of the different combinations of interfering signals is to model as far as possible the range of possible scenarios with interfering signals of different bandwidths that may be encountered inside and outside the base-station and UE receiver operating band. While the types of requierments are very similar between base station and UE, the signal levels are different, since the interference scenarios for the base station and UE are very different. There is also no UE requirement corresponding to the base-station *In-Channel Selectivity* (ICS) requirement.

The following requirements are defined for LTE base station and UE, starting from interferers with large frequency separation and going close in (see also Figure 17.10). In all cases where the interfering signal is an LTE signal, it has the same bandwidth as the wanted signal, but at most 5 MHz.

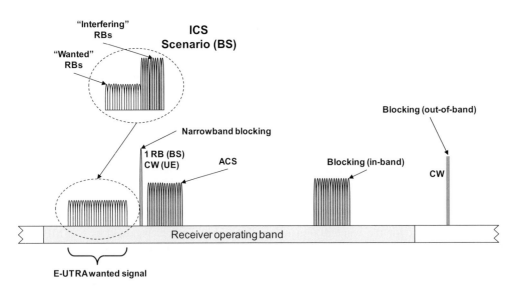

FIGURE 17.10

Base-station and UE requirements for receiver susceptibility to interfering signals in terms of blocking, ACS, narrowband blocking, and in-channel selectivity (BS only).

- *Blocking*. This corresponds to the scenario with strong interfering signals received outside the operating band (out-of-band blocking) or inside the operating band (in-band blocking), but not adjacent to the wanted signal. In-band blocking includes interferers in the first 20 MHz outside the band for the base station and the first 15 MHz for the UE. The scenarios are modeled with a *Continuous Wave* (CW) signal for the out-of-band case and an LTE signal for the in-band case. There are additional (optional) base-station blocking requirements for the scenario when the base station is co-sited with another base station in a different operating band. For the UE, a fixed number of *exceptions* are allowed from the out-of-band blocking requirement, for each assigned frequency channel and at the respective *spurious response frequencies*. At those frequencies, the UE must comply with the more relaxed spurious response requirement.
- *Adjacent channel selectivity*. The ACS scenario is a strong signal in the channel adjacent to the wanted signal and is closely related to the corresponding ACLR requirement (see also the discussion in Section 17.9.3). The adjacent interferer is an LTE signal. For the UE, the ACS is specified for two cases with a lower and a higher signal level.
- *Narrowband blocking*. The scenario is an adjacent strong narrowband interferer, which in the requirement is modeled as a single resource block LTE signal for the base station and a CW signal for the UE.
- *In-channel selectivity* (ICS). The scenario is multiple received signals of different received power levels inside the channel bandwidth, where the performance of the weaker "wanted" signal is verified in the presence of the stronger "interfering" signal. ICS is only specified for the base station.
- *Receiver intermodulation*. The scenario is *two* interfering signals near to the wanted signal, where the interferers are one CW and one LTE signal (not shown in Figure 17.10). The interferers are

placed in frequency in such a way that the main intermodulation product falls inside the wanted signal's channel bandwidth. There is also a *narrowband intermodulation* requirement for the base station where the CW signal is very close to the wanted signal and the LTE interferer is a single RB signal.

For all requirements except in-channel selectivity, the wanted signal uses the same reference channel as in the corresponding reference sensitivity requirement. With the interference added, the same 95% relative throughput is met as for the reference channel, but at a "desensitized" higher wanted signal level.

Performance

18.1 PERFORMANCE ASSESSMENT

Computer simulation of mobile systems is a very powerful tool to assess the system performance. The "real-life" performance can, of course, be measured and evaluated in the field for an already deployed system and such values represent a valid example of performance for a certain system configuration. But there are several advantages with computer simulations:

- Evaluation can be made of system concepts that are not deployed or still under development.
- There is full control of the environment, including propagation parameters, traffic, system layout, etc., and full traceability of all parameters affecting the result.
- Well-controlled "experiments" comparing similar system concepts or parts of concepts can be done under repeatable conditions.

In spite of the advantages, simulation results obviously do not give a full picture of the performance of a system. It is impossible to model all aspects of the mobile environment and to properly model the behavior of all components in a system. Still, a very good picture of system performance can be obtained and it can often be used to find the potential limits for performance. Because of the difficulty in modeling all relevant aspects, relative capacity measures for introducing features will be more accurate than absolute capacity numbers, if a good model is introduced of the feature in question.

The capacity of a system is difficult to assess without comparisons, since a system performance number in itself does not provide very much information. It is when set in relation to how other systems can perform that the number becomes interesting. Since making comparisons is an important component in assessing performance, it also makes performance numbers a highly contentious issue. One always has to watch out for "apples and oranges" comparisons, since the system performance depends on so many parameters. If parameters are not properly selected to give comparable conditions for two systems, the performance numbers will not be comparable either.

In this context, it is also essential to take into account that system performance and capacity will be feature dependent. Many features such as MIMO and advanced antenna techniques that have been introduced in 3G (HSPA) and 4G (LTE) systems are very similar between systems. If a certain feature is a viable option for several systems that are evaluated in parallel, the feature should be included in the evaluation for all systems.

Any simulated performance number should be viewed in the context that real radio network performance will depend on many parameters that are difficult to control or model, including:

- The mobile environment, including channel conditions, angular spreads, clutter type, terminal speeds, indoor/outdoor usage, and coverage holes.
- User-related behavior, such as voice activity, traffic distribution, and service distribution.
- System tuning of service quality and network quality.
- Deployment aspects such as site types, antenna heights and types, and frequency planning.
- A number of additional parameters that are usually not modeled, such as signaling capacity and performance, and measurement quality.

There is no single universal measure of performance for a telecommunications system. Indeed, end-users (subscribers) and system operators define good performance quite differently. On the one hand, end-users want to experience the highest possible level of quality. On the other hand, operators want to derive maximum revenue, for example by squeezing as many users as possible into the system. Performance-enhancing features can improve perceived quality of service (end-user viewpoint) or system performance (operator viewpoint). The good news, however, is that LTE and its evolution has the potential to do both. Compared to 3G systems, LTE yield better data rates and shorter delay. That is, LTE and its evolution can greatly improve both the service experience (end-user viewpoint) and the system capacity (operator viewpoint).

18.1.1 **End-User Perspective of Performance**

Users of circuit-switched services are assured of a fixed data rate. The quality of service in the context of speech or video telephony services is defined by perceived speech or video quality. Superior-quality services have fewer bit errors in the received signal.

By contrast, users who download a web page or movie clip via packet data describe quality of service in terms of the delay they experience from the time they start the download until the web page or movie clip is displayed. Best-effort services do not guarantee a fixed data rate. Instead, users are allocated whatever data rate is available under present conditions. This is a general property of packet-switched networks – that is, network resources are not reserved for each user. Given that the delay increases with the size of the object to be downloaded, absolute delay is not a fair measure of quality of service.

Performance of a packet data service in cellular systems can be characterized by several different measures depending on the perspective taken (see Figure 18.1). A single user in a radio network experiencing good radio conditions may enjoy the *peak data rate* of the radio interface. A user will, however, normally share radio resources with other users. If radio conditions are less than optimal or there is interference from other users, the radio-interface data rate will be less than the peak data rate. In addition, some data packets might be lost, in which case the missing data must be retransmitted, further reducing the effective data rate as seen from higher protocol layers. Furthermore, the effective data rate diminishes even further as the distance from the cell increases (due to poorer radio conditions at cell edges). The data rate experienced above the MAC layer, after sharing the channel with other users, is denoted *user throughput*.

The *Transmission Control Protocol* (TCP) – the protocol at the transport layer – is commonly used together with IP traffic. However, due to the so-called slow-start algorithm, which is sensitive to latency in the network, it is especially prone to cause delay for small files. The slow-start algorithm is meant to ensure that the packet transmission rate from the source does not exceed the capability of network nodes and interfaces.

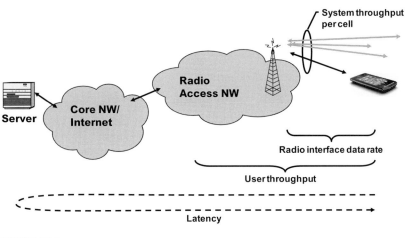

FIGURE 18.1

Definitions of data rates for performance.

Network *latency*, which in principle is a measure of the time it takes for a packet to travel from a client to a server and back again, has a direct impact on performance with TCP. Therefore, an important design objective for LTE has been to reduce network latency. One other quality-related criterion (end-user viewpoint) relates to the setup time for initiating, for example, a web-browsing session.

18.1.2 Operator Perspective

Radio resources need to be shared when multiple users are in the network. As a result, all data must be queued before it can be transmitted, which restricts the effective data rate to each user. Notwithstanding this fact, by scheduling radio resources, operators can improve system throughput or the total number of bits per second transmitted over the radio interface. A common measure of system performance is "spectral efficiency", which is the system throughput per MHz of spectrum in each cell of the system.

LTE employs intelligent scheduling methods to optimize performance, from both end-user and operator viewpoints.

An important performance measure for operators is the number of active users who can be connected simultaneously. Given that system resources are limited, there will thus be a trade-off between number of active users and perceived quality of service in terms of user throughput.

18.2 PERFORMANCE IN TERMS OF PEAK DATA RATES AND LATENCY

As described in Chapter 1, LTE has been developed in a process where design targets for performance parameters play an important role. One target is for the peak data rate over the radio interface. The original design targets for the first release of LTE are documented in 3GPP TR 25.913 [10]. The target capability when operating in a 20 MHz spectrum allocation is a peak data rate of 100 Mbit/s in

Table 18.1 LTE Peak Spectral Efficiency [92]

Peak Spectral Efficiency	ITU Requirement (bit/s/Hz)	LTE Fulfillment			
		Release 8		Release 10	
		FDD	TDD	FDD	TDD
Downlink	15	15.3	15.0	30.6	30.0
Uplink	6.75	4.2	4.0	16.8	16.0

the downlink and 50 Mbit/s in the uplink. The numbers assume two receive antennas in the terminal for the downlink capability and one transmit antenna for the uplink capability. These target numbers are exceeded by a good margin by the peak data rate capability of the specified LTE standard. LTE release 8 supports peak data rates of 300 Mbit/s in the downlink and 75 Mbit/s in the uplink by using spatial multiplexing of four layers (4×4 MIMO) in the downlink and 64QAM in both downlink and uplink. With the assumptions in the design targets – that is, spatial multiplexing of two layers – the downlink peak data rate is 150 Mbit/s, which is still significantly higher than the target.

The design targets for LTE release 10 ("LTE-Advanced") are documented in 3GPP TR 36.913 [10], based on the targets set by ITU-R [65]. There is no absolute peak data rate target expressed for LTE release 10; it is instead expressed relative to the channel bandwidth as a peak spectral efficiency, with targets of 15 bit/s/Hz for downlink and 6.75 bit/s/MHz for uplink [65]. LTE release 10 exceeds those numbers by a good margin, as shown in Table 18.1. The assumptions for deriving the peak spectral efficiency numbers is a deployment with 20 MHz channel bandwidth, 8×8 MIMO in the downlink, and 4×4 MIMO in the uplink.

The ITU-R requirement for downlink peak spectral efficiency is in fact fulfilled already by LTE release 8, assuming 4×4 MIMO in the downlink [92].

The ITU-R target for *latency* in [65] is set with a different definition than in Figure 18.1. Instead of a two-way *latency*, there is a 10 ms target for the *one-way latency* in both downlink and uplink. The one-way latency is defined as the one-way transit time between a packet being available at the IP layer in the user base station and the availability of this packet at the IP layer in the user terminal, and vice versa. This is achieved with a good margin for LTE, where the latency achieved is 4 ms for LTE FDD and 4.9 ms for LTE TDD.

18.3 PERFORMANCE EVALUATION OF LTE-ADVANCED

An essential part of the submission of LTE release 10 to ITU-R as a candidate for IMT-Advanced was the performance evaluation of the spectral efficiency. The technical requirements are set by ITU-R Report M.2134 [65] and the detailed evaluation methodology is described in ITU-R Report M.2135 [93].

For the work on LTE release 10, 3GPP performed a large simulation campaign with input from several 3GPP members. The resulting performance numbers formed an essential part of the submission of LTE-Advanced as a candidate for IMT-Advanced and is reported in detail in 3GPP TR 36.912 [92]. In addition to the ITU-R evaluation criteria, an additional test environment with higher

Table 18.2 Test Environment and Deployment Parameters for the Evaluation

	Test Environment			
	Indoor	**Microcellular**	**Base Coverage, Urban**	**High Speed**
Deployment scenario	Indoor hotspot	Urban micro	Urban macro	Rural macro
Channel model	InH	UMi	UMa	RMa
Carrier frequency (GHz)	3.4	2.5	2.0	0.8
Inter-site distance (m)	60	200	500	1732
Terminal speed (km/h)	3	3	30	120
User distribution	100% indoor	50% indoor, 50% outdoor	100% outdoor	100% outdoor
BS antenna height (m)	6	10	25	35
BS antenna gain (dBi)	0	17	17	17
BS output power (dBm/20 MHz)	21	44	49	49
UE output power (dBm)	21	24	24	24

performance targets than the ones set by ITU-R was defined in 3GPP. Performance numbers for this additional environment are also reported in [92].

An evaluation based on the ITU-R methodology and assumptions is presented in [94] and is also outlined below. The evaluation is based on time-dynamic system simulations, where users are dropped independently with uniform distribution over a simulated LTE system with a large number of cells. The simulations include overhead for control channels and time-dynamic models of feedback channels. Further details are given below. With the level of details applied to model the protocols and the simulation assumptions used for the evaluation described below, the technology potential of LTE release 10 is demonstrated, in addition to showing that the IMT-Advanced performance requirements are exceeded.

Detailed evaluations of LTE release 8 in comparison to earlier 3GPP releases of HSPA can be found in [95,96].

18.3.1 **Models and Assumptions**

This section presents the models and assumptions used for the evaluation in [94]. A summary of the test environments is given in Table 18.2 and the assumptions for the LTE-Advanced system characteristics are given in Table 18.3.

The evaluation is performed in four deployment scenarios, each corresponding to a different test environment defined by ITU-R [93]:

- *Indoor Hotspot*. Deployment scenario for the *indoor* environment, having isolated office cells or hotspots for stationary or pedestrian users, with high user density and high user throughput.
- *Urban Micro*. Deployment scenario for the *microcellular* environment, having small cells with outdoor and outdoor-to-indoor coverage for pedestrian and slow vehicular users, provided by

Table 18.3 LTE-Advanced System Characteristics for the evaluation in [94]

General Characteristics	
Duplex method assumptions	FDD
	TDD: Configuration 1, DwPTS/GP/UpPTS length set to 12/1/1 OFDM symbols (see Chapter 9)
Spectrum allocation	10 MHz DL + 10 MHz UL for FDD,
	10 MHz for TDD
	Double bandwidth for InH case
Antenna configuration at base station (BS)	Vertically co-polarized
	Antennas with 4 (InH) or 0.5 (UMi, UMa, RMa) wavelengths separation
Antenna configuration at terminal (UE)	Vertically co-polarized
	0.5 wavelengths separation
Network synchronization	Synchronized, not explicitly utilized other than for avoiding UE–UE and BS–BS interference for TDD
Detailed Radio-Interface Characteristics and Models	
Scheduler	DL: Proportional fair in time and frequency
	UL: Quality-based frequency-domain multiplexing
Downlink transmission scheme	InH: Transmission mode 4; closed-loop codebook-based precoded adaptive rank spatial multiplexing
	UMi, UMa, RMa: Transmission mode 5; coordinated beam-forming (within site) with MU-MIMO
Uplink transmission scheme	1 Tx, 4Rx antennas, no MU-MIMO
Receiver type	Minimum Mean Squared Error (MMSE) in DL and UL
Uplink power control	Open loop with fractional path-loss compensation, parameters chosen according to the deployment scenario. Effective noise rise below 10 dB
Hybrid-ARQ scheme	Incremental redundancy, synchronous, adaptive
Link adaptation	Non-ideal, based on delayed feedback
Channel estimation	Non-ideal channel estimation
	Non-ideal channel-state reports in downlink; CQI error per resource block is $N(0,1)$ dB, error-free feedback of the reports, 6 ms reporting delay, 5 ms reporting periodicity
	Uplink quality estimated from PUSCH, 6 ms delay, 20 ms sounding period
Feedback channel errors	Error-free, but quantized and delayed
Control channel overhead	DL: 3 OFDM symbols per subframe
	UL: 4 resource blocks
	Overhead for common control channels (synchronization, broadcast and random access; ~1% for 10 MHz) has not been deducted

below rooftop outdoor base stations. It has high traffic per unit area and will be interference limited.

- *Urban Macro.* Deployment scenario for the *base coverage urban* environment, having large cells and continuous coverage for pedestrians up to fast vehicular users, provided by above rooftop outdoor base stations. It will be interference limited and has mostly non-line-of-sight propagation conditions.
- *Rural Macro.* Deployment scenario for the *high speed* environment, having large cells and continuous coverage for high-speed vehicular and train users. It will be noise and/or interference limited.

Each deployment scenario described in [93] corresponds to a set of deployment parameters, including detailed propagation models for line-of-sight, non-line-of-sight, and/or outdoor-to-indoor propagation. The channel model is based on the so-called WINNER II channel model [97], which is a Spatial Channel Model (SCM) applicable for MIMO simulations [98]. Further details of the test environments and deployment parameters are given in Table 18.2.

The simulation methodology is "time dynamic", where a system of base stations and terminals is simulated over a limited time frame (20–100 seconds). This is repeated to create a number of samples to reach sufficient statistical confidence. For each simulation, terminals are randomly positioned over a model of a radio network and the radio channel between each base station and terminal antenna pair is simulated according to the propagation and fading models. A full buffer traffic model with an average of 10 users per cell is assumed. This results in the system operating at the maximum 100% system load.

Based on the channel realizations and the active interferers, a *signal-to-interference-and-noise ratio* (SINR) is calculated for each terminal (or base-station) receive antenna. The SINR values are then mapped to *block error probability* for the modulation and coding scheme employed for each user. MIMO and the modulation-and-coding scheme according to the LTE standard are selected based on delayed feedback. Retransmissions are explicitly modeled and the user throughput for each active user i will be the number of correctly received bits χ_i (above the MAC layer) divided by the simulated time T. The distribution of user throughput between users is used as a basis for assessing end-user quality. The served traffic per cell is calculated from the total number of received bits χ_i for all users, averaged over all cells and divided by the simulated time T. Statistics are collected from each simulation run and then new terminals are randomly positioned for the next sample.

18.3.2 Evaluation Criteria

ITU-R defines two requirements related to the efficiency of the radio interface for evaluating the performance of the IMT-Advanced candidate radio-interface technologies (RITs) [65]. The first is *cell spectral efficiency*, defining the operator perspective, and the second is *cell-edge spectral efficiency*, defining the end-user perspective.

The *cell spectral efficiency* is the aggregated throughput over all users, averaged over all cells and divided by the channel bandwidth. The measure relates to the system throughput in Figure 18.1 and is a measure of the maximum total "capacity" available in the system to be shared between users; it is measured in bits/s/Hz/cell.

The cell spectral efficiency η is defined as:

$$\eta = \frac{\sum_{i=1}^{N} \chi_i}{T \cdot \omega \cdot M}, \tag{18.1}$$

Table 18.4 ITU-R Requirements for IMT-Advanced Spectral Efficiency

Test Environment and Corresponding Deployment Scenario	Cell Spectral Efficiency (bit/s/Hz/cell)		Cell-Edge User Spectral Efficiency (bit/s/Hz)	
	Downlink	Uplink	Downlink	Uplink
Indoor (InH)	3	2.25	0.1	0.07
Microcellular (UMi)	2.6	1.8	0.075	0.05
Base coverage, urban (UMa)	2.2	1.4	0.06	0.03
High speed (RMa)	1.1	0.7	0.4	0.015

where χ_i denotes the number of correctly received bits for user i in a system with N users and M cells, ω is the channel bandwidth, and T is the time over which the data bits are received.

The *cell-edge user spectral efficiency* is based on the distribution between users of the *normalized user throughput* (see also Figure 18.1), which is defined as the average user throughput over a certain period of time divided by the channel bandwidth, and is measured in bit/s/Hz. The cell-edge user spectral efficiency is defined as the 5% point of the cumulative distribution function (CDF) of the normalized user throughput. It is thus a measure of the end-user perceived "quality of service" for the 5% of the users with the lowest user throughput.

The normalized user throughput for user i is defined as:

$$\gamma_i = \frac{\chi_i}{T_i \cdot \omega}, \tag{18.2}$$

where T_i is the active session time for user i.

The requirements defined by ITU-R [65] for cell spectral efficiency and cell-edge user spectral efficiency are listed in Table 18.4. The requirement levels are shown as dashed lines together with the simulated performance values in Figures 18.2 and 18.5.

18.3.3 Performance Numbers for FDD

Simulations for FDD and TDD and for all test environments were submitted for the ITU-R evaluation and are also presented here. The numbers here, however, are not identical to the numbers put forward to ITU-R in [92], but will be seen as one sample of results from the evaluation in [94].

For FDD, the cell spectral efficiency and cell-edge spectral efficiency are shown in Figure 18.2. All results exceed the ITU-R performance requirements in Table 18.4, in some cases by a very good margin. For the indoor scenario and the downlink, this is achieved using an uncorrelated antenna configuration together with single-user MIMO. For the other scenarios, a correlated antenna setup and intra-site coordinated beam-forming (CBF) with MU-MIMO are used in the downlink. A quite basic uplink configuration not utilizing MIMO is used for all environments. It should be noted that the cell-edge spectral efficiency is highly dependent on the system load and that the ITU-R simulations are performed with maximum load and a full buffer model. An example of LTE performance numbers including lower and varying loads can be found in [96].

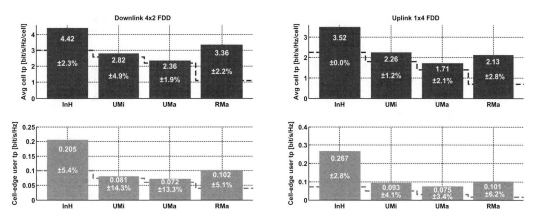

FIGURE 18.2

FDD cell spectral efficiency and cell-edge user spectral efficiency, compared with ITU-R requirements (downlink and uplink).

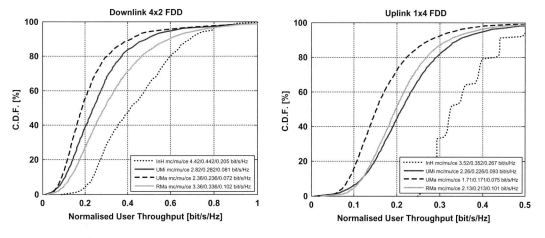

FIGURE 18.3

FDD normalized user throughput distributions (downlink and uplink).

The normalized user throughput distributions are given in Figure 18.3 and Signal-to-Noise-and-Interference Ratio (SINR) distributions are shown in Figure 18.4. The figures demonstrate the impact of the very advantageous SINR distributions achieved in the indoor hotspot scenario and to some extent also for rural macro, resulting in higher spectral efficiency numbers in those environments.

18.3.4 Performance Numbers for TDD

For TDD, the cell spectral efficiency and cell-edge spectral efficiency are shown in Figure 18.5, while normalized user throughput distributions are illustrated in Figure 18.6. There is a small drop in the

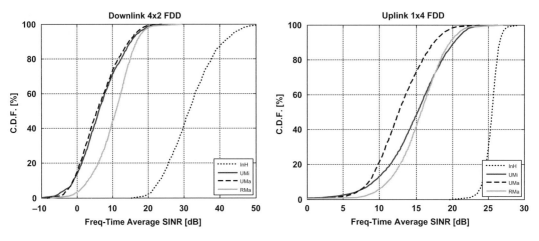

FIGURE 18.4

FDD SINR distributions (downlink and uplink).

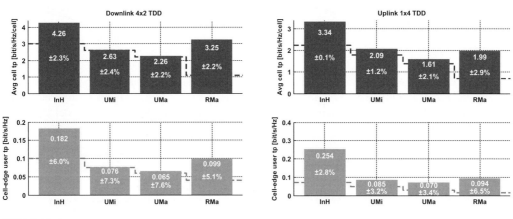

FIGURE 18.5

TDD cell spectral efficiency and cell-edge user spectral efficiency, compared with ITU-R requirements (downlink and uplink).

spectral efficiency numbers compared to FDD, but all results exceed the ITU-R performance requirements. The difference compared to FDD is due to a higher overhead from the guard period between uplink and downlink. The average delay between making measurements in the terminal and receiving the measurement result at the base station is also increased compared to FDD, due to the TDD time-domain structure. This has some impact on the performance of scheduling and link adaptation. The channel reciprocity is not utilized in the simulated TDD scheme.

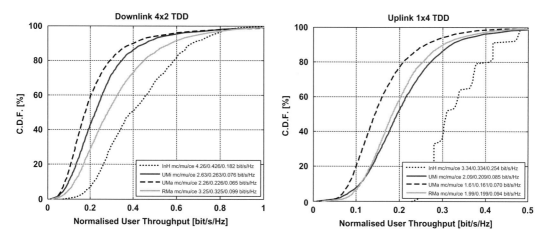

FIGURE 18.6

TDD normalized user throughput distributions (downlink and uplink).

18.4 CONCLUSION

The simulation results presented in this chapter demonstrate the high potential of LTE/LTE-Advanced in terms of both overall spectral efficiency to the benefit of operators and high cell-edge performance to the benefit of the end-user. The performance is assessed in four different test environments. The results for FDD and TDD in downlink and uplink all exceed the requirements set up by ITU-R for the evaluation of IMT-Advanced candidates. In addition, the peak radio-interface data rates and the latency achieved by LTE meet the ITU-R requirements.

Other Wireless Communications Systems

Most mobile and cellular systems go through a continuous development of the underlying technology and of the features, services, and performance supported. The direction of development is from the baseline "narrowband" systems of the 1990s supporting mainly voice services to today's "wideband" systems that target a much wider set of services, including broadband wireless data.

The fundamental technologies introduced in the systems to support new and better performing services all fall within the scope of Chapters 2–6 of this book. When the problems to solve are fundamentally the same, the solutions engineered tend to be quite similar. Wideband transmission, higher-order modulation, fast scheduling, advanced receivers, multi-carrier, OFDM, MIMO, etc. have been added as continuous developments and, at other times, as more revolutionary steps for the different technologies.

The application of these different technical solutions to LTE has been discussed in the previous chapters of this book. A similar development is occurring for the other technologies in the IMT-2000 family. This chapter gives a brief overview of HSPA and GSM, developed in 3GPP, 1xEV-DO, developed in 3GPP2, and 806.16 m/WiMAX, developed in IEEE in conjunction with WiMAX forum.

19.1 HSPA

Wideband Code-Division Multiple Access (WCDMA), also known as *Universal Terrestrial Radio Access* (UTRA), is the dominant third-generation cellular technology, developed by 3GPP with the first version of the specifications released in 1999. WCDMA is based on code-division multiple access using a chip rate of 3.84 Mchip/s, is originally designed for 5 MHz of spectrum allocation, and provides wide-area coverage of data rates[1] up to 384 kbit/s in both uplink and downlink. Circuit-switched as well as packet-switched services are supported.

High-Speed Packet Access (HSPA) is an evolution of the WCDMA radio interface, adding significantly enhanced support for packet data by applying technologies such as shared-channel transmission, channel-dependent scheduling, link adaptation, and hybrid ARQ with soft combining. The first version of HSPA appeared in release 5 of the WCDMA standard in 2002, but the specifications have

---

[1] The specifications support peak data rates of up to 2 Mbit/s in both uplink and downlink, but practical implementations typically support up to 384 kbit/s only.

been updated following a similar release scheme as the LTE specifications. Later versions of HSPA include features such as spatial multiplexing, carrier aggregation, and broadcast support. Figure 19.1 outlines the evolutionary steps of HSPA, which are also summarized below:

- Release 5 was the first release of HSPA, introducing the basic set of features (shared-channel transmission, channel-dependent scheduling, link adaptation, and hybrid ARQ with soft combining) in the downlink. Peak data rates up to 14 Mbit/s in the downlink were supported.
- Release 6 provided the basic set of features for uplink transmissions, resulting in an uplink peak data rate of 5.7 Mbit/s.
- Release 7 brought spatial multiplexing of two layers to the downlink, as well as downlink 64QAM and uplink 16QAM, to HSPA. Other enhancements in release 7 included continuous packet connectivity to provide better DRX/DTX capabilities and enhanced CELL_FACH to reduce the latency associated with state switching.
- Release 8 supports simultaneous usage of spatial multiplexing and 64QAM modulation in the downlink.
- Release 9 introduced carrier aggregation in a similar way as later done for LTE, thereby increasing the HSPA bandwidth to 10 MHz.
- Release 10 further increased the bandwidth supported by HSPA to 20 MHz by using four carriers in the downlink. The downlink peak data rate is 168 Mbit/s using roughly 20 MHz of spectrum.

As seen from the list above, the performance of HSPA has increased considerably in less than 10 years since its first incarnation in release 5. The fact that HSPA is an *evolution* and builds upon the basic WCDMA structure sets some constraints on what it is possible to introduce compared to a clean-slate design such as LTE, but also provides the possibility to gradually improve the performance of an already deployed network. This is an important aspect from an operator's perspective. The evolution is still ongoing and HSPA can, in many aspects, provide performance similar to that of LTE.

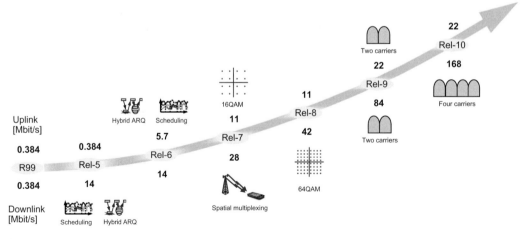

FIGURE 19.1

Evolution of HSPA.

HSPA exists in both FDD and TDD versions but, unlike LTE, the two duplex schemes are designed more or less independently of each other, resulting in significant differences in the physical-layer design. In the following, a brief overview of the most important technology components in the FDD version of HSPA is provided, as well as a short summary of the TDD version. For a more detailed description of HSPA, see [96].

19.1.1 **Architecture**

The WCDMA radio-access network architecture, upon which HSPA builds, is different from LTE, with the main difference being the existence of a *Radio Network Controller* (RNC), motivated by the support of macro diversity in WCDMA. The functional split between the RNC and the NodeB[2] is such that the NodeB handles most of the physical-layer processing, while the RNC handles the higher-layer protocols and all the radio-resource management. The WCDMA/HSPA radio-access network architecture is illustrated in Figure 19.2.

HSPA to a large extent relies on the same basic features as LTE, such as channel-dependent scheduling, hybrid ARQ with soft combining, rate adaptation, etc. These features to a large extent rely on rapid adaptation to the instantaneous radio conditions and hence need to be located in the NodeB, while the higher-layer protocols reside in the RNC. At the same time, an important design objective of HSPA was, as far as possible, to retain the original WCDMA functional split between layers and nodes. Minimization of the architectural changes is desirable as it simplifies introduction of HSPA in already deployed networks and also secures operation in environments where not all cells have been upgraded with HSPA functionality. Therefore, HSPA introduced new MAC sublayers in the NodeB, to handle scheduling, rate adaptation, and hybrid-ARQ retransmissions.

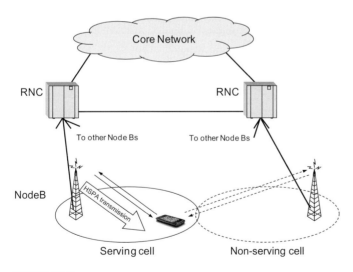

FIGURE 19.2

WCDMA/HSPA architecture.

---

[2] In WCDMA/HSPA, the term NodeB is used instead of eNodeB.

Each terminal using HSPA will receive data from one cell, the *serving cell*, which is also the cell responsible for scheduling. Unlike LTE, uplink soft handover is explicitly supported, in which case the uplink data transmission will be received in multiple cells and the terminal will receive power control commands from multiple cells.

19.1.2 Channel-Dependent Scheduling

Channel-dependent scheduling is used in HSPA for similar reasons as in LTE, namely to rapidly adapt to varying traffic and radio-channel conditions. The basic operation is similar to LTE, with scheduling decisions taken by the NodeB once per 2 ms TTI, but the difference in the uplink multiple-access scheme between HSPA and LTE has particularly affected the scheduling design.

The downlink transmission scheme in HSPA is based on reserving a certain fraction of the total downlink radio resources available within a cell, channelization codes and transmission power in the case of WCDMA/HSPA, and dynamically scheduling these resources across the users. The shared resource, illustrated in Figure 19.3, consists of a set of channelization codes of spreading factor 16. The dynamic allocation of the code resource for transmission to a specific user is done on a 2 ms TTI basis (see lower part of Figure 19.3). The reasons for a short TTI are similar as for LTE, namely to follow fast channel variations and to reduce the overall latency. From a scheduling perspective, using a set of orthogonal codes is similar to using a set of orthogonal frequency resources in LTE, and hence the downlink scheduling strategy for HSPA is similar to that of LTE. Channelization codes not used for HSPA data transmission are used for other downlink channels, for example control signaling and non-HSPA services.

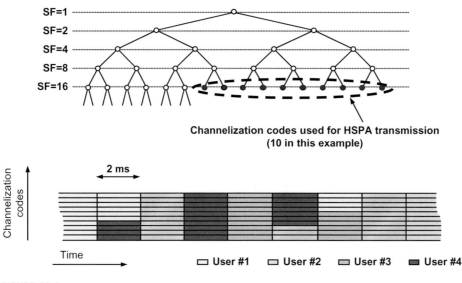

FIGURE 19.3

Shared resource in HSPA downlink – channelization codes.

The uplink transmission scheme, on the other hand, is inherently non-orthogonal and this motivates a different scheduling mechanism than the one used for the orthogonal LTE uplink. The higher the data rate a terminal is using, the higher the received power at the NodeB must be to maintain the E_b/N_0 required for successful demodulation. By increasing the transmission power, the terminal can transmit at a higher data rate. However, due to the non-orthogonal uplink, the received power from one terminal represents interference when receiving from other terminals due to the use of non-orthogonal spreading codes in the uplink. Hence, the shared resource in the uplink is basically the amount of tolerable interference in the cell. If the interference level is too high, some transmissions in the cell, most notably control channels and non-scheduled uplink transmissions, may not be received properly. On the other hand, an interference level that is too low may indicate that terminals are artificially throttled and the full system capacity is not being exploited. Therefore, the task of the uplink scheduler is to give users with data to transmit permission to use as high a data rate as possible without exceeding the maximum tolerable interference level in the cell.

The uplink scheduling framework for HSPA is based on *scheduling grants* sent by the NodeB scheduler to control the terminal transmission activity and *scheduling requests* sent by the terminals to request resources. The scheduling grants essentially control the maximum allowed transmission power the terminal may use; a larger grant implies the terminal may use a higher data rate but also contributes more to the interference level in the cell. Both multi-bit absolute grants, directly setting the grant level in the terminal to the desired level, as well as single-bit relative grants increasing/decreasing the grant level in the terminal are supported. Within the limits set by the scheduling grants, the terminals are allowed to autonomously select the data rate to use. A terminal not utilizing all its granted resources will transmit at a lower power, thereby reducing the intra-cell interference. Hence, shared resources not utilized by one mobile terminal can be exploited by another mobile terminal through statistical multiplexing.

For very high data rates, a non-orthogonal transmission scheme may result in relatively large amounts of interference from high-rate transmissions to low-rate transmissions. HSPA therefore makes it possible to also separate data transmissions in the time domain in the uplink, where terminals are allowed to transmit in certain 2 ms TTIs only. By separating high-rate and low-rate terminals in the time domain, the interference problem can be alleviated.

The fact that HSPA supports uplink macro diversity also has an impact on the scheduling framework. Although the scheduling grants are provided to the terminal by the serving cell only, there is the chance that neighboring cells may affect the grant level in the terminal. In essence, this is an overload indicator, common to multiple terminals, which a cell can set in order to ask terminals in the soft handover region to lower their data rates, thereby reducing the interference level in the non-serving cell.

Different data rates in the uplink as well as in the downlink are supported in a similar way as in LTE by supporting different modulation orders, up to 16QAM in the uplink and up to 64QAM in the downlink, in combination with rate matching to adjust the coding rate. For historical reasons, a 10 ms TTI is supported in the uplink in addition to the 2 ms TTI.

19.1.3 Hybrid ARQ with Soft Combining

Fast hybrid ARQ with soft combining in HSPA uses the same structure with multiple, parallel hybrid-ARQ processes as in LTE. Retransmissions typically take place 12 ms after the initial transmission in the downlink. For the uplink, the corresponding figure is 16 ms for a 2 ms TTI (40 ms for a 10 ms TTI).

Since the RLC is located in the RNC, while the hybrid-ARQ functionality is located in the NodeB, a separate reordering mechanism is located in the NodeB to ensure in-sequence delivery of received transport blocks to the RLC (in LTE, reordering is an integral part of the RLC).

19.1.4 Control-Plane Latency Reductions

Starting in release 7, mechanisms to reduce the overall latency were introduced in HSPA, more specifically *Continuous Packet Connectivity* (CPC) and *Enhanced CELL_FACH*, to match the bursty nature of packet-data traffic.

In WCDMA/HSPA, a terminal can be in one of four different states, URA_PCH, CELL_PCH, CELL_FACH, and CELL_DCH, ranging from low activity, low power consumption, to high activity and higher power consumption. Thus, from a delay perspective it is preferable to keep the terminal in CELL_DCH, while from an interference and power-consumption perspective, one of the paging states is preferred. RRC signaling, originating from the RNC, is used to move the terminal between the different states.

To improve the packet-data support in HSPA, a set of features known as *Continuous Packet Connectivity* (CPC) was introduced in release 7. CPC consists of three building blocks:

1. *Discontinuous transmission* (DTX), to reduce the uplink interference and thereby increase the uplink capacity, as well as to save battery power. Without DTX, the terminal is constantly transmitting on a control channel in the uplink while being in the CELL_DCH state.
2. *Discontinuous reception* (DRX), to allow the terminal to periodically switch off the receiver circuitry and save battery power.
3. *HS-SCCH-less operation*, to reduce the downlink control signaling overhead for small amounts of data, as will be the case for services such as voice-over IP.

The intention with these features is to provide an "always-on" experience for the end-user by keeping the terminal in CELL_DCH for a longer time and avoiding frequent state changes to the low-activity states, as well as improving the capacity for services such as voice-over IP (Figure 19.4).

The purpose of CPC is, as discussed above, to provide an "always-on" user experience by keeping the terminal in the active state CELL_DCH while still providing mechanisms for reduced power consumption. However, eventually the terminal will be switched to CELL_FACH if there has been no transmission activity for a certain period of time. Once the terminal is in CELL_FACH, signaling on the *Forward Access Channel* (FACH), a low-rate common downlink transport channel, is required to

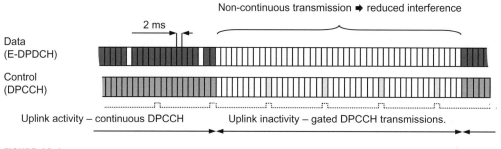

FIGURE 19.4

Simplified illustration of DTX in continuous packet connectivity.

move the terminal to CELL_DCH prior to any data exchange using the HSPA features taking place. The physical resources to which the FACH is mapped is semi-statically configured by the RNC and, to maximize the resources available for HS-DSCH and other downlink channels, the amount of resources (and thus the FACH data rate) is typically kept small, of the order of a few tens of kbit/s.

To reduce the latency associated with state changes, release 7 improved the performance by allowing HSPA to be used also in the CELL_FACH state. This is often referred to as *Enhanced CELL_FACH Operation*. Using the high-rate HSPA transmission scheme also in CELL_FACH allows for a significant reduction in the delays associated with switching to the CELL_DCH state. Instead of using a low-rate semi-statically allocated control channel, the signaling from the network to the terminal can be carried using high-rate HSPA channels. This can result in a significant reduction in call-setup delay and a corresponding improvement in user perception. In release 8, this is further improved by also supporting HSPA functionality in the uplink when being in the CELL_FACH state.

19.1.5 Spatial Multiplexing

Spatial multiplexing in HSPA supports up to two layers in the downlink. Two transport blocks, one per layer, are used in multi-stream transmission and HSPA-MIMO is thus a multi-codeword scheme (see Chapter 5 for a discussion on single- vs. multi-codeword schemes), allowing for a successive-interference-cancellation receiver in the terminal.

Similar to LTE, precoding is used prior to transmitting the two layers. Using precoding is beneficial for several reasons, especially in the case of single-stream transmission. In this case, precoding provides both *diversity gain* and *array gain* as both transmit antennas are used and the weights are selected such that the signals from the two antennas add coherently at the receiver. This results in a higher received carrier-to-interference ratio than in the absence of precoding, thus increasing the coverage for a certain data rate. Furthermore, if a separate power amplifier is used for each physical antenna, precoding ensures that both of them are also used in the case of single-layer transmissions, thereby increasing the total transmit power.

19.1.6 Carrier Aggregation

In later releases of HSPA, carrier aggregation is supported with up to four carriers in the downlink and two carriers in the uplink. A similar structure as for LTE has been adopted, where each carrier in principle is independently processed.

19.1.7 UTRA TDD

In parallel to the development of WCDMA and its evolution to HSPA, 3GPP also works on UTRA TDD. Although the higher layers are virtually identical between FDD and TDD, the physical-layer design is, for historical reasons, quite different. Furthermore, there are three versions of the TDD specifications, with the chip rate being one distinguishing factor.

Initially, TDD supported 3.84 Mchip/s only, but later 1.28 and 7.68 Mchip/s versions were added. The 3.84 and 7.68 versions share many features, while the 1.28 version, also known as *Time-Division-Synchronous CDMA* (TD-SCDMA), is substantially different from the other two. Of the three versions, TD-SCDMA is the only version deployed on a larger scale, mainly in China, with the other two versions being limited to niche applications.

TD-SCDMA was primarily developed in China as an industry standard within CCSA. In release 4, TD-SCDMA was introduced into 3GPP as an alternative to the 3.84 Mchip/s TDD version. The main differences from WCDMA/HSPA are, apart from the use of TDD instead of FDD, the lower chip rate and resulting (approximate) 1.6 MHz carrier bandwidth, the optional 8PSK higher-order modulation, a different 5 ms time-slot structure, and beam-forming support with eight antennas.

The HSPA enhancements to TD-SCDMA are similar to those applied to FDD, such as the use of 16QAM higher-order modulation and the use of hybrid ARQ. Some main differences are that, due to the TDD frame structure, timing relations are very different and a TTI of 5 ms is used. In general, UTRA TDD development is completed at a later stage in the 3GPP standardization process compared to FDD.

Some features found in TD-SCDMA originate from the work within the CCSA [99], including:

- *Multi-frequency operation.* In this mode, multiple 1.28 Mchip/s carriers are supported in one cell, but BCH is only sent on one carrier called *primary frequency*, in order to decrease inter-cell interference. The carrier on the primary frequency contains all common channels while traffic channels can be on both primary and secondary carriers. Each terminal still operates on a single 1.6 MHz carrier.
- *Beam-forming improvement.* Angle-of-arrival parameters for beam-forming can be signaled in the RACH and FACH data frames.
- *Multi-carrier HSPA.* In a cell using multi-carrier HSPA, data can be transmitted to a terminal on more than one carrier. There is a UE capability defined for receiving up to six carriers.

As discussed in Chapter 9, the LTE TDD mode has been designed to ease coexistence with and migration from TD-SCDMA.

19.2 GSM/EDGE

Worldwide GSM deployment started in 1992 and GSM quickly became the most widely deployed cellular standard in the world, with more than 2.5 billion subscribers today. A major evolutionary step for GSM was initiated in late 1996, at the same time as UTRA development started in ETSI. This step sprung out of the study of higher-order modulation for a 3G TDMA system that was performed within the European FRAMES project (see also Chapter 1).

The enhancement of GSM is called EDGE (*Enhanced Data Rates for GSM Evolution*). The EDGE development was initially focused on higher end-user data rates by introducing higher-order modulation in GSM, both for circuit-switched services and the GPRS packet-switched services. During the continued work, focus moved to the enhancement of GPRS (EGPRS), where other advanced radio-interface components were added, including link adaptation, hybrid ARQ with soft combining, and advanced scheduling techniques, as described in Chapter 6 of this book. In this way GSM became the first cellular standard to add such enhancements,[3] later followed by WCDMA/HSPA, CDMA2000, and other technologies. GSM/EDGE is, however, a more narrowband technology than WCDMA/HSPA and CDMA2000, implying that the peak data rates achievable are not as high. HSPA and CDMA2000 have,

---

[3]The first cellular technology to deploy higher-order modulation was iDEN, which uses "M-16QAM". It is, however, a proprietary technology and as such is not standardized.

however, added a time-division structure to make full use of advanced scheduling techniques for high-rate data services. Being a TDMA system, EDGE already has a time-division structure.

19.2.1 Objectives for GSM/EDGE Evolution

An evolved GSM/EDGE standard is developed in 3GPP based on a feasibility study for an "Evolved GSM/EDGE Radio Access Network (GERAN)" [100]. The objective is to improve service performance for GERAN for a range of services, including interactive best-effort services as well as conversational services including *voice-over IP*.

The Evolved GERAN is compatible with the existing GSM/EDGE in terms of frequency planning and coexistence with legacy terminals. It also reuses the existing network architecture (Figure 19.5) and has minimum impact on the BTS, BSC, and core network hardware. For the standardization of Evolved GERAN, a number of performance targets were set [100]:

- *Improved spectrum efficiency.* The target is a 50% increase in an interference-limited scenario (measured in kbit/s/MHz/cell for data or Erlang/MHz/cell for voice).
- *Increased peak data rates.* The target is a 100% increase in both downlink and uplink.
- *Improved coverage for voice and data.* The target is a sensitivity increase of 3 dB in downlink (noise-limited scenario).
- *Improved service availability.* The target is a 50% increase in mean data rate for uplink and downlink at cell edges (when cells are planned for voice).
- *Reduced latency.* The target is a *round-trip time* (RTT) of less than 450 ms for initial access and less than 100 ms after initial access (in non-ideal radio conditions, counting from the terminal to the GGSN and back, as shown in Figure 19.5).

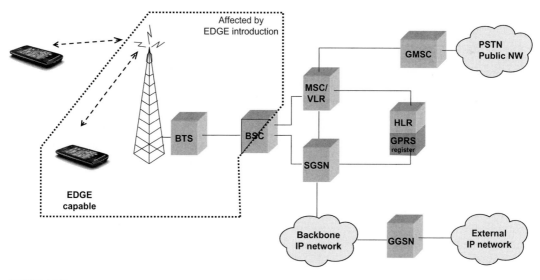

FIGURE 19.5

GSM/EDGE network structure.

A number of alternative solutions were studied in 3GPP to achieve the performance and compatibility targets listed above. These are not fundamentally different from the solutions selected for LTE or other technologies and are all described in a more general context in Chapters 2–6 of this book. The technologies chosen for standardization in 3GPP for GERAN Evolution were:

- Dual-antenna terminals
- Multi-carrier EDGE
- Reduced TTI and fast feedback
- Improved modulation and coding
- Higher symbol rates.

The specific application of these technologies to GERAN is discussed below. The solutions are included in release 7 of the GSM/EDGE specifications.

19.2.2 Dual-Antenna Terminals

As was shown in Chapter 5, multiple receiver antennas are an effective means against multipath fading and to provide an improved signal-to-noise ratio through "energy gain" when combining the antenna signals. There will thus be improvements for both interference-limited and noise-limited scenarios. There are also possibilities for interference cancellation through multiple antennas. While multiple antennas have an implementation impact for the terminal, there is no impact on the base-station hardware or software.

For GSM/EDGE, a dual-antenna solution called *Mobile Station Receive Diversity* (MSRD) is standardized. Analysis shows that a dual-antenna solution in GSM terminals can give a substantial coverage improvement of up to 6 dB. In addition, the dual-antenna terminals could potentially handle almost 10 dB more interference [79,101].

19.2.3 Multi-Carrier EDGE

The GSM radio interface is based on a TDMA structure with eight time slots and 200 kHz carrier bandwidth. To increase data rates, today's GPRS and EDGE terminals can use multiple time slots for transmission, reception, or both. Using 8PSK modulation (Figure 19.6) and assigning the maximum of eight time slots, the GSM/EDGE standard gives a theoretical peak data rate of close to 480 kbit/s. However, from a design and complexity point of view, it is best to avoid simultaneous transmission and reception. Today's terminals typically receive on a maximum of five time slots because they must also transmit (on at least one time slot) as well as measure the signal strength of neighboring cells.

To increase data rates further, multiple carriers for the downlink and the uplink can be introduced, similar to what is discussed in Chapter 2. This straightforward enhancement increases peak and mean data rates in proportion to the number of carriers employed. For example, given two carriers with eight time slots each, the peak data rate will be close to 1 Mbit/s. Using dual carriers can be seen as a straightforward extension of the multi-slot principle, allowing a multi-slot configuration to span more than one carrier. The limiting factor in this case is the complexity and cost of the terminal, which must have either multiple transmitters and receivers or a wideband transmitter and receiver. The use of multiple carriers has only a minor impact on base transceiver stations. A dual-carrier downlink solution is standardized in release 7 of the GSM/EDGE standard.

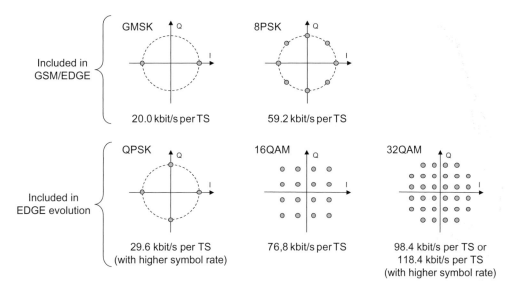

Included in
GSM/EDGE

GMSK

20.0 kbit/s per TS

8PSK

59.2 kbit/s per TS

Included in
EDGE evolution

QPSK

29.6 kbit/s per TS
(with higher symbol rate)

16QAM

76,8 kbit/s per TS

32QAM

98.4 kbit/s per TS or
118.4 kbit/s per TS
(with higher symbol rate)

FIGURE 19.6

Existing and new modulation schemes for GSM/EDGE. The highest specified radio-interface data rate using GPRS is shown for each scheme [101]. Note that the view of the nonlinear binary GMSK scheme is simplified in the figure.

19.2.4 Reduced TTI and Fast Feedback

Latency is usually defined as the round-trip time over the radio-access network. It has a major influence on user experience, and conversational services such as voice and video telephony especially require low latency, but services such as web browsing and e-mail download are also improved considerably. A major parameter in the radio interface that has an impact on latency is the *Transmission Time Interval* (TTI), as discussed for LTE in Chapter 8. It is difficult to substantially improve latency without reducing the TTI.

The *round-trip time* (RTT) in GSM/EDGE networks can be 150 ms [101], including network delays but not retransmissions over the radio interface. Radio blocks are transmitted interleaved over four consecutive bursts on one assigned time slot over 20 ms. In GSM/EDGE evolution, a reduced TTI with radio blocks interleaved over four consecutive bursts assigned on *two time slots* in 10 ms is introduced. This reduction of the TTI from 20 to 10 ms can reduce the round-trip time from 150 to less than 100 ms [101]. Multiplexing of users with the two different interleaving schemes is also possible on the same RF carrier.

Faster responses to incorrectly received radio blocks speed up the retransmission of radio blocks and can also help in reducing the latency. In GSM/EDGE evolution, a fast ACK/NACK reporting mechanism is introduced to reduce time for the network to realize that a block is lost. ACK/NACK reports can also "piggy-back" on user data, which reduces the overhead. Combined with the shorter TTI, the total time to retransmit lost radio blocks is thus reduced significantly and the throughput is increased.

19.2.5 Improved Modulation and Coding

The main evolutionary step taken for GSM when EDGE was introduced was the higher-order modulation to enhance the data rates over the radio interface – Enhanced Data Rates for GSM Evolution. 16QAM was considered for EDGE, but 8PSK was finally chosen as a smooth evolutionary step that gives three bits per modulated symbol instead of one. This increases the peak data rate from approximately 20 to 60 kbit/s per time slot.

Figure 19.6 shows the further steps that are standardized for Evolved EDGE, giving 4 bits/symbol for 16QAM and 5 bits/symbol for 32QAM. Since the signal points in the new schemes move more closely together, they are more susceptible to interference. With more bits per symbol, however, the higher data rates allow for more robust channel coding, which can more than compensate for the increased susceptibility to interference. QPSK modulation with 2 bits/symbol is also included in Evolved GSM/EDGE, but used only with the higher symbol rate described below.

Both GSM and EDGE use convolutional codes, while Turbo codes are also standardized for Evolved EDGE. Decoding Turbo codes is more complex than regular convolutional codes. However, since Turbo codes are already used today for WCDMA and HSPA, and many terminals support both GSM/EDGE and WCDMA/HSPA, the decoding circuitry for Turbo codes already exists in the terminals and can be reused for EDGE.

Turbo codes perform well for large code-block sizes, making them better suited for higher data rate EDGE channels using 8PSK or higher-order modulation. It is estimated that, compared to the existing EDGE scheme with 8PSK modulation, the combination of Turbo codes and 16QAM improves the user data rates 30–40% for the median user in a system [100]. Turbo codes are standardized for the downlink in release 7 of GSM/EDGE.

19.2.6 Higher Symbol Rates

The modulation can also be improved by simply increasing the modulation symbol rate. As a part of GERAN Evolution in 3GPP, the combination of higher-order modulation and a 20% increase in symbol rate has been standardized [102]. For both uplink and downlink, two "levels" of terminal capabilities are defined, called Levels A and B. Each level defines a set of modulation schemes, where Level B also applies the higher symbol rate for some modulation schemes. Table 19.1 shows the combinations that are defined.

The higher symbol rate of 325 ksymbol/s will operate with the same nominal carrier bandwidth and carrier raster as legacy GERAN having 271 ksymbols/s, which puts some requirements on the

Table 19.1 Combinations of Modulation Schemes and Symbol Rates in GSM/EDGE Evolution

Terminal Capability	Symbol Rate (ksymbols/s)	Modulation Schemes
Uplink Level A	271	GMSK, 8PSK, 16QAM
Uplink Level B	271	GMSK
	325	QPSK, 16QAM, 32QAM
Downlink Level A	271	GMSK, 8PSK, 16QAM, 32QAM
Downlink Level B	271	GMSK
	325	QPSK, 16QAM, 32QAM

transmitter filter. The amount of transmitter filtering has to be weighed against the receiver complexity and performance, and the amount of interference in adjacent channels. An alternative wider Tx filter has been standardized for the uplink. A wider Tx filter is also studied for the downlink.

19.2.7 **Voice Service over Adaptive Multi-User Channels**

The focus of GSM/EDGE evolution has been improvements of the data services through higher spectral efficiency and increased peak data rates. At the same time, voice is still a predominant service for many operators and an improvement of voice performance is also desirable. Earlier improvements of voice capacity have mainly been in the voice coding area, where half-rate voice coding was a major step taken in earlier releases. There are eight time slots in the GSM TDMA frame structure, with one full-rate or two half-rate users occupying each slot.

For release 9 of the GERAN specification, multi-user channels on each time slot were investigated and a new mode called *Voice services over Adaptive Multi-user channels* (VAMOS) was introduced [103]. In VAMOS mode, *two* users are multiplexed on the same physical resource – that is, on the same frequency and TDMA time slot. The two users form a *VAMOS pair*, each user being mapped on one *VAMOS subchannel*. In the uplink, each subchannel is modulated with GSMK. In the downlink, a new *Adaptive Quadrature Phase-Shift Keying* (AQPSK) modulation scheme is used jointly for the two subchannels. The two subchannels in a VAMOS pair have different training sequences to facilitate joint demodulation of the two subchannels.

The AQPSK modulation scheme shown in Figure 19.7 is used in the downlink when both subchannels in a VAMOS pair have a burst scheduled for transmission. The bits from one subchannel in the VAMOS pair are mapped on the first bit that forms the QPSK constellation (corresponding to I) and the bits from the second user subchannel are mapped on the second bit (corresponding to Q). Depending on the choice of the parameter α, the power ratio between the two subchannels mapped on I and Q can be controlled, thereby providing partially independent power control of the two subchannels.

Use of the VAMOS mode was studied in detail by GERAN [103], with the conclusion that there is no harmful degradation of voice services for legacy users.

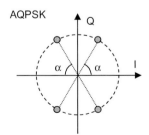

FIGURE 19.7

The AQPSK modulation scheme used in the downlink
for two users forming a VAMOS pair.

19.3 CDMA2000 AND HRPD/1x EV-DO

CDMA2000 evolved as a cellular standard under the name IS-95 and later became a part of the IMT-2000 family of technologies described in Chapter 1. When it became a more global IMT-2000 technology, the name was changed to CDMA2000 and the specification work was moved from the *US Telecommunications Industry Association* (TIA) to 3GPP2. Being a sister organization to 3GPP, 3GPP2 is responsible for CDMA2000 specifications.

The CDMA2000 standard is going through an evolution similar to that of WCDMA/HSPA. In the different evolutionary steps, the same gradual shift of focus from voice and circuit-switched data to best-effort data and broadband data can be seen as for WCDMA/HSPA. The basic principles used are also very similar to those found in HSPA.

The evolutionary steps of CDMA2000 are shown in Figure 19.8. After the CDMA2000 1x standard was formed as an input to ITU for IMT-2000, two parallel evolutionary tracks were initiated for better support of data services. The first one was EV-DO (*Evolution-Data Only*),[4] which has continued to be the main track, as further described below. It is also called HRPD (*High-Rate Packet Data*). A parallel track was EV-DV (*Evolution for integrated Data and Voice*), developed to give parallel support of data and circuit-switched services on the same carrier. It is not used at the moment and has not been developed further within 3GPP2.

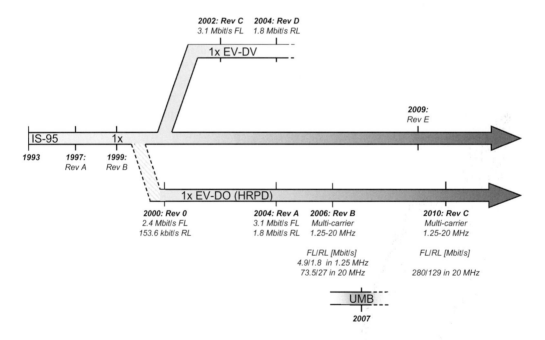

FIGURE 19.8

The evolution from IS-95 to CDMA2000 1x and 1x EV-DO.

---

[4]The abbreviation "DO" is also interpreted as "Data Optimized", since the EV-DO carrier targets data services.

The figure also illustrates UMB (*Ultra Mobile Broadband*), an OFDM-based standard including support for multi-antenna transmission and channel bandwidths up to 20 MHz. It is similar to LTE in terms of the technologies and features used, with one main difference that UMB uses OFDM in the uplink while LTE uses single-carrier modulation. UMB is not backwards compatible with CDMA2000. At the moment UMB is not used, and has not been developed further within 3GPP2, but some features from UMB, most notably the OFDM-based multi-antenna schemes, have subsequently been the basis for the corresponding features in EV-DO Rev C.

19.3.1 CDMA2000 1x

The CDMA2000 standard, which originally supported both a single-carrier (1x) and a multi-carrier (3x) mode, was adopted by ITU-R under the name IMT-2000 CDMA *Multi-Carrier* (MC) [2]. It offers several improvements over the earlier versions of IS-95 that give better spectral efficiency and higher data rates. The most important aspect from a 3G evolution perspective, however, is that 1x has been a platform for further evolution of packet-data modes, as shown in Figure 19.8. While 1x EV-DV is not undergoing any new development, 1x EV-DO is deployed today and is going through several new evolutionary steps in Rev 0, Rev A, Rev B, and Rev C that are described further below. While all steps up to Rev A are fundamentally based on direct sequence spread spectrum and a 1.25 MHz carrier bandwidth, Rev B and Rev C diverge from this by including wider carrier bandwidths and for Rev C also OFDM operation.

The 3x mode of CDMA2000 was never deployed in its original form, but was still an essential component of the submission of CDMA2000 to the ITU-R. However, multi-carrier CDMA is today once again included in the CDMA2000 evolution through EV-DO Rev B.

The 1x version of CDMA2000 (that is, not the EV-DO/HRPD track in Figure 19.8) has evolved into Rev E (the letters C and D were used for the EV-DV versions). Compared to Rev B, several small enhancements have been introduced in order to increase the voice capacity. Receiver requirements have also been defined to facilitate terminals simultaneously receiving circuit-switched voice call on one CDMA2000 carrier and packet data on EV-DO carriers.

19.3.2 1x EV-DO Rev 0

EV-DO Rev 0 defines a new uplink and downlink structure for CDMA2000 1x, where DO originally implied "*Data Only*". The reason is that an EV-DO carrier has a structure optimized for data that does not support the voice and circuit-switched data services of a CDMA2000 1x carrier. In this way, the whole carrier works as a shared downlink resource for data transmission. An operator would deploy an additional carrier for EV-DO, thereby separating circuit-switched and packet-switched connections on different carriers. A drawback is reduced flexibility in that there cannot be a simultaneous packet-data and legacy circuit-switched service (such as voice) to the same user on one carrier. 1x EV-DO has later also been named HRPD (*High-Rate Packet Data*).

With EV-DO Rev 0, a peak data rate of 2.4 Mbit/s is supported in the downlink on a 1.25 MHz carrier. There are several components in EV-DO Rev 0, several of which have similarities with the HSPA evolution described in Section 19.1:

- *Shared-channel transmission.* EV-DO Rev 0 has a *Time-Division Multiplexing* (TDM) downlink, with transmission to only one user at the time with the full power of the Base Station (BS). This makes the downlink a resource that is shared between the users in the time domain only. This is similar to the "shared-channel transmission" for HSPA, with the difference that HSPA can also share the downlink in the code domain, primarily with non-HSPA users.

- *Channel-dependent scheduling.* The adaptive data scheduler of EV-DO Rev 0 takes into account fairness, queue sizes, and measures of the channel state. It thereby exploits multi-user diversity in fading channels in a way similar to the channel-dependent scheduling used in HSPA.
- *Short TTI.* The TTI is reduced from 20 ms in CDMA2000 to 1.67 ms in EV-DO Rev 0. This is important to enable fast channel-dependent scheduling and rapid retransmissions, also resulting in lower latency. The TTI for EV-DO Rev 0 is thereby of the same order as the 2 ms TTI used for HSPA.
- *Rate control.* EV-DO Rev 0 employs rate control through adaptive modulation and coding, thereby maximizing the throughput for a given channel condition. This is similar to HSPA, but an EV-DO base station follows the rate request from the mobile, while in HSPA the feedback is a recommendation and the NodeB takes the final decision.
- *Higher-order modulation.* EV-DO Rev 0 supports 16QAM modulation in the downlink, which is similar to the first HSPA release.
- *Hybrid ARQ.* The EV-DO Hybrid-ARQ scheme is similar to the scheme used for HSPA.
- *Virtual SOHO.* EV-DO Rev 0 does not use soft handover in the downlink like CDMA2000. Instead, the terminal supports "virtual soft handover" via adaptive server selection initiated by the terminal, which can be seen as fast cell selection within the "active set" of base stations. These server changes may result in some packet transmission delays [104].
- *Receive diversity in the mobile.* EV-DO Rev 0 has terminal performance numbers specified assuming receive diversity, similar to performance requirements specified for HSPA advanced receivers.

19.3.3 1x EV-DO Rev A

The next step in the evolution of CDMA2000 is 1x EV-DO Rev A. The focus is on an uplink improvement similar to the enhanced uplink of HSPA, but it also includes an updated downlink, a more advanced quality-of-service handling, and an add-on multicast mode [104].

The downlink of EV-DO Rev A is based on the EV-DO Rev 0 downlink, with the following differences:

- *Higher peak rates.* EV-DO Rev A downlink supports 3.1 Mbit/s as compared to the 2.4 Mbit/s of EV-DO Rev 0. Rev A also offers a finer quantization of data rates.
- *Shorter packets.* New transmission formats for EV-DO Rev A enable 128-, 256-, and 512-bit packets. This, together with new multi-user packets, where data to multiple terminals share the same packet in the downlink, improves support for lower-rate, delay-sensitive services.

The major enhancements in EV-DO Rev A compared to Rev 0 are in the uplink. This results in a more packet-oriented uplink with higher capacity and data rates. Peak uplink data rates of up to 1.8 Mbit/s are supported:

- *Higher-order modulation.* In addition to BPSK modulation in EV-DO Rev 0, the uplink physical layer of Rev A supports QPSK and optionally 8PSK modulation.
- *Hybrid ARQ.* Improved performance is achieved through an uplink hybrid ARQ scheme, similar to HSPA.
- *Reduced latency.* The use of smaller packet sizes and a shorter TTI enables a reduced latency of up to 50% compared to EV-DO Rev 0 [105].

- *Capacity/latency trade-off.* Each packet can be transmitted in one of two possible transmission modes: *LoLat* gives low latency through a higher power level, ensuring that the packet is received within the latency target, while *HiCap* gives higher total capacity by allowing for more retransmissions and lower transmit power levels. A similar effect can be obtained in HSPA through different HARQ profiles [96].

19.3.4 1x EV-DO Rev B

The next step for the 1x EV-DO series of standards is Rev B, which enables higher data rates by aggregation of multiple carriers. This fundamental way of increasing the data rate is also discussed in Chapter 2. Rev B permits up to sixteen 1.25 MHz carriers to be aggregated, forming a 20 MHz wide system [106], and giving a theoretical peak data rate of up to 46.5 Mbit/s in the downlink. For reasons of cost, size, and battery life, Rev B devices will most likely support up to three carriers [105], giving a peak downlink data rate of 9.3 Mbit/s.

The lower layers of the radio interface are similar to and compatible with those in Rev A, making it possible for both single-carrier Rev 0 and Rev A terminals to function on a Rev B network that supports multi-carrier operation.

Carriers do not have to be symmetrically allocated in the uplink and the downlink. For asymmetric applications such as file downloads, a larger number of carriers can be set up for downlink than for uplink. This reduces the amount of overhead for the uplink transmission. One uplink channel can carry feedback information for multiple forward link channels operating on multiple downlink carriers to a single terminal in asymmetric mode.

19.3.5 1x EV-DO Rev C

The latest step in the evolution of 1x EV-DO is Rev C, which introduces both smaller software-based enhancements as well as larger enhancements requiring hardware modifications.

The major new features in Rev C are inclusion of OFDM and support for spatial multiplexing. OFDM is supported in the forward link only. A subcarrier spacing of 6.8 kHz is used, resulting in an OFDM symbol duration of 162.76 µs including a cyclic prefix of 16.28 µs. OFDM- and CDMA-based transmissions are time multiplexed to support Rev C and legacy terminals on the same carrier.

Spatial multiplexing supports up to four transmission layers in the forward link. Forward-link spatial multiplexing is only supported in conjunction with OFDM. In the reverse link, which only supports CDMA-based transmission, up to two layers are supported together with codebook-based precoding under control of the base station. Higher-order modulation in the reverse link has been increased to 64QAM.

19.4 IEEE 802.16E, MOBILE WiMAX AND 802.16 m

The IEEE 802.16 family of radio-access technologies is developed by IEEE, initially targeting fixed wireless access applications but later being extended to also support mobility, thus providing a radio-access technology that can be used to offer mobile-data/broadband services.

The initial 802.16 specification was designed for line-of-sight operation in the 10–60 GHz frequency range. The next revision, 802.16a, introduced support for non-line-of-sight operation with focus on the 2–11 GHz frequency range, but still targeting the fixed-wireless-access application. Support for mobility was introduced as part of IEEE 802.16e, finalized in December 2005. 802.16e is the basis for all currently commercial mobile-data networks based on the 802.16 technology. Part of 802.16e has also, at a relatively late stage (2007), been approved by the ITU as an additional IMT-2000 technology in parallel with WCDMA/HSPA, TD-SCDMA, and CDMA2000/1xEV-DO. The 802.16e-based IMT-2000 technology is only supporting TDD operation and is, in the ITU, referred to as *IMT-2000 OFDMA TDD WAN*.

The latest step taken within the IEEE 802.16 community is the development of the 16 m version of the specification (IEEE 802.16 m), a key goal of which was to ensure compliance with the ITU requirements on IMT-Advanced [65]. In October 2010, 802.16 m was approved by the ITU as a second IMT-Advanced-compliant technology under the name *WirelessMAN-Advanced*, in parallel with LTE/LTE-Advanced.

The IEEE 802.16 specifications cover only the physical and MAC layers of the overall protocol stack, with no definition of higher layers. Furthermore, the 802.16 specifications contain an extremely large set of different features, often serving essentially the same purpose but doing it in different ways, including even multiple alternatives for the basic physical-layer transmission scheme. Including all these options and alternative features in a mobile terminal is not feasible. Thus, it is generally accepted that the 802.16 specifications in themselves do not provide a complete and implementable specification for a mobile-communication system.

The *WiMAX forum* is an industry-led, non-profit cooperation formed to promote and certify compatibility and interoperability of 802.16-based products. As part of this, the WiMAX forum selects the set of features, from the full set of features specified by the IEEE, that should be included in any equipment that conforms to the different *WiMAX System Profiles*. The first such profile, System Profile Release 1.0, was published in 2007, with the second profile, System Profile Release 1.5, finalized in 2009. System Profile Release 2.0 corresponds to the 802.16 m specifications currently under development.

The system profiles for 802.16e developed by the WiMAX forum are also, jointly, often referred to as Mobile WiMAX or just WiMAX.

19.4.1 IEEE 802.16e and Mobile WiMAX

Although IEEE 802.16 includes several alternatives for the basic physical-layer transmission scheme, including both OFDM and single-carrier transmission, Mobile WiMAX is based on OFDM transmission supporting different bandwidths for both uplink and downlink. In contrast to LTE, which is based on a single numerology for all transmission bandwidths, different bandwidths of 802.16e may correspond to partly different numerology, including different subcarrier spacing. However, the main focus of Mobile WiMAX has been on the 5 and 10 MHz bandwidths,[5] both corresponding to a subcarrier spacing $\Delta f \approx 10.94$ kHz and only differing in the number of subcarriers within the transmission bandwidth. Note that, although wider bandwidths are supported by the 802.16e specifications, 10 MHz is the maximum transmission bandwidth supported by the current Mobile WiMAX profiles.

802.16e includes support for both TDD and FDD operation, including the possibility for half-duplex FDD (see Chapter 9 for a discussion on half-duplex FDD in the context of LTE). However, WiMAX has mainly focused on the TDD mode of operation. In particular, TDD was the only duplex operation supported by the first WiMAX release (1.0). Also, as already mentioned, the WiMAX/802.16e-based

---

[5]There are additional bandwidths supported by Mobile WiMAX, more specifically 3.5, 7.5, and 8.75 MHz.

IMT-2000 standard only supports TDD operation. This overview will therefore focus on the TDD mode of operation for Mobile WiMAX.

As illustrated in Figure 19.9, in the time domain, a 5 ms frame[6] is divided into two parts, a downlink part and an uplink part, together consisting of 48 OFDM symbols.[7] With one symbol used as preamble, there is a maximum of 47 symbols available for downlink and uplink transmission. These symbols can be assigned for downlink and uplink transmission respectively, by means of a number of defined downlink/uplink configurations, ranging from 32:12 (35 downlink symbols and 12 uplink symbols) to 26:21 (26 downlink symbols and 21 uplink symbols). This only leaves a very small time available for downlink/uplink switching. If additional switching time is needed, for example to handle large cells, one can simply discard a number of symbols at the end of the downlink part of the frame.

Downlink control signaling for scheduling and hybrid-ARQ operation is transmitted at the beginning of the downlink part. Hence, unlike LTE, where such control signaling can be provided in each downlink subframe, in 802.16e such control signaling can be provided once per 5 ms frame only. Hybrid ARQ with soft combining is also supported both in 802.16 and by Mobile WiMAX. However, in contrast to HSPA and LTE, the first Mobile WiMAX releases only support Chase combining and not incremental redundancy.

Similar to LTE, 802.16e and Mobile WiMAX supports QPSK, 16QAM, and 64QAM[8] data modulation, providing spectral efficiency up to 6 bit/s/symbol. Also similar to LTE, adaptive modulation and coding ("link adaptation") can be used to adjust the modulation scheme and coding rate, and thus the data rate, to match the instantaneous channel conditions. The 802.16e specifications support a range of channel coding schemes, including both Turbo codes, similar to HSPA and LTE, and so-called LDPC codes [53]. However, LDPC codes are not supported in current Mobile WiMAX releases.

802.16e allows for mapping of data to the basic OFDM time–frequency grid in different ways to achieve diversity either by means of resource allocations that spread the coded information in the frequency domain, or optimization for channel-dependent scheduling and link adaptation by concentrating the transmission in the frequency domain. The different resource assignment schemes can be used

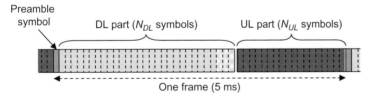

FIGURE 19.9

Basic frame structure for IEEE802.16e/Mobile WiMAX ($N_{DL}:N_{UL} = 29:18$).

---

[6]Additional frame lengths are specified in 802.16e. However, 5 ms is the only frame length included in the Mobile WiMAX profiles.
[7]This assumes a cyclic-prefix length of approximately 11.4 μs. The 802.16e specifications include both larger and smaller cyclic-prefix lengths, corresponding to somewhat less/more OFDM symbols within a 5 ms frame. However, these cyclic-prefix lengths are not part of the Mobile WiMAX system profiles.
[8]64QAM for the uplink is not supported/mandatory in the first Mobile WiMAX releases.

within the same subframe by dividing the subframe into different time-domain regions or "zones". Uplink channel-dependent scheduling was not part of the first release of Mobile WiMAX but was added in later releases.

A wide range of multi-antenna transmission schemes are included in 802.16e. Already since the first release (1.0), Mobile WiMAX has included a subset of these, more specifically open-loop spatial multiplexing. Later releases have introduced additional multi-antenna transmission techniques for both downlink and uplink.

19.4.2 IEEE 802.16 m – WiMAX for IMT-Advanced

The work on 802.16m was initiated with the aim to extend the family of 802.16 radio-access technologies and ensure fulfillment of the IMT-Advanced requirements defined by the ITU.

In contrast to LTE-Advanced being an evolution of LTE release 8/9, 802.16m is not a direct evolution of 802.16e, adding new complementary features to extend the performance and capabilities of the radio-access technology. Rather, 802.16m is, in many respects, a new radio-access technology although retaining several of the basic characteristics of 802.16e, including the basic OFDM numerology. This also implies that 802.16e and 802.16m can coexist on the same carrier by means of time multiplexing the two radio-access technologies within the 802.16e 5ms frame structure. In contrast, earlier releases of LTE are included as integrated parts of LTE release 10, providing a more straightforward backwards compatibility and support of legacy terminals, for example.

Technology wise, 802.16m introduces many features similar to LTE release 10, including the use of multi-carrier transmission (carrier aggregation) for bandwidths beyond 20MHz and support for relaying functionality.

802.16m also introduces substantially shorter subframes of length roughly 0.6ms to reduce hybrid-ARQ round-trip time and, in general, allow for reduced latency over the radio interface.

Regarding the resource assignment, 802.16m does not support the resource mapping schemes specified for 802.16e.[9] Rather, 802.16m introduces *physical resource units* consisting of a number of frequency-contiguous subcarriers during one subframe, very similar to the LTE resource blocks. Actually, the number of subcarriers in a resource unit equals 18 which, together with the subcarrier spacing of 10.94kHz, results in a resource unit with a bandwidth that is very similar to the LTE resource-block bandwidth of 180kHz. Above the physical resource units are *logical resource units* that can be distributed or localized, where distributed resource units are used for cases when frequency diversity is used. Once again, this is very similar to LTE localized and distributed resource mapping, as described in Chapter 10.

Taking into account the many similarities between LTE and 802.16m, it is not surprising that performance evaluations indicate similar performance of the two radio-interface technologies. Thus, similar to LTE, 802.16m also fulfils all the requirements for IMT-Advanced as defined by the ITU.

As already mentioned, 802.16m was approved by the ITU in October 2010, as a second IMT-Advanced-compliant technology under the name WirelessMAN-Advanced. Thus, similar to LTE, IEEE 802.16m can also be seen as a fully compliant 4G technology.

---

[9]They can, of course, be supported on a 802.16m carrier as part of 802.16e/802.16m time multiplexing.

19.5 SUMMARY

The IMT-2000 technologies and the other technologies introduced above are developed in different standardization bodies, but all show many commonalities. The reason is that they target the same type of application and operate under similar conditions and scenarios. The fundamental constraints for achieving high data rates, good quality of service, and system performance will require that a set of efficient tools are applied to reach those targets.

The technologies described in Chapters 2–6 of this book form a set of such tools and it turns out that many of these are applied across most of the technologies discussed here. To cater for high data rates, different ways to transmit over wider bandwidth is employed, such as single- and multi-carrier transmission including OFDM, often with the addition of higher-order modulation. Multi-antenna techniques are employed to exploit the "spatial domain", using techniques such as receive and transmit diversity, beam-forming, and multi-layer transmission. Most of the schemes also employ dynamic link adaptation and scheduling to exploit variations in channel conditions. In addition, coding schemes such as Turbo codes are combined with advanced retransmission schemes such as hybrid ARQ.

As mentioned above, one reason that solutions become similar between systems is that they target similar problems for the systems. It is also true to some extent that some technologies and their corresponding abbreviations go in and out of "fashion". Most 2G systems were developed using TDMA, while many 3G systems are based on CDMA and the evolution towards 4G is based on OFDM. Another reason for this stepwise shift of technologies is of course that as technology develops, more complex implementations are made possible. A closer look at many of the evolved wireless communication systems of today also shows that they often combine multiple techniques from previous steps, and are built on a mix of TDMA, OFDM, and spread spectrum components.

Final Thoughts

The focus of this book has been on the LTE radio-access technology, from the first version in release 8, via release 9 to release 10. LTE release 10 is the latest version of the LTE radio-access specifications, sometimes also referred to as *LTE-Advanced* and approved by the ITU as the first *IMT-Advanced* technology.

The basic technologies described in Chapters 2–6 served as a foundation for the detailed discussion and explanation of the LTE radio-access technology in the subsequent chapters, providing information about the LTE radio-network architecture, the basic transmission structure, and all the protocols and procedures needed for a complete radio-access network.

The capabilities and system performance of LTE were illustrated in Chapter 18 and it was shown that all the requirements defined by the ITU for IMT-Advanced were fulfilled, in many cases by a significant margin.

At the same time as ITU-R is completing the work on the formal IMT-Advanced recommendations, activities are initiated to look at what will come beyond IMT-Advanced in the 2012–2020 time frame. Several regional workshops on *"IMT for the 2010–2020 Decade"* are planned for 2011 and a new report looking at global broadband wireless services and marketplace for IMT has been initiated within ITU-R WP5D. The basis for the continued work is that future mobile broadband consumers will expect similar user experience and access to similar services and applications in the mobile environment as for fixed broadband, with the added convenience of mobility. Studies within ITU-R will include long-term market forecasts and take into account new user behavior, specifically considering personal computers and smart phones.

20.1 WHERE TO GO IN THE FUTURE?

The evolution of LTE does not end with LTE release 10. Rather LTE will continue to evolve into release 11, release 12, and so on, with each new release bringing additional capabilities and further enhanced system performance into the LTE radio-access technology. Not only will the additional capabilities provide better performance in existing applications, they may also open up for, or even be motivated by, new application areas. Examples hereof are home automation, smart transportation, security, and e-books, but the list is continuously growing as additional applications benefiting from mobile connectivity are emerging. Similarly, new ways of deploying cellular networks, for example more extensive use of massive beam-forming or ubiquitous access to optical fibers in the backhaul, may call for enhancements in the radio interface.

4G LTE/LTE-Advanced for Mobile Broadband.
© 2011 Erik Dahlman, Stefan Parkvall & Johan Sköld. Published by Elsevier Ltd. All rights reserved.

Initial 3GPP discussions on release 11 have already started, with actual work expected to be initiated in early 2011, aiming for completion of this release in mid-2012. In the following, some possible areas for the future evolution of LTE will be briefly discussed. Some of these areas may happen in release 11 or 12, while others may be more futuristic.

20.1.1 Advanced Multi-Cell Coordination

Coordinated Multi-Point transmission and reception (CoMP) refers to a wide range of different techniques with the common denominator being the dynamic coordination of transmission and/or reception at multiple geographically separated sites with the aim to enhance system performance and end-user service quality. Many different coordination schemes of very different characteristics fall under the joint umbrella of CoMP, ranging from dynamic inter-cell scheduling coordination to joint transmission/reception at multiple sites. In the former case, CoMP can, to a large extent, be seen as an extension of the inter-cell interference coordination that is already part of LTE, discussed in Chapter 13.

Joint reception means that the signals received at multiple sites are jointly processed for enhanced reception performance. Maximum-ratio combining and interference-rejection combining are examples of schemes that can be used to combine the uplink transmission received at multiple points. This is, in many respects, similar to softer handover used within a site, for example in WCDMA/HSPA-based systems, but extended to multiple sites.

Joint transmission implies that data is transmitted to a mobile terminal jointly from several sites, thereby not only reducing the interference but also increasing the received power. The transmission from the sites can also take the instantaneous channel conditions at the different terminals into account to enhance the received signal strength, while at the same time reducing the interference between different transmissions.

In general, both joint reception and transmission pose high requirements on low latency in the communication between the network node involved in the joint processing and the different antennas involved in the reception/transmission. Hence, in practice, it can be expected that the different sites may be connected in the form of a *centralized RAN* (C-RAN) deployment for example, as briefly discussed in Chapter 13.

20.1.2 Network Energy Efficiency

Low energy consumption for mobile terminals has been an important requirement ever since the emergence of hand-held terminals roughly 20 years ago. The driving force has been the reduction in battery size and improved battery time. Today, reduced energy consumption *also in the radio-access network* is receiving increased attention for several reasons:

- The cost of energy is a far from negligible part of the overall operational cost for the operator. Thus, reduced energy consumption is one component in the everlasting quest for reduced operating costs for network operators.
- In some rural areas, it may not even be possible to connect the base station to the electrical grid. With sufficiently low energy consumption, reasonably sized solar panels could be used as power source, instead of the diesel generators commonly used today.

- In today's world, where energy consumption and the related climate impact is seen as one of the great challenges for the future, the mobile industry should lead the way for reduced energy consumption.

Regarding the third point, it is important to bear in mind though that the entire ICT (*Information and Communication Technologies*) sector today (2010) contributes roughly 2% of the overall world energy consumption and the contribution of the mobile-communication sector is just a fraction of that. At the same time, the emergence of mobile broadband communication opens up tremendous opportunities for reducing the global energy consumption in general by, for example, virtual meetings replacing traveling to physical meetings.

Still, low energy consumption in mobile networks is important. In particular, taking into account large future traffic increases, where traffic growth of several hundred times up to perhaps 1000 times can be expected in a longer-term perspective, low energy consumption of the mobile networks will most likely be as important a performance metric as capacity, data rates and latency, and must be treated as such.

Reducing the energy consumption of mobile networks is, to a large extent, an implementation issue. However, it is important to ensure that the basic principles of the radio-access technology allows for low energy consumption. One key characteristic is that the energy consumption should scale with the use of the radio resources, not only on average but also on a short-term basis. That is, during points in time where there is no traffic, the energy consumption should be reduced to an absolute minimum, for example by avoiding all but absolutely necessary transmissions. LTE already provides several tools that can be used for this. Nevertheless, the future evolution of LTE should further strive for minimizing transmission of signals strictly not needed.

This will be even more important as the network becomes more and more dense with more and more network nodes. In such network deployments, which may typically be heterogeneous networks as discussed in Chapter 13, pico cells may often be deployed to provide high data rates, rather then being needed for capacity reasons. Thus, the load per cell may be relatively low and each pico cell may often be more or less "empty", further stressing the importance of very low energy consumption when idle.

20.1.3 Machine-Type Communication

Similar to earlier mobile-communication systems, LTE has been designed with data services in mind and much effort has been put into developing techniques for providing high data rates and low latencies for services such as file downloading and web browsing. However, with the increased availability of mobile broadband, connectivity has also become a realistic option for machine-type communication. Machine-type communication spans a wide range of applications, from massive deployment of low-cost battery-powered sensors to remote-controlled utility meters, to surveillance cameras. Many of these applications can be handled by LTE already; communicating with a surveillance camera, for example, is not significantly different from uploading a file, in which case high data rates are paramount. However, other applications may not require transmission of large amounts of data or low latency, but rather pose challenges in terms of a vast amount of devices connecting to the network.

According to some sources [107], 50 billion connections, most of them machine-type communications, can be expected in the year 2020. Handling such a large number of devices is likely to be a challenge mainly for the core network, but improvements in the area of connection setup and power-efficient handling of control signaling in the radio-access network may be of interest.

20.1.4 New ways of Using Spectrum

Spectrum has been, is, and will continue to be a scarce resource for the mobile-communication industry. In particular, in light of the continuous increase in the data rates requested and the corresponding need for wider bandwidths, it is expected that spectrum will remain a scarce resource.

Historically, up until now, the mobile industry has relied on spectrum dedicated for mobile communication and licensed to a certain operator. This will also clearly be the main track for the future. However, in situations where licensed spectrum is not available, other possibilities for increasing the spectrum availability are of interest. This could include the use of unlicensed spectrum, or secondary spectrum primarily used for other communication services, as a complement to operation in the licensed spectrum. Broadcast spectrum not used (in some areas) is often referred to as "white space" [108]. Related to this is the concept of *cognitive radio* [109,110] – that is, radio-access-related functionality related to "smart" selection of spectrum usage. However, the applicability of cognitive radio to cellular communication is a relatively new area and further studies are required to assess the feasibility and impact of such usage.

20.1.5 Direct Device-to-Device Communication

One possible longer-term evolution path would be to extend the LTE radio-access technology with support for direct device-to-device communication – that is, make it possible for two mobile terminals to communicate directly with each other without going via the network. This has been studied in academia for some years, for example as part of the European research project WINNER [111]. Scenarios where the possibility for direct device-to-device communication could be of interest include situations where no network infrastructure is available, for example national security and public safety (NSPS), and situations where network infrastructure is available but when communication directly between the terminals could be more efficient, either in terms of requiring a smaller amount of radio resources for the same communication quality or allowing for improved communication quality. In the latter case, one could envision different degrees of network interaction in the device-to-device communication. However, this area is still in its infancy and scenarios where device-to-device communication offers benefits and how to interact with infrastructure-based communication need to be better understood.

Related to device-to-device communication is the possibility to also use LTE as a radio-access technology in the home, for example a digital camera connecting to the TV. Although such connectivity is possible to some extent today by using (semi-proprietary) technology, basing this type of communication on LTE could be attractive. When outside the home, the devices can communicate with the cellular LTE network, but while at home they can seamlessly connect to the local LTE network covering the home. Furthermore, for devices that are to be used at home only, operation without a SIM card and interaction with an external core network should be possible. Adding this type of functionality to LTE is most likely a relatively simple task.

20.2 **CONCLUDING REMARKS**

The list of technology areas above should be seen as examples. Some of these technologies will most likely be part of future releases of LTE, while others may not happen at all. There may also be other technologies, not listed above and maybe yet to be discovered, that could be of interest for LTE evolution. What is clear though is that LTE is a very flexible platform that can evolve in different directions to meet the future needs of wireless communication. Given the size of the LTE ecosystem, such evolution is a very attractive path for future wireless communication.

References

[1] ITU-R, International mobile telecommunications-2000 (IMT-2000), Recommendation ITU-R M.687-2, February 1997.

[2] ITU-R, Detailed specifications of the radio interfaces of international mobile telecommunications-2000 (IMT-2000), Recommendation ITU-R M.1457-9, May 2010.

[3] ITU-R, Principles for the process of development of IMT-advanced, Resolution ITU-R 57, October 2007.

[4] ITU-R, Framework and overall objectives of the future development of IMT-2000 and systems beyond IMT-2000, Recommendation ITU-R M.1645, June 2003.

[5] ITU-R, Invitation for submission of proposals for candidate radio interface technologies for the terrestrial components of the radio interface(s) for IMT-advanced and invitation to participate in their subsequent evaluation, ITU-R SG5, Circular Letter 5/LCCE/2, March 2008.

[6] ITU-R, ITU paves way for next-generation 4G mobile technologies; ITU-R IMT-advanced 4G standards to usher new era of mobile broadband communications, ITU Press Release, 21 October 2010.

[7] ITU-R, Working document towards a preliminary draft new recommendation ITU-R M.[IMT.RSPEC] ITU-R WP5D, Contribution 870, Attachment 5.11.

[8] ITU-R, Frequency arrangements for implementation of the terrestrial component of international mobile telecommunications-2000 (IMT-2000) in the bands 806–960 MHz, 1710–2025 MHz, 2110–2200 MHz and 2500–2690 MHz, Recommendation ITU-R M.1036-3, July 2007.

[9] M. Olsson, S. Sultana, S. Rommer, L. Frid, C. Mulligan, SAE and the Evolved Packet Core – Driving the Mobile Broadband Revolution, Academic Press, 2009.

[10] 3GPP, 3rd generation partnership project; Technical specification group radio access network; Requirements for Evolved UTRA (E-UTRA) and Evolved UTRAN (E-UTRAN) (Release 7), 3GPP TR 25.913.

[11] C.E. Shannon, A mathematical theory of communication, Bell System Tech. J. 27 (July and October 1948) 379–423, 623–656.

[12] J.G. Proakis, Digital Communications, McGraw-Hill, New York, 2001.

[13] G. Bottomley, T. Ottosson, Y.-P. Eric Wang, A generalized RAKE receiver for interference suppression, IEEE J. Sel. Area Comm. 18 (8) (August 2000) 1536–1545.

[14] WiMAX Forum, Mobile WiMAX – Part I: A technical overview and performance evaluation, White Paper, August 2006.

[15] ETSI, Digital video broadcasting (DVB): Framing structure, channel coding and modulation for digital terrestrial television, ETS EN 300 744 v. 1.1.2.

[16] NTT DoCoMo et al., Views on OFDM parameter set for evolved UTRA downlink, Tdoc R1-050386, 3GPP TSG-RAN WG 1, Athens, Greece, 9–13 May 2005.

[17] L. Hanzo, M. Munster, B.J. Choi, and T. Keller, OFDM and MC-CDMA for broadband multiuser communications, WLANs, and broadcasting, Wiley-IEEE Press, Chichester, UK, ISBN 0-48085879.

[18] R. van Nee, R. Prasad, OFDM for Wireless Multimedia Communications, Artech House Publishers, London, January 2000.

[19] J. Tellado and J.M. Cioffi, PAR reduction in multi-carrier transmission systems, ANSI T1E1.4/97-367.

[20] P.V. Etvalt, Peak to average power reduction for OFDM schemes by selective scrambling, Electron. Lett. 32 (21) (October 1996) 1963–1964.

[21] W. Zirwas, Single frequency network concepts for cellular OFDM radio systems, International OFDM Workshop, Hamburg, Germany, September 2000.

[22] A. Oppenheim, R.W. Schafer, Digital Signal Processing, Prentice-Hall International, ISBN 0-13-214107-8 01.

[23] S. Haykin, Adaptive Filter Theory, Prentice-Hall International, NJ, USA, 1986. ISBN 0-13-004052-5 025.

[24] D. Falconer, S.L. Ariyavisitakul, A. Benyamin-Seeyar, B. Eidson, Frequency domain equalization for single-carrier broadband wireless systems, IEEE Commun. Mag. 40 (4) (April 2002) 58–66.

[25] G. Forney, Maximum likelihood sequence estimation of digital sequences in the presence of intersymbol interference, IEEE T. Inform. Theory IT-18 (May 1972) 363–378.

[26] G. Forney, The viterbi algorithm, Proceedings of the IEEE, Piscataway, NJ, USA, Vol. 61, No. 3, March, 1973, 268–278.

[27] Motorola, Comparison of PAR and Cubic Metric for Power De-rating, Tdoc R1-040642, 3GPP TSG-RAN WG1, May 2004.

[28] U. Sorger, I. De Broeck, M. Schnell, IFDMA – A new spread-spectrum multiple-access scheme, Proceedings of the ICC'98, Atlanta, GA, USA, June 1998, pp. 1267–1272.

[29] W. Lee, Mobile Communications Engineering, McGraw-Hill, New York, USA, ISBN 0-07-037039-7.

[30] J. Karlsson, J. Heinegard, Interference rejection combining for GSM, Proceedings of the 5th IEEE International Conference on Universal Personal Communications, Cambridge, MA, USA, 1996, pp. 433–437.

[31] L.C. Godara, Applications of antenna arrays to mobile communications, Part I: Beam-forming and direction-of-arrival considerations, Proceedings of the IEEE, Vol. 85, No. 7, July, 1997, 1029–1030.

[32] L.C. Godara, Applications of antenna arrays to mobile communications, Part II: beam-forming and direction-of-arrival considerations, Proceedings of the IEEE, Piscataway, Vol. 85, No. 7, July, 1997, 1031–1060.

[33] A. Huebner, F. Schuehlein, M. Bossert, E. Costa, H. Haas, A simple space–frequency coding scheme with cyclic delay diversity for OFDM, Proceedings of the 5th European Personal Mobile Communications Conference, Glasgow, Scotland, April 2000, pp. 106–110.

[34] 3GPP, 3rd generation partnership project; technical specification group radio access network; Physical Channels and Mapping of Transport Channels onto Physical Channels (FDD), 3GP TS 25.211.

[35] V. Tarokh, N. Seshadri, A. Calderbank, Space–time block codes from orthogonal design, IEEE T. Inform. Theory 45 (5) (July 1999) 1456–1467.

[36] A. Hottinen, O. Tirkkonen, and R. Wichman, Multi-Antenna Transceiver Techniques for 3 G and Beyond, John Wiley, Chichester, UK, ISBN 0470 84542 2.

[37] C. Kambiz, L. Krasny, Capacity-achieving transmitter and receiver pairs for dispersive MISO channels, IEEE T. Commun. 42 (April 1994) 1431–1440.

[38] R. Horn and C. Johnson, Matrix analysis, Proceedings of the 36th Asilomar Conference on Signals, Systems and Computers, Pacific Grove, CA, USA, November 2002.

[39] K.J. Kim, Channel estimation and data detection algorithms for MIMO–OFDM systems, Proceedings of the 36th Asilomar Conference on Signals, Systems and Computers, Pacific Grove, CA, USA, November 2002.

[40] M.K. Varanasi and T. Guess, Optimum decision feedback multi-user equalization with successive decoding achieves the total capacity of the gaussian multiple-access channel, Proceedings of the Asilomar conference on Signals, Systems, and Computers, Monterey, CA, November 1997.

[41] S. Grant, J.-F. Cheng, L. Krasny, K. Molnar, Y.-P.E Wang, Per-antenna-rate-control (PARC) in frequency selective fading with SIC-GRAKE receiver, 60th IEEE Vehicular Technology Conference, Los Angeles, CA, USA 2, September 2004, 1458–1462.

[42] S.T. Chung, A.J. Goldsmith, Degrees of freedom in adaptive modulation: A unified view, IEEE T. Commun. 49 (9) (September 2001) 1561–1571.

[43] A.J. Goldsmith, P. Varaiya, Capacity of fading channels with channel side information, IEEE T. Inform. Theory 43 (November 1997) 1986–1992.

[44] R. Knopp, P.A. Humblet, Information capacity and power control in single-cell multi-user communications, Proceedings of the IEEE International Conference on Communications, Seattle, WA, USA, Vol. 1, 1995, 331–335.

[45] D. Tse, Optimal power allocation over parallel gaussian broadcast channels, Proceedings of the International Symposium on Information Theory, Ulm, Germany, June 1997, p. 7.

[46] M.L. Honig and U. Madhow, Hybrid intra-cell TDMA/inter-cell CDMA with inter-cell interference suppression for wireless networks, Proceedings of the IEEE Vehicular Technology Conference, Secaucus, NJ, USA, 1993, pp. 309–312.

[47] S. Ramakrishna, J.M. Holtzman, A scheme for throughput maximization in a dual-class CDMA system, IEEE J. Sel. Area Comm. 16 (6) (1998) 830–844.

[48] S.-J. Oh, K.M. Wasserman, Optimality of greedy power control and variable spreading gain in multi-class CDMA mobile networks, Proceedings of the AMC/IEEE MobiComp, Seattle, WA, USA, 1999, pp. 102–112.

[49] J.M. Holtzman, CDMA forward link waterfilling power control, Proceedings of the IEEE Vehicular Technology Conference, Tokyo, Japan, Vol. 3, May, 2000, 1663–1667.

[50] J.M. Holtzman, Asymptotic analysis of proportional fair algorithm, Proceedings of the IEEE Conference on Personal Indoor and Mobile Radio Communications, San Diego, CA, USA, Vol. 2, 2001, 33–37.

[51] P. Viswanath, D. Tse, R. Laroia, Opportunistic beamforming using dumb antennas, IEEE T. Inform. Theory 48 (6) (2002) 1277–1294.

[52] S. Lin and D. Costello, Error Control Coding, Prentice-Hall, Upper Saddle River, NJ, USA.

[53] C. Schlegel, Trellis and Turbo Coding, Wiley–IEEE Press, Chichester, UK, March 2004.

[54] J.M. Wozencraft, M. Horstein, Digitalised Communication Over Two-way Channels, Fourth London Symposium on Information Theory, London, UK, September 1960.

[55] D. Chase, Code combining – a maximum-likelihood decoding approach for combining and arbitrary number of noisy packets, IEEE T. Commun. 33 (May 1985) 385–393.

[56] M.B. Pursley, S.D. Sandberg, Incremental-redundancy transmission for meteor-burst communications, IEEE T. Commun. 39 (May 1991) 689–702.

[57] S.B. Wicker, M. Bartz, Type-I hybrid ARQ protocols using punctured MDS codes, IEEE T. Commun. 42 (April 1994) 1431–1440.

[58] J.-F. Cheng, Coding performance of hybrid ARQ schemes, IEEE T. Commun. 54 (June 2006) 1017–1029.

[59] P. Frenger, S. Parkvall, and E. Dahlman, Performance comparison of HARQ with chase combining and incremental redundancy for HSDPA, Proceedings of the IEEE Vehicular Technology Conference, Atlantic City, NJ, USA, October 2001, pp. 1829–1833.

[60] J. Hagenauer, Rate-compatible punctured convolutional codes (RCPC codes) and their applications, IEEE T. Commun. 36 (April 1988) 389–400.

[61] 3GPP, 3rd generation partnership project; Technical specification group radio access network; Physical Channels and Modulation (Release 8), 3GPP TS 36.211.

[62] 3GPP, 3rd generation partnership project; Technical specification group radio access network; Multiplexing and Channel Coding (Release 8), 3GPP TS 36.212.

[63] 3GPP, 3rd generation partnership project; Technical specification group radio access network; Physical Layer Procedures (Release 8), 3GPP TS 36.213.

[64] 3GPP, 3rd generation partnership project; Technical specification group radio access network; Physical Layer – Measurements (Release 8), 3GPP TS 36.214.

[65] ITU-R, Requirements related to technical performance for IMT-Advanced radio interface(s), Report ITU-R M.2134, 2008.

[66] 3GPP, 3rd generation partnership project; Technical specification group radio access network; Requirements for further advancements for Evolved Universal Terrestrial Radio Access (E-UTRA) (LTE-Advanced) (Release 9), 3GPP TR 36.913.

[67] 3GPP, 3rd generation partnership project; Technical specification group radio access network; Evolved universal terrestrial radio access (E-UTRA); User Equipment (UE) Radio Access Capabilities, 3GPP TS 36.306.

[68] IETF, Robust header compression (ROHC): Framework and four profiles: RTP, UDP, ESP, and Uncompressed, RFC 3095.

[69] J. Sun, O.Y. Takeshita, Interleavers for turbo codes using permutation polynomials over integer rings, IEEE T. Inform. Theory 51 (1) (January 2005) 101–119.

[70] O.Y. Takeshita, On maximum contention-free interleavers and permutation polynomials over integer rings, IEEE T. Inform. Theory 52 (3) (March 2006) 1249–1253.

[71] D.C. Chu, Polyphase codes with good periodic correlation properties, IEEE T. Inform. Theory 18 (4) (July 1972) 531–532.

[72] J. Padhye, V. Firoiu, D.F. Towsley, J.F. Kurose, Modelling TCP reno performance: A simple model and its empirical validation, ACM/IEEE T. Network. 8 (2) (2000) 133–145.

[73] CRAN International Workshop, CRAN – Road Towards Green Radio Access Network, China Mobile, 23 April 2010.

[74] 3GPP, 3rd generation partnership project; Technical specification group radio access network; Evolved universal terrestrial radio access (E-UTRA) and evolved universal terrestrial radio access network (E-UTRAN); User equipment (UE) conformance specification; Radio Transmission and Reception (Part 1, 2 and 3), 3GPP TS 36.521.

[75] 3GPP, Evolved universal terrestrial radio access (E-UTRA); Physical layer for relaying operation, 3GPP TS 36.216.

[76] 3GPP, 3rd generation partnership project; Technical specification group radio access network; Evolved universal terrestrial radio access (E-UTRA); User Equipment (UE) radio transmission and reception, 3GPP TS 36.101.

[77] APT Wireless Forum, APT Report on Harmonised Frequency Arrangements for The Band 698–806 MHz, APT/AWF/REP-14, September 2010 Edition.

[78] 3GPP, 3rd generation partnership project; Technical specification group radio access network; UMTS-LTE 3500 MHz Work Item Technical Report (Release 10), 3GPP TR 37.801.

[79] 3GPP, 3rd generation partnership project; Technical specification group radio access network; Feasibility Study for Evolved Universal Terrestrial Radio Access (UTRA) and Universal Terrestrial Radio Access Network (UTRAN) (Release 7), 3GPP TR 25.912.

[80] 3GPP, E-UTRA, UTRA and GSM/EDGE; Multi-Standard Radio (MSR) Base Station (BS) Radio Transmission and Reception, 3GPP TR 37.104.

[81] 3GPP, E-UTRA, UTRA and GSM/EDGE; Multi-Standard Radio (MSR) Base Station (BS) Conformance Testing, 3GPP TR 37.141.

[82] 3GPP, 3rd generation partnership project; Technical specification group radio access network; Evolved universal terrestrial radio access (E-UTRA); User Equipment (UE) Radio Transmission and Reception, 3GPP TR 36.803.

[83] 3GPP, 3rd generation partnership project; Technical specification group radio access network; Evolved universal terrestrial radio access (E-UTRA); Base Station (BS) Radio Transmission and Reception, 3GPP TR 36.804.

[84] 3GPP, Evolved universal terrestrial radio access (E-UTRA); User Equipment (UE) Radio transmission and reception, 3GPP TR 36.807.

[85] 3GPP, Evolved universal terrestrial radio access (E-UTRA); Carrier Aggregation Base Station (BS) Radio transmission and reception, 3GPP TR 36.808.

[86] 3GPP, 3rd generation partnership project; Technical specification group radio access network; Evolved universal terrestrial radio access (E-UTRA); Base Station (BS) Radio transmission and reception, 3GPP TS 36.104.

[87] 3GPP, 3rd generation partnership project; Technical specification group radio access network; Evolved universal terrestrial radio access (E-UTRA); Base Station (BS) conformance testing, 3GPP TS 36.141.

[88] FCC, Title 47 of the Code of Federal Regulations (CFR), Federal Communications Commission.

[89] 3GPP, 3rd generation partnership project; Technical specification group radio access network; Evolved universal terrestrial radio access (E-UTRA); Radio Frequency (RF) system scenarios, 3GPP TR 36.942.

[90] ITU-R, Unwanted Emissions in the Spurious Domain, Recommendation ITU-R SM.329-10, February 2003.

[91] 3GPP, 3rd generation partnership project; Technical specification group radio access network; Radio Frequency (RF) system scenarios, 3GPP TR 25.942.

[92] 3GPP, 3rd generation partnership project; Technical specification group radio access network; Feasibility study for further advancements for E-UTRA (LTE-Advanced) (Release 9), 3GPP TR 36.912.

[93] ITU-R, Guidelines for Evaluation of Radio Interface Technologies for IMT-Advanced, Report ITU-R M.2135-1, December 2009.

[94] A. Furuskär, Performance evaluations of LTE-advanced – the 3GPP ITU proposal, 12th International Symposium on Wireless Personal Multimedia Communications (WPMC 2009), Invited Paper.

[95] E. Dahlman, H. Ekström, A. Furuskär, Y. Jading, J. Karlsson, M. Lundevall, S. Parkvall, The 3G long-term evolution – radio interface concepts and performance evaluation, 63rd Vehicular Technology Conference, VTC 2006 – Spring, Vol. 1, pp. 137–141, IEEE, 2006.

[96] E. Dahlman, S. Parkvall, J. Sköld, P. Beming, 3G Evolution – HSPA and LTE for Mobile Broadband, second ed., Academic Press, 2008.

[97] WINNER, IST-WINNER II Deliverable 1.1.2 v. 1.2. WINNER II Channel Models, IST-WINNER2, Tech. Rep., 2007.

[98] 3GPP, 3rd generation partnership project; Technical specification group radio access network; Spatial Channel Model for Multiple Input Multiple Output (MIMO) Simulations (Release 9), 3GPP TR 25.996.

[99] Alcatel Shanghai Bell, CATT, CMCC, RITT, Spectrum Communications, TD-TECH, ZTE, Introduce TD-SCDMA Industry Standard in CCSA to 3GPP, Document R4-071394, 3GPP TSG-RAN WG4 meeting #44, Athens, Greece, August 2007.

[100] 3GPP, 3rd generation partnership project; Technical specification group GSM/EDGE radio access network; Feasibility Study for Evolved GSM/EDGE Radio Access Network (GERAN) (Release 7), 3GPP TR 45.912.

[101] H. Axelsson, P. Björkén, P. de Bruin, S. Eriksson, H. Persson, GSM/EDGE continued evolution, Ericsson Rev. (01) (2006) 20–29 Telefonaktiebolaget LM Ericsson, Stockholm, Sweden.

[102] 3GPP, Updated New WID on Higher Uplink Performance for GERAN Evolution (HUGE), Tdoc GP-061901, 3GPP TSG GERAN #31, Denver, USA, 4–8 September 2006.

[103] 3GPP, Circuit Switched Voice Capacity Evolution for GSM/EDGE Radio Access Network (GERAN), 3GPP TR 45.914.

[104] N. Bhushan, C. Lott, P. Black, R. Attar, Y.-C. Jou, M. Fan, et al., CDMA2000 1xEV-DO revision A: A physical layer and MAC layer overview, IEEE Commun. Mag. (February 2006) 75–87.

[105] M.W. Thelander, The 3G Evolution: Taking CDMA2000 into the Next Decade, Signal Research Group, LLC, White Paper developed for the CDMA Development Group, October 2005.

[106] R. Attar, Evolution of CDMA2000 cellular networks: Multicarrier EV-DO, IEEE Commun. Mag. (March 2006) 46–53.

[107] M. Alendal, Operators need an ecosystem to support 50 billion connections, Ericsson Bus. Rev. (3) (2010).

[108] M. Nekovee, A survey of cognitive radio access to TV white spaces, Int. J. Digi. Multimed. Broadcast. (2010) 1–11.

[109] J. Mitola, Cognitive Radio – An Integrated Agent Architecture for Software Defined Radio, Royal Institute of Technology (KTH), Sweden, May 2000.

[110] S. Haykin, Cognitive radio: Brain-empowered wireless communications, IEEE J. Sel. Area. Comm. 23 (2) (February 2005) 201–220.

[111] Wireless World Initiative New Radio, Eurescom, 2006, https://www.ist-winner.org.

Index

W

WCDMA, 3, 389–395
White space, 414
WiMAX, 405–408

X

X2 Interface, 111

Z

Zadoff-Chu sequences, 212